Lecture Notes in Computer Science 4777

Commenced Publication in 1973
Founding and Former Series Editors:
Gerhard Goos, Juris Hartmanis, and Jan van Leeuwen

Subhash Bhalla (Ed.)

Databases in Networked Information Systems

5th International Workshop, DNIS 2007
Aizu-Wakamatsu, Japan, October 17-19, 2007
Proceedings

 Springer

Volume Editor

Subhash Bhalla
University of Aizu
Department of Computer Software, Database Systems Laboratory
Tsuruga, Ikki Machi, Aizu-Wakamatsu, Fukushima 965-8580, Japan
E-mail: bhalla@u-aizu.ac.jp

Library of Congress Control Number: 2007936616

CR Subject Classification (1998): H.2, H.3, H.4, H.5, C.2

LNCS Sublibrary: SL 3 – Information Systems and Application, incl. Internet/Web
and HCI

ISSN   0302-9743
ISBN-10   3-540-75511-X Springer Berlin Heidelberg New York
ISBN-13   978-3-540-75511-1 Springer Berlin Heidelberg New York

Springer is a part of Springer Science+Business Media

springer.com

© Springer-Verlag Berlin Heidelberg 2007

Typesetting: Camera-ready by author, data conversion by Scientific Publishing Services, Chennai, India
Printed on acid-free paper   SPIN: 12170821   06/3180   5 4 3 2 1 0

Preface

Information systems in transportation, health-care and public utility services depend on computing infrastructure. Many research efforts are being made in related areas, such as wireless computing, data warehousing, stream processing, text-based computing, and information accesse by Web users. Government agencies in many countries plan to launch facilities in education, health-care and information support as e-government initiatives. In this context, Web content and query interface adaptation for ubiquitous access to data resources has become an active research field. A number of new opportunities have evolved in design and modeling based on the new computing needs of users. Database systems play a central role in supporting networked information systems for access and storage management aspects.

The fifth international workshop on Databases in Networked Information Systems (DNIS 2007) was held on October 17–19, 2007 at the University of Aizu in Japan. The workshop program included research contributions and invited contributions. A view of research activity in geospatial decision making and related research issues was provided by the session on this topic. The invited talk was given by Dr. Cyrus Shahabi. The workshop session on Web Data Management Systems included invited papers by Prof. Masahito Hirakawa and Dr. Gao Cong. The sessions on Networked Information Systems included the invited contributions by Prof. Divyakant Agrawal and Prof. Krithi Ramamritham. I would like to thank the members of the Program Committee for their support and all the authors who submitted their papers to DNIS 2007.

The sponsoring organizations and the organizing committee deserve praise for the support they provided. A number of individuals contributed to the success of the workshop. I thank Dr. Umeshwar Dayal, Dr. Meichun Hsu, Prof. D. Agrawal, Prof. Krithi Ramamritham, and Dr. Cyrus Shahabi for providing continuous support and encouragement.

The workshop received invaluable support from the University of Aizu. In this context, I would like to thank Prof. Tsunoyama, President of the University of Aizu. I would also like to thank Prof. Sedukhin, Head of Department of Computer Software, for making the support available. I express my gratitude to the members and chairman of the International Affairs Committee, for supporting the workshop proposal. Many thanks also to the faculty members at the university for their cooperation and support.

August 2007   Subhash Bhalla

Organization

DNIS 2007 was organized by the Database Systems Laboratory, University of Aizu, Aizu-Wakamatsu City, Fukushima, (PO 965-8580) Japan.

Executive Committee

Program Chair   S. Bhalla, University of Aizu, Japan
Organizing Chair   Takanori Kuroiwa, University of Aizu, Japan
Executive Chair   N. Bianchi-Berthouze, University College London, UK

Programme Committee

D. Agrawal, University of California, USA
F. Andres, National Center for Science Information Systems, Japan
N. Berthouze, University of Aizu, Japan
S. Bhalla, University of Aizu, Japan
P.C.P. Bhatt, Indian Institute of Information Technology, Bangalore, India
L. Capretz, University of Western Ontario, Canada
M. Capretz, University of Western Ontario, Canada
G. Cong, Microsoft Research, China
U. Dayal, Hewlett-Packard Laboratories, USA
W.I. Grosky, University of Michigan-Dearborn, USA
R. Gupta, University of Aizu, Japan
J. Herder, University of Applied Sciences, Fachhochschule Düsseldorf,
    Germany
M. Hirakawa, Shimane University, Japan
S. Jajodia, George Mason University, USA
Q. Jin, University of Aizu, Japan
H. Kubota, Kyoto University, Japan
A. Kumar, Pennsylvania State University, USA
T. Kuroiwa, University of Aizu, Japan
J. Li, University of Tsukuba, Japan
A. Mondal, Institute of Industrial Sciences, University of Tokyo, Japan
K. Myszkowski, Max-Planck-Institut für Informatik, Germany
M. Nakano, Institute of Industrial Sciences, University of Tokyo, Japan
A. Pasko, Hosei University, Tokyo, Japan
L. Pichl, University of Aizu, Japan
K. Ramamritham, Indian Institute of Technology, Bombay, India
P.K. Reddy, International Institute of Information Technology, Hyderabad,
    India

C. Shahabi, University of Southern California, USA
M. Sifer, Sydney University, Australia
L.T. Yang, St. Francis Xavier University, Canada

Sponsoring Institution

International Affairs Committee, University of Aizu,
Aizu-Wakamatsu City, Fukushima P.O. 965-8580, Japan

Table of Contents

Geospatial Decision-Making

Web Data Management Systems

Location Privacy in Geospatial Decision-Making*

Cyrus Shahabi and Ali Khoshgozaran

University of Southern California
Department of Computer Science
Information Laboratory (InfoLab)
Los Angeles, CA, 90089-0781
{shahabi,jafkhosh}@usc.edu

Abstract. Geospatial data-sets are becoming commonplace in many application domains, especially in the area of decision-making. Current state-of-the-art in geospatial systems either lack the ease-of-use and efficiency or sophisticated querying and analysis features needed by these applications. To address these shortcomings, we have been working on a generic and scalable geospatial decision making system dubbed *GeoDec*. In this paper, we first discuss many of the new features of GeoDec, particularly its spatial querying utilities. Next, we argue that in some applications, a user of GeoDec may not want to reveal the location of the query and/or its result set to the GeoDec server to preserve his/her privacy. Hence, for GeoDec to remain applicable in these scenarios, it should be able to evaluate the spatial queries without knowing the locations of the query and/or results. Towards this end, we present our novel space-encoding approach which would enable the GeoDec server to evaluate the spatial queries *blindly*.

1 Introduction

Geospatial information, in the form of traditional maps, have been used for several centuries for decision-making tasks. The oldest map is known to be from 2500 B.C. of a city near Babylon. In the past forty years, the field of Geospatial Information Systems (GIS), with ESRI leading the industry, has been increasing the role of geospatial information in decision-making tasks by allowing their digital manipulation. However, it was not until the last couple of years that the power of digital geospatial information has been brought to mass population through online map services such as Yahoo! maps [Yah] and most recently Google Earth [Goo]. Nowadays, you cannot see a news story without a screen-shot of Google Earth.

While the GIS industry targets the two ends of user population, the expert and the naive, there is a large group of users in the middle who are satisfied with neither the bare navigation/visualization features of Google Earth-like applications, nor the unscalable and non-generic solutions of GIS utilities. The reason for the lack of a middle-ground

* This research has been funded in part by NSF grants EEC-9529152 (IMSC ERC), IIS-0238560 (PECASE), IIS-0324955 (ITR), and unrestricted cash gifts from Google and Microsoft. Any opinions, findings, and conclusions or recommendations expressed in this material are those of the author(s) and do not necessarily reflect the views of the National Science Foundation.

S. Bhalla (Ed.): DNIS 2007, LNCS 4777, pp. 1–15, 2007.
© Springer-Verlag Berlin Heidelberg 2007

solution is the hard fundamental challenges in building a system that is generic, easy-to-use and scalable while at the same time can perform complex geospatial and temporal analysis efficiently.

Towards this end, in April 2005, we embarked on a multidisciplinary project, called Geospatial Decision-Making (GeoDec) [SCC+06]. GeoDec aims to enable geospatial decision-making for users in a variety of geographic application domains, including urban planning, emergency response, military intelligence, simulation and training. Since then, we have developed an end-to-end system that allows navigation through a 3D model of a location (e.g., a city) and enables users to issue queries and retrieve information as they navigate about the area. GeoDec helps the user to be immersed in an information-rich environment which facilitates his/her decision-making. The immersion in GeoDec system is the result of applying relevant techniques developed independently in the fields of databases, artificial intelligence, computer graphics and computer vision to the problem described above. In particular, the system seamlessly integrates satellite imagery, accurate 3D models, textures, video streams, road vector data, maps and point data for a specific geographic location. In addition, users can intuitively interact with the system using a glove-based interface and a large screen to issue a variety of spatial queries.

This paper has two main parts. In the first part, we report on some of our new developments in the GeoDec project. In the second part, we focus on one of the new challenges we face in any geospatial decision-making system including GeoDec: *Location Privacy*. The problem is that some users of geospatial decision-making systems may not want to reveal their locations or the locations of their query results to the system's location server. The challenge is how to provide all the spatial query features of GeoDec to the users without revealing the users locations to the GeoDec server. Here, we define some general metrics to evaluate any location privacy scheme and then discuss our recently proposed privacy model based on space encoding.

2   Geodec: Enabling Geographical Decision-Making

In previous work [SCC+06], we developed a multidisciplinary project, called GeoDec, for geospatial decision-making. The goal of GeoDec is to provide the needed information for decision makers in a variety of geographic application domains, including location-based services, urban planning, emergency response, military intelligence, simulation and training. The system, like Google Earth, supports the navigation through a 3D model of a geographical location (e.g., a city). In addition, with GeoDec one can also issue queries and retrieve information as he/she navigates about an area. In particular, the system seamlessly integrates satellite imagery, accurate 3D models, textures and video streams, road vector data, maps, point data, and temporal data for a specific geographic location (see Figure 1).

2.1   Geodec Architecture

Figure 2 depicts the 3-tier architecture that is used in GeoDec. This architecture consists of a data tier, an integration tier, and a presentation tier. The data tier focuses on

Fig. 1. A Screen-shot of a Geographic Location in GeoDec's User Interface

the efficient storing and indexing of geospatial data. The integration tier provides the ability to efficiently query and integrate heterogeneous geospatial sources. And the presentation tier provides a uniform representation so that query results can be visualized in commercial systems (such as Microsoft Virtual Earth [Vir] or Google Earth [Goo]) or in more specialized user interfaces.

I. Query Formulation and Visualization (Negaah): Negaah is a visualization interface for GeoDec that allows a user to navigate the 3D environment in real-time, and submit some customized queries on geospatial data based on a user-defined selection area. The user can selectively query and display different layers of information, and move forward or backwards in time. Negaah also supports formulating more sophisticated spatial queries such as KNN query for points of interest and the line of sight query around the selected query point.

II. Spatio-Temporal Query Middleware (Jooya): All the queries in Negaah are directed to GeoDec's information mediator/spatio-temporal database component through a middleware layer, Jooya. Jooya offers a universal way of specifying the type of query (e.g., nearest neighbor, range, shortest path, etc.), as well as its parameters and retrieves the results back in a unified format. Depending on the user's query type, Jooya either sends a query to our Spatio-temporal Database Manager (Darya), or to Prometheus [TAK04], our information mediator (which provides a uniform query interface to a set of web sources that contain information about a geographic location). Examples of queries sent to Darya are queries for infrequently updated or bulky data such as road network data, video metadata, and 3D building models. On the other hand, Jooya uses Prometheus to query highly dynamic data such as live traffic information obtained via a variety of web sources. Jooya returns results in a standard format (specifically, Google Earth's KML format) and this architecture enables any visualization layer that can support KML to sit on top of Jooya for its integrated query and access needs (Figure 2). Jooya, however, would need to include a customized query-interface blade for each new GUI. This enables Jooya to act as an interface for web based geospatial GUIs such as Google Earth as well.

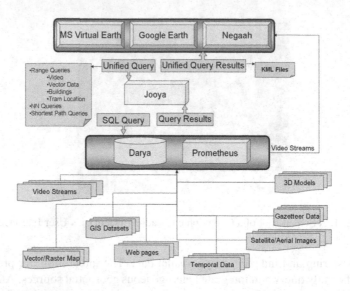

Fig. 2. Geodec Architecture

III. Spatio-Temporal Database Manager (Darya): As part of GeoDec's back-end, we have also developed a Spatio-temporal Database Manager (Darya). This module is in charge of managing spatio-temporal data stored in a database, a task that includes data modeling, storage and retrieval (querying). For storage and transmission, our main focus is now on bulky vector data such as road networks. Mainly Darya embeds a multi resolution vector data compression technique which effectively compresses the result of query windows sent to it, taking into account the client's display resolution. Our vector compression approach enables Darya to efficiently store and transfer bulky data. Furthermore Darya stores several heterogeneous data sources such as satellite/areal imagery, temporal data, GIS gazetteer data, 3D models, video streams, vector/raster maps etc. which allows the query middleware to integrate several data sources and create a compelling and information-rich representation of the geographic area.

One of the key benefits of the above architecture is its scalability. As discussed above, separating these three layers allows Geodec to act as a tool for rapid construction of an information-rich geographic space. The steps to reach that goal include adding the necessary spatio-temporal data to Darya (e.g., satellite imagery, vector data, etc.) and implementing the necessary query-interface blades in Jooya. Once the additions are made, Negaah (or any other proprietary user interface that can communicate with Jooya) will allow users to interactively work with and query the recently added geographic area.

2.2 GeoDec Querying Features

The incorporation of several data sources at the database layer, blended with the integration modules at the middle layer allows Geodec to respond to a wide range of sophisticated queries. In this section we briefly study these queries.

Geodec's query formulation and visualization interface allows users to construct different types of spatial queries for their area of interest. In order to facilitate the process of decision-making Geodec allows users to query Darya for any spatial data relevant to a *specific event* which is of interest to the user. This way, a single generic event query will be translated into several heterogeneous low-level queries by the middle layer and are sent to Darya. Jooya then receives the different pieces of query results returned from the spatio-temporal database layer and compiles the results into different information layers (such as 3D models, moving object trajectories, video feeds, vector data, etc.) for Negaah. When a user notices an event taking place in a geographic area, he uses Geodec to specify different pieces of information he has about the event and what he is expecting to obtain from the system. For instance he uses a bounding box to specify the region of interest while navigating in the 3D model. He can also specify a time duration in which the event has taken place and any other possible piece of information about the event. The user then selects different layers of information each of which corresponding to one (or more) spatial queries. The query results are then provided to the user in different layers to allow him to view/hide each dimension of result returned. We now review several types of spatial datasets and their associated queries to show how an event query is resolved by several spatial queries, using the architecture described in Section 2.1.

3D Building Models: The spatio-temporal database layers store the 3D building information of the geographic area texture mapped with satellite imagery of the area. Each building is associated with a timeframe of its existence. This way we can move back and forth in time and see the changes in an area caused by constructing/demolishing buildings. Once a user specifies the bounding box and timeframe for each event, we can reconstruct the actual 3D space for user's area of interest at the time the event had taken place.

Multi-resolution Vector Data Storage and Retrieval: Darya stores vector data (e.g., point data, spatial extents and road-network data) obtained from Navteq [Nav] for the entire United States. One challenge with storage and access of vector data is their large size. For example, the Santa Monica Blvd. in Los Angeles area, which is only 1.115 miles long, consists of 234 line segments in the Navteq street data set. Since storage is cheap, there is no reason to reduce the size of these data just to save space. However, transferring these bulky data over the network and then rendering it is a slow process. This is especially a waste of resources if the display resolution and zoom level (in case of GeoDec GUIs) is not fine-grained enough to require all the details. Hence, during an offline process, we utilize our multi-resolution vector data compression scheme [KKSS06] to construct different levels of detail for this data so that later, during the query time, Darya can minimize the communication/rendering overhead by dynamically choosing the right level of detail based on user's query so that we only retrieve and transmit what is absolutely needed for the display. The vector data is then superimposed on top of the areal image for the geographic location and are added to the 3D model of the area.

Moving Object Storage/Retrieval: Another important dataset which is maintained by Darya is the moving object data. For Geodec, we are tracking and storing the location of several moving objects (such as university trams) which are equipped with GPS devices. Therefore when a user specifies a moving object of interest for an event query, we can construct the trajectory of moving object locations for the query timeframe and allow the user to move back and forth in time to see the exact location of the moving object at any given time in past. It also allows users to track the live location of all moving objects in an area of interest as well. The trajectory of the selected moving object will then be added as a separate layer to the query result.

Fig. 3. The Result of a Line-of-Sight Query in GeoDec

Point Data Integration: Darya incorporates gazetteer data (such as building names, parcel information, points of interest, etc.) into its repository to provide users with a wide range of location-based services and spatial queries. Users can use Geodec to issue KNN or range queries associated with each event by specifying a query point in the 3D space. This allows users to view several points of interest deemed related to the event query result.

Video Feeds: Darya also maintains a database of video streams for several surveillance cameras installed in the different locations of an area. Therefore while responding to an event query, Darya queries its video server and retrieves any video surveillance feed associated with the event that might be of interest to the user and sends the video information (e.g., video feed, camera location and coverage area) to the GUI. In order to find the relevant camera feeds, Darya performs a spatio-temporal join between the event and the video feeds. The available video feeds are then highlighted in the area to be viewed by the user.

Line of Sight: In order to further enhance the capabilities of Geodec in dealing with dynamic data related to an event, Geodec allows users to associate a line of sight query to any moving object associated to an event. This way the line of sight module is blended with the moving object module to allow users to see a color coded 3D view of their area of interest based on whether the points in space are visible from the location of the moving object or not. As the user moves back and forth in time (i.e., the location of the moving object changes), Negaah dynamically updates the coloring of the 3D space (see Figure 3).

As described above, blending different *modalities* of information together, Jooya constructs an immersive result set for the user which is highly interactive. Ultimately, Geodec allows users to save their event queries and their associated data layers for future use.

3   Location Privacy

As discussed in Section 2, one of the main utilities of GeoDec is its rich spatial querying facility. An important requirement for successful evaluation of spatial queries is to know the exact location of the query point(s). With location-based services, for example, the query location is usually the location of the user operating a portable device such as a cell-phone, a PDA, a car navigation system or a laptop. With the advent of inexpensive GPS devices, many of these portable devices can incorporate (and already have incorporated) GPS systems. Hence, the location of the user (or query point) can be accurately identified and reported to a location service provider. In particular, the user of GeoDec can only install its GUI, e.g., Negaah, on his/her PDA or laptop and then use it to issue customized spatial queries based on his/her location. The problem here is that the user would then need to reveal his/her location to other layers of GeoDec, such as Jooya and Darya, which are installed on other servers and maintained by perhaps untrusted entities. This has major privacy implications to the extent that some users would prefer to avoid the potential benefits, not to compromise their location privacy. Therefore, it is essential for Geodec to be able to offer two modes of operations where it can both resolve a user's query with strict location privacy requirements blindly (i.e., without knowing their identity, location or query result), and operate in a normal non-private mode (i.e., in the original 2D space).

Recently, we proposed a fundamentally new approach to evaluate spatial queries, without revealing the original location of the user [KS07]. We term this novel method of performing spatial queries as the *"blind evaluation"* of spatial queries. The main idea behind our approach is to transform the original space and query location to a different space in which the results of the queries remain unchanged. Hence, the location server can perform the queries in the transformed space without acquiring any knowledge on the identities and locations of either the query point(s) or the result set. We formulate our transformation approach into a framework analogous to that of the conventional encryption schemes. That is, we separate the transformation *algorithm* from the transformation *key* for our space transformation approach. Hence, each client, having access to both the transformation key and algorithm, can apply the transformation on its location (i.e., encrypting its location) without requiring a trusted third party. The un-trusted location server, however, not having access to the transformation key, cannot acquire the original locations of query points and other points in the space even if the transformation algorithm is known. Meanwhile, by standing on the shoulders of the encryption giants, we benefit from all the techniques developed in the past two decades for encryption-key management, maintenance, distribution and security.

Current encryption schemes [IW06] cannot maintain the space distance property and hence cannot efficiently evaluate spatial queries in the transformed space. On the other hand, the currently proposed location privacy approaches [GG03, GL04, MCA06, Mok06, BWJ05, GL, CBP, BS03] rely on an intermediate third-party. Our approach is a fundamentally new encryption/tranformation methodology that maintains the distance properties of space and hence is both efficient and free of the reliance on a third party intermediator.

Let us formally define our hypothesis.

MAIN HYPOTHESIS. There exists a one-way transformation that can encode the 2-D space of static and dynamic objects in which spatial queries can be evaluated privately, accurately and efficiently.   ❏

Towards this end, here, we first clearly define the terms used in our hypothesis: non-reversible transformation, static and dynamic objects, and spatial queries. Subsequently, we define our privacy, accuracy and efficiency metrics. Next, we briefly discuss our space encoding approach based on Hilbert Curve [Hil91] and its vulnerability issues. In a recent publication [KS07], we conducted some preliminary evaluation of our approach assuming *static* objects and KNN queries. We briefly summarize the main observations in Sec. 3.7. Our ultimate goal is to extend our approach to support any spatial query on both static and *dynamic* objects.

3.1 Definitions of Terms in the Main Hypothesis

A transformation is *one-way* if it can be easily calculated in one direction (i.e., the forward direction) and is computationally impossible to calculate in the other (i.e., backward) direction [Sti02]. The process of transforming the original space with such a one-way mapping can be viewed as *encrypting* the elements of the 2-D space. With this view, in order to make decryption possible and efficient the function has to allow fast computation of its inverse given some extra knowledge, termed *trapdoor* [Sch84]. In practice, many one-way transformations may be reversible even without the knowledge of the trapdoor but the process must be too complex (equivalent to exhaustive try) to make such transformation computationally secure.

A *static* object is defined by a point or a polygon in 2-D space, e.g., with latitude and longitude, and its position does not change over time, e.g., location of a restaurant. A *dynamic* object is the same as a static one except that its position changes over time, e.g., a moving vehicle.

Spatial queries can be divided into two main classes. The first class of spatial queries consists of nearest-neighbor (NN) query and its variations. These queries search for data objects that minimize a distance-based function with reference to one or more query objects (e.g., points). Examples are K Nearest Neighbor (KNN) [RKV95, HS99], Reverse k Nearest Neighbor (RkNN) [KM00, SAA00, TPL04], k Aggregate Nearest

Neighbor (kANN) [PTMH05] and skyline queries [BKS01,PTFS05,SS06]. The second class is the spatial range queries. This includes identifying a range, as a circle with a center point and a radius, a rectangle with a corner point and width and height, or other polygon shapes with a list of points (i.e., vertices).

To evaluate all these spatial queries blindly, we mainly need to hide the location of the point or points in the query and response. Hereafter, without the loss of generality we define our metrics (Secs. 3.2 to 3.4) and discuss our space transformation approach (Secs. 3.5 and 3.6) assuming the K Nearest-Neighbor query (KNN), where we look for all the k closest points to a query point. The same approach and metrics can be used for other types of spatial queries with minor modifications. However, to evaluate our approach given the metrics, we need to study each query type individually. So far, as discussed in Sec. 3.7, we have only studied and evaluated our approach assuming KNN queries. In future, we plan to extend and evaluate our approach for other types of spatial queries. For now, we start by a formal definition of KNN.

Given a set of static objects $S = (o_1, o_2, \ldots, o_n)$ in 2-D space, the KNN query with respect to query point q finds a set $S' \subset S$ of K objects where for any object $o' \in S'$ and $o \in S - S'$, $D(o', q) \leq D(o, q)$ where D is the Euclidean distance function. In a typical KNN query scenario, the static objects represent points of interest (POI) and the query points represent user locations.

3.2 Privacy Metrics

We now formally define our privacy metrics with which we evaluate our proposed approach.

Metric 1. u-anonymity: While resolving a KNN query, the user issuing the query should be indistinguishable among the entire set of users. That is, for each query Q, $P(Q) = \frac{1}{M}$ where $P(Q)$ is the probability that query Q is issued by a user u_i and M is the total number of users. Note that satisfying this metric ensures the server does not know which user queried from a point q_i; however, we also need to ensure that the server does not know which point the query Q is issued from. This requirement is captured in Metric 2.

Metirc 2. a-anonymity: While resolving a KNN query, the location of the query point should not be revealed. That is, for each query Q, $P'(Q) = \frac{1}{area(A)}$, where A is the entire region covering all the objects in S, and $P'(Q)$ is the probability that query Q was issued by a user located at any point inside A.

Note that Metrics 1 and 2 impose much stronger privacy requirements than the commonly used K-anonymity [SS, GL, GG03, KGMP06, BS03, GL04, Mok06, MCA06], in which a user is indistinguishable among K other users or his location is blurred in a cloaked region R. The above metrics for location privacy are free of factors such as K and R. They are in fact identical to an extreme case of setting $R = A$ for spatial cloaking, and an extreme case of setting $K = M$ for K-anonymity.

Metric 3. Result set anonymity: The location of all points of interest in the result set should be kept secret from the location server. More precisely $\dot{P}(o_j) = k/n$ for $j = 1 \ldots n$ where $\dot{P}(o_j)$ is the probability that o_j is a member of the result set of size k for query Q and n is the total number of POI's.

Definition 1. *Blind evaluation of KNN*: We say a KNN query is blindly evaluated if the *u-anonymity, a-anonymity* and *result set anonymity* constraints defined above are all satisfied. In blind evaluation of KNN, the identity and location of the query point as well as the result set should not be revealed.

We term our approach *blind evaluation of KNN queries* because it attempts to prevent any leak of information to essentially blind the server from acquiring information about a user's location. The following example shows how the above properties should be satisfied in a typical KNN query. Suppose a user asks for his 3 closest gas-stations. In this case a *malicious entity* should acquire neither the location of the user (i.e., a-anonymity) nor its identity (i.e., u-anonymity) nor the actual location or identity of any of the 3 closest gas stations in the response set (i.e., result set anonymity) while the user should receive the actual points of interest matching his query.

3.3 Accuracy Metrics

Definition 2. Suppose the actual result of a KNN query, issued by a user located at point Q is $R = (o_1, o_2, \ldots, o_K)$, and it is approximated by a transformation T as $R' = (o'_1, o'_2, \ldots, o'_K)$. T is KNN-invariant if it yields acceptable values for the following two metrics:

Metric 1: The *Resemblance*, denoted by α, defined as

$$\alpha = \frac{|R \cap R'|}{|R|} \tag{1}$$

where $|R|$ denotes the size of a set R. In fact α measures what percentage of the points in the actual query result set R are included in the approximated result set R'.

Metric 2: The *Displacement*, denoted by β, defined as

$$\beta = \frac{1}{K}(\sum_{i=1}^{K} ||Q - o'_i|| - \sum_{i=1}^{K} ||Q - o_i||) \tag{2}$$

where $||Q - o_i||$ is the Euclidean distance between the query point Q and o_i. Therefore β measures how *closely* R is approximated by R' on average. Obviously, since R is the ground truth, $\beta \geq 0$.

Although there is no fixed threshold for acceptable α and β values, depending on the application and the scenario, certain values may or may not be considered satisfactory. In [KS07], we evaluated our approach against these two metrics and showed that it is an accurate enough KNN-invariant transformation.

3.4 Efficiency Metrics

Our main efficiency metrics are the well-known, widely practiced *query response time* and *server throughput*. Due to lack of space and popularity of these metrics, we do not discuss them further. However, in general, a space encoding technique that results in $O(n)$ computation complexity (where n is the total number of points in space) to answer spatial queries is unacceptable. Using hierarchical index structures, e.g., R-Trees, the computation complexity can usually be reduced to logarithmic in non-encrypted spaces. Hence, ideally we would like to achieve such complexity in the encrypted space.

3.5   Space Transformation Using Dual Hilbert Curves (STUDHC)

Introduced in 1890 by an Italian mathematician G. Peano [Sag94], space filling curves belong to a family of curves which pass through all points in space without crossing themselves. The important property of these curves is that they retain the *proximity* and *neighboring* aspects of the data. Consequently, points which lie close to one another in the original space mostly remain close to each other in the transformed space. One of the most popular members of this class is Hilbert curves [Hil91] since several studies show the superior clustering and distance preserving properties of these curves [LK01, Jag90, FR89, MvJFS01].

Similar to [MvJFS01] we define H_d^N for $N \geq 1$ and $d \geq 2$, as the Nth order Hilbert curve for a d-dimensional space. H_d^N is therefore a linear ordering which maps a d-dimensional integer space $[0, 2^N - 1]^d$ into an integer set $[0, 2^{Nd} - 1]$ as follows: $H = \ell(P)$ for $H \in [0, 2^{Nd} - 1]$, where P is the coordinate of each point in the d-dimensional space. We call the output of this function its *H-value*. Note that it is possible for two or more points to have the same H-value in a given curve.

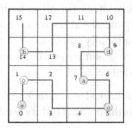

Fig. 4. A H_2^2 Pass of the 2-D Space

As mentioned above, our motivating application is location privacy and therefore we are particularly interested in 2-D space and thus only deal with 2-D curves ($d = 2$). Therefore $H = \ell(X, Y)$ where X and Y are the coordinates of each point in the 2-D space. Figure 4 illustrates a sample scenario showing how a Hilbert curve can be used to transform a 2-D space into H-values. In this example, points of interest (POI) are traversed by a second order Hilbert curve and are *indexed* based on the order they are visited by the curve (i.e., H in the above formula). Therefore, in our example the points a, b, c, d, e are represented by their H-values 7, 14, 5, 9 and 0, respectively. Depending on the desired resolution, more fine-grained curves can be recursively constructed.

An important property of a Hilbert curve that makes it a very suitable tool for our proposed scheme is that ℓ becomes a one-way function if the curve parameters are not known. These parameters, which collectively form a *key* for this one-way transformation, include the curve's starting point (X_0, Y_0), curve orientation θ, curve order N and curve scale factor Γ. We term this key, *Space Decryption Key* or SDK where $SDK = \{X_0, Y_0, \theta, N, \Gamma\}$.

In [KS07], we showed that two Hilbert curves, termed *dual curve*, where one is a 90 degree rotation of the other one, would actually result in a more accurate encryption of space without compromising its resilience.

Use of Hilbert curves to protect users location information is also suggested by [KGMP06]. However, [KGMP06] bears significant difference to our work in several aspects. First, it uses Hilbert curves to construct the anonymization of spatial region and to make a user k-anonymous. It does not use it as a space encryption algorithm. Basically, it uses $k-1$ closest H-values around a user to come up with the k-anonymity set. Second, it uses an anonymizer between the users and untrusted location server to blur user locations. This is what we are trying to avoid. Third, it satisfies the k-anonymity and does not achieve a-anonymity and u-anonymity metrics proposed in Sec. 3.3. Finally, it does not transform points-of-interests and users locations into another space in order to preserve location privacy. Instead, it utilizes efficient cloaking techniques to compute spatial queries for a region that includes user location.

3.6 Vulnerability of STUDHC

STUDHC is a One-Way Transformation: A malicious entity, not knowing our transformation key, has to exhaustively check for all combinations of curve parameters to find the right curve by comparing the H-values for all points of interest. As we show in Theorem 1, we make it computationally impossible to reverse the transformation and get back the original points. Even a nominal error in approximating curve parameters will generate a completely different set of H-values.

Theorem 1. The complexity of a brute-force attack to find the transformation key discussed above is $O(2^{4p})$ where p is the number of bits used to discretisize each parameter.

Proof. Please refer to [KS07] for the proof.   ☐

Key Management, Maintenance, Distribution and Security

One advantage of building our model based on encryption schemes is that once we have the concept of the *encryption key*, we can immediately benefit from all the techniques developed and matured in the past two decades in managing keys. For example, one concern might be what happens if the space transformation key is compromised. That is, malicious users can exist in the system trying to subscribe to the service and acquire the key pairs to share them with the location server. Similar to all other encryption keys used widely in mobile devices, our transformation key pair is not accessible by users and is kept in modules in charge of decryption inside their devices' hardware. These devices are all tamper-proof and hence the above scenario will not happen.

Similar arguments can be made for issues in securely distributing, maintaining and updating the keys. We consider all issues related to key management beyond the scope of this paper as they are already being investigated actively by the encryption community and several practical techniques are already adapted by the industry. Note that there is a huge difference between having an *off-line* trusted entity that manages and maintains keys and a trusted *on-line* intermediator that intercepts all communications. While the former is an integral component of many encryption schemes, the latter is a major security flaw. Furthermore, our trusted entity does not need to know user locations/identity simply because there is no need for it to anonymize such data.

Reverse Engineering and Use of Known Landmarks: One of the classic known attacks to unknown transformations is through the use of known landmarks. In the most powerful form of this attack, an attacker subscribes to the service as a client and conspires with the untrusted location server to probe himself with known landmarks. However, note that in our proposed scheme the clients cannot get to know the value of the key they are using to encrypt their locations (e.g., by making their communication devices tamper proof) and thus they do not get to know their encrypted location in the transformed space (in order to share it with the location server). Furthermore, since the clients communicate with the location server through pseudonyms, the server cannot trace back a received query point (such as the one sent from the attacker) to a client to infer its location in the original space. Therefore, no matter how many landmarks are used, this attack will not reveal the key to the attacker.

3.7 Preliminary Evaluation: Privacy, Accuracy and Efficiency of STUDHC

In [KS07], we reported on our preliminary evaluation of STUDHC assuming KNN queries and static objects. In this section, we briefly discuss the main observations.

Theorem 2. Using an $H_2{}^N$ Hilbert curve to encode the space satisfies the *a-anonymity, u-anonymity* and *result set anonymity* properties defined in Sec. 3.2.

Proof. Please refer to [KS07] for the proof.   ❑

In [KS07], we also performed several experiments with real-world datasets to evaluate the effectiveness of our approach. We showed that our proposed technique achieves a very close approximation of performing KNN queries in the original space by generating a result set whose elements on average have less than 0.08 mile displacement to the elements of the actual result set in a 26 mile by 26 mile area containing more than 10000 restaurants. We also showed that a malicious attacker gains almost no useful knowledge about the parameters of our encoding techniques, even when significant knowledge about the key is compromised. Hence, a nominal displacement error in approximating only one of the key parameters, (a meter displacement in a 670 sq-mile area) will result in no useful information to compromise our encryption.

Finally, in [KS07] we showed that the KNNs computational complexity in our scheme is $O(K \times \frac{2^{2N}}{n})$ where N, the curve order, is a small constant. Moreover, since only the K closest points are sent back to the client, the communication complexity becomes $O(K)$. These are much lower than $O(n)$ complexity but we believe the computational complexity can still be improved further.

4 Conclusion and Future Directions

This paper consisted of two main parts. In the first part, we reported on the new extensions and developments of a system that we built in the past two years to enable geospatial decision making, dubbed *GeoDec*. In particular, we focused on several new spatial querying capabilities of GeoDec and discussed the importance of these spatial queries in decision making applications. In the second part, we argued that for many of the spatial queries supported by systems such as GeoDec, it is critical to preserve the privacy of the locations of both the query point and the result set. Subsequently, we introduced novel privacy metrics to be met in order for a system to preserve location privacy. We then discussed our space-encoding approach to location privacy and showed that our approach meets the defined metrics. As part of our future plan, we intend to extend our location privacy approach to support dynamic/moving objects as well as other types of spatial queries.

References

[BKS01]   Börzsönyi, S., Kossmann, D., Stocker, K.: The Skyline Operator. In: Proceedings of ICDE 2001, pp. 421–430 (2001)

[BS03]   Beresford, A.R., Stajano, F.: Location privacy in pervasive computing. IEEE Pervasive Computing 2(1), 46–55 (2003)

[BWJ05]   Bettini, C., Wang, X.S., Jajodia, S.: Protecting privacy against location-based personal identification. In: Jonker, W., Petković, M. (eds.) SDM 2005. LNCS, vol. 3674, pp. 185–199. Springer, Heidelberg (2005)

[CBP]   Cheng, R., Bertino, E., Prabhakar, S.: Preserving user location privacy in mobile data management infrastructures. In: Danezis, G., Golle, P. (eds.) PET 2006. LNCS, vol. 4258, Springer, Heidelberg (2006)

[FR89]   Faloutsos, C., Roseman, S.: Fractals for secondary key retrieval. In: PODS 1989: Proceedings of the eighth ACM SIGACT-SIGMOD-SIGART symposium on Principles of database systems, pp. 247–252. ACM Press, New York (1989)

[GG03]   Gruteser, M., Grunwald, D.: Anonymous usage of location-based services through spatial and temporal cloaking. In: MobiSys. USENIX (2003)

[GL]   Gedik, B., Liu, L.: A customizable k-anonymity model for protecting location privacy

[GL04]   Gruteser, M., Liu, X.: Protecting privacy in continuous location-tracking applications. IEEE Security & Privacy 2(2), 28–34 (2004)

[Goo]   Google earth, http://earth.google.com

[Hil91]   Hilbert, D.: Uber die stetige abbildung einer linie auf ein flachenstuck. In: Math. Ann. vol. 38, pp. 459–460 (1891)

[HS99]   Hjaltason, G.R., Samet, H.: Distance Browsing in Spatial Databases. TODS, ACM Transactions on Database Systems 24(2), 265–318 (1999)

[IW06]   Indyk, P., Woodruff, D.P.: Polylogarithmic private approximations and efficient matching. In: Theory of Cryptography, Third Theory of Cryptography Conference, New York, USA, pp. 245–264 (2006)

[Jag90]   Jagadish, H.V.: Linear clustering of objects with multiple atributes. In: Proceedings of the 1990 ACM SIGMOD International Conference on Management of Data, pp. 332–342. ACM Press, Atlantic City, NJ (1990)

[KGMP06]   Kalnis, P., Ghinita, G., Mouratidis, K., Papadias, D.: Preserving anonymity in location based services. A Technical Report TRB6/06, National University of Singapore (2006)

[KKSS06]   Khoshgozaran, A., Khodaei, A., Sharifzadeh, M., Shahabi, C.: A multi-resolution compression scheme for efficient window queries over road network databases. In: Perner, P. (ed.) ICDM 2006. LNCS (LNAI), vol. 4065, Springer, Heidelberg (2006)

[KM00]   Korn, F., Muthukrishnan, S.: Influence sets based on reverse nearest neighbor queries. In: Proceedings of the 2000 ACM SIGMOD international conference on Management of data, pp. 201–212. ACM Press, New York (2000)

[KS07]   Khoshgozaran, A., Shahabi, C.: Blind Evaluation of Nearest Neighbor Queries
             Using Space Transformation to Preserve Location Privacy. In: Proceedings of the
             10th International Symposium on Spatial and Temporal Databases (SSTD 2007)
             (July 2007)
[LK01]   Lawder, J.K., King, P.J.H.: Querying multi-dimensional data indexed using the
             hilbert space-filling curve. SIGMOD Record 30(1), 19–24 (2001)
[MCA06]   Mokbel, M.F., Chow, C.-Y., Aref, W.G.: The new casper: Query processing for
             location services without compromising privacy. In: Proceedings of the 32nd In-
             ternational Conference on Very Large Data Bases, pp. 763–774. ACM, Seoul,
             Korea (2006)
[Mok06]   Mokbel, M.F.: Towards privacy-aware location-based database servers. In:
             Barga, R.S., Zhou, X. (eds.) ICDE Workshops, p. 93. IEEE Computer Society,
             Los Alamitos (2006)
[MvJFS01]   Bongki Moon, H.V.J., Faloutsos, C., Saltz, J.H.: Analysis of the clustering prop-
             erties of the hilbert space-filling curve. IEEE Transactions on Knowledge and
             Data Engineering 13(1), 124–141 (2001)
[Nav]   Navteq, http://www.navteq.com
[PTFS05]   Papadias, D., Tao, Y., Fu, G., Seeger, B.: Progressive Skyline Computation in
             Database Systems. ACM Trans. Database Syst. 30(1), 41–82 (2005)
[PTMH05]   Papadias, D., Tao, Y., Mouratidis, K., Hui, C.: Dimitris Papadias, Yufei Tao, Kyr-
             iakos Mouratidis, and Chun Kit Hui. ACM Trans. Database Syst. 30(2), 529–576
             (2005)
[RKV95]   Roussopoulos, N., Kelley, S., Vincent, F.: Nearest Neighbor Queries. In: Pro-
             ceedings of the 1995 ACM SIGMOD International Conference on Management
             of Data, San Jose, California, May 22-25, 1995, pp. 71–79. ACM Press, New
             York (1995)
[SAA00]   Stanoi, I., Agrawal, D., El Abbadi, A.: Reverse nearest neighbor queries for dy-
             namic databases. In: ACM SIGMOD Workshop on Research Issues in Data Min-
             ing and Knowledge Discovery, pp. 44–53. ACM Press, New York (2000)
[Sag94]   Sagan, H.: Space-Filling Curves. Springer, Heidelberg (1994)
[SCC⁺06]   Shahabi, C., Chiang, Y.-Y., Chung, K., Huang, K.-C., Khoshgozaran-Haghighi,
             J., Knoblock, C., Lee, S., Neumann, U., Nevatia, R., Rihan, A., Thakkar, S.,
             You, S.: Geodec: Enabling geospatial decision making. In: IEEE International
             Conference on Multimedia & Expo(ICME), IEEE Computer Society Press, Los
             Alamitos (2006)
[Sch84]   Schroeder, M.R.: Number Theory in Science and Communication. Springer, Hei-
             delberg (1984)
[SS]   Samarati, P., Sweeney, L.: Protecting privacy when disclosing information: k-
             anonymity and its enforcement through generalization and suppression.
[SS06]   Sharifzadeh, M., Shahabi, C.: The Spatial Skyline Queries. In: VLDB 2006.
             Proceedings of the 32nd International Conference on Very Large Data Bases
             (September 2006)
[Sti02]   Stinson, D.R.: Cryptography, Theory and Practice. Chapman & Hall/CRC (2002)
[TAK04]   Thakkar, S., Ambite, J.L., Knoblock, C.A.: A data integration approach to au-
             tomatically composing and optimizing web services. In: ICAPS Workshop on
             Planning and Scheduling for Web and Grid Services (2004)
[TPL04]   Tao, Y., Papadias, D., Lian, X.: Reverse kNN Search in Arbitrary Dimensionality.
             In: Proceedings of the Thirtieth International Conference on Very Large Data
             Bases, pp. 744–755. Morgan Kaufmann, San Francisco (2004)
[Vir]   Microsoft virtual earth. http://maps.live.com/
[Yah]   Yahoo! maps. http://maps.yahoo.com/

Geometrical Information Fusion
from WWW and Its Related Information

Yasuhiko Morimoto

Hiroshima University
1-7-1 Kagamiyama, Higashi-Hiroshima, 739-8521, Japan
morimoto@mis.hiroshima-u.ac.jp
http://www.morimo.com/morimo-ken

Abstract. We considered a spatial data mining method that extracts spatial knowledge by computing geometrical patterns from web pages and access log files of HTTP servers. There are many web pages that contain location information such as addresses, postal codes, and telephone numbers. We can collect such web pages by web-crawling programs. For each page determined to contain location information, we apply geocoding techniques to compute geographic coordinates, such as latitude-longitude pairs. Next, we augment the location information with keyword descriptors extracted from the web page contents. We then apply spatial data mining techniques on the augmented location information. In addition, we can use hyperlinks and access log files to find linkage between pages with location information to derive spatial knowledge.

Keywords: Spatial Data Mining, Web Mining, Co-Location Pattern, Graph Clustering.

1 Introduction

With rapid growth of capability of collecting and storing information, we can easily construct very large information repository. The world-wide web becomes one of the largest information sources of any kind. Many people have been interested in finding valuable knowledge from the WWW. We often call such technologies "web mining". Web mining is utilized in various kinds of business such as internet search engines, recommendation systems, P4P (pay for performance) advertisement, SEM/SEO (search engine marketing / search engine optimization) and so on.

In the WWW, there are many web pages that contain location information such as addresses, postal codes, telephone numbers and so on. Commercial web pages typically contain address and contact telephone numbers as well as descriptions of the products and services offered. The introductions of news articles appearing on web pages often state the locations where the events took place, or where they were reported from. It is natural to assume that associations exist between the general contents of a web page and the specific location information it contains.

S. Bhalla (Ed.): DNIS 2007, LNCS 4777, pp. 16–32, 2007.

In addition to the location information, most of web pages contain hyperlinks. A link of a page to another page implies some kinds of relationships between the two pages. Access log files of HTTP servers contains information of the referrer page of each page access, which tells us where a visitor of the page comes from and which pages in the HTTP server site are viewed together in each visit. Hyperlinks of web pages and access patterns of users which surf along with hyperlinks can be considered as a weighted directed graph of web pages, where each web page is associated with some locations.

This paper addresses the problem of geometrical information fusion from WWW and its related information. By using spatial mining and graph mining algorithms, we extract spatial knowledge from the web.

2   Geospatial Association

Table 1 shows an example of a *geospatial association* extracted from web pages by means of a web crawling program.

Table 1. Geospatial Association

ID	Coordinates	Concepts
705	(37.260 -121.919)	('diet' ···)
6062	(37.890 -122.259)	('diet' ···)
1401	(37.381 -122.125)	('qualification' 'duty' ···)
1858	(37.772 -122.414)	('genome' ···)
1858	(37.772 -122.414)	('biotech' ···)
··· ···		···

In the geospatial association format, the ID is the unique identification number of a web page, the Coordinates values are geographic coordinates (latitude-longitude pairs), and the Concepts values are lists of representative keywords for the web page. Each record of the table implies the existence of a web page whose contents relate to these concepts as well as including information referring to a location.

Figure 1 gives a conceptual overview of how our system extracts geospatial information and associates it with web pages. For each web page, we parse the source code of the page and try to find geospatial information (G1). We then translate the geospatial information into coordinate values such as latitude-longitude pairs (G2).

Simultaneously, we extract concepts from the collection of web pages. Figure 2 provides an overview of the extraction process. From each page, we eliminate HTML tags and extract keywords such as names, terms, and abbreviations (C1). From the set of extracted keywords, we create a matrix of document vectors, where each vector corresponds to an individual web page, and each vector attribute corresponds to a keyword from the total set of extracted keywords (C2).

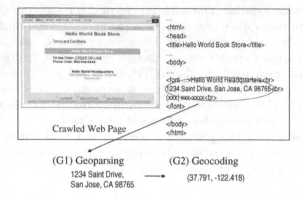

(G1) Geoparsing

1234 Saint Drive,
San Jose, CA 98765

(G2) Geocoding

(37.791, -122.418)

Fig. 1. Geospatial Information Extraction

We use the *term frequency inverse document frequency* (`tf-idf`) model for representing extracted keywords [1]. Next, we reduce the number of columns of the `tf-tdf` matrix in order to lower computational costs (**C3**). After this dimensional reduction, clusters of web pages are produced by means of a clustering algorithm applied to the reduced-dimensional matrix (**C4**). Finally, we label each cluster with several significant keywords indicating concepts associated with its constituent web pages (**C5**).

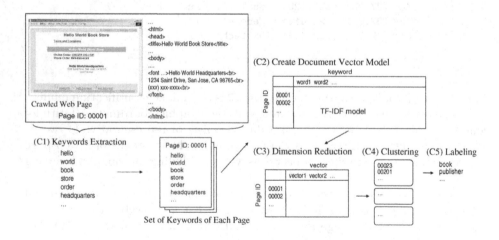

Fig. 2. Concepts Extraction

2.1 Geospatial Information Extraction

The process of recognizing geographic context is referred to as *geoparsing*, and the process of assigning geographic coordinates is known as *geocoding*. In the field of GIS, various geoparsing and geocoding techniques have been explored and are utilized. In [2] and [3], geoparsing and geocoding techniques are presented that are especially well-suited for web pages. From a large collection of web pages produced by a web crawler, we select those pages containing location information that can be identified using geoparsing techniques, and create lists of the format shown in Table 2, consisting of a web page ID (Page ID), geospatial coordinate values (Coordinates), the URL (URL), and the title of the page (Title).

Table 2. Geocoded Web Pages

ID	Coordinates	URL	Title
1	37.79,-122.41	www.···	Museum of ···
2	37.64,-122.42	www.···	San Jose Book
2	37.78,-122.39	www.···	San Jose Book
···		···	···

The format of postal addresses varies greatly from one country to another. Moreover, within a given country a variety of expressions may be used. However, within many countries, recognition of postal addresses and zip (postal) codes from natural language text data is well-established. For example, within the United States, we can geocode such location information using the product "Tiger/Zip+4" available with the TIGER dataset [4].

In addition to explicit location information such as addresses, web pages contain other types of information from which we can infer location. Such implicit information can also be utilized to make the lists in Table 2 and to increase the accuracy of location. Phone numbers are an important example of such implicit location information, since they are organized according to geographic principles. IP addresses, hostnames, routing information, geographic feature names, and hyperlinks can also be utilized to infer location [3].

Note that some web pages refer to more than one explicitly recognizable location. Such web pages can therefore be assigned more than one set of coordinate values.

2.2 Vector Space Modeling

In addition to the location information, we extract keywords, such as names, terms, and abbreviations, for each web page. Then, we use vector space modeling (VSM) for representing a collection of web pages. In the VSM, each web page is represented by a document vector consisting of a set of weighted keywords.

The weights of the entries in each document vector are determined according to the *term frequency inverse document frequency* (tf-idf) model. The weight of the i-th keyword in the j-th document, denoted by $a(i, j)$, is a function of the keyword's term frequency $tf_{i,j}$ and the document frequency df_i as expressed by the following formula:

$$a(i,j) = \begin{cases} (1 + tf_{i,j}) \ \log_2 \dfrac{N}{df_i}, & \text{if } tf_{i,j} \geq 1, \\ \\ 0, & \text{if } tf_{i,j} \geq 1. \end{cases}$$

where $tf_{i,j}$ is defined as the number of occurrences of the i-th keyword w_i in the j-th document d_j, and df_i is the number of documents in which the keyword w_i appears. Once each $a(i,j)$ has been determined, a data set consisting of M web pages spanning N keywords attributes can be represented by an M-by-N matrix $\mathbf{A} = (a_{i,j})$.

Since the number of keywords, N, is too large to handle, we apply a singular value decomposition technique to reduce the dimension of the M-by-N matrix into an M-by-n matrix, formed by taking the n ($n \ll N$) eigenvectors corresponding to the largest n eigenvalues of the original matrix [5].

We empirically determined that clusters of web pages can be efficiently computed using values of n of approximately 200. Accordingly, we use this value of n for the target reduced dimension.

2.3   Clustering Based on Vector Space

We, then, compute clusters of web pages from the M-by-n reduced-dimensional matrix. Though the number of web pages M tends to be very large, we used a scalable and effective clustering method suitable for clustering large sets of text data.

In general, a web page may contain several topics in its contents, and thus may contribute to several concepts. Any clustering method based on text data drawn from web pages should take into account the following desiderata:

- An individual data element need not be assigned to exactly one cluster. It could belong to many clusters, or none.
- Clusters should be allowed to overlap to a limited extent. However, no two clusters should have a high degree of overlap unless one contains the other as a subcluster.
- Cluster members should be mutually well-associated. Chains of association whose endpoints consist of dissimilar members should be discouraged.
- Cluster members should be well-differentiated from the non-members closest to them. However, entirely unrelated elements should have little or no influence on the cluster formation.

We measure the level of mutual association within clusters using *shared neighbor* information, as introduced by Jarvis and Patrick [6]. Shared neighbor information and its variants have been used in the merge criterion of several agglomerative clustering methods [7,8]. We made clusters of web pages based on the shared neighbor information using a method proposed by Houle [9,10].

2.4   Cluster Labeling

We assigned labels to a cluster based on a ranked list of keywords that occur most frequently within the web pages of the cluster, in accordance with the keyword weighting strategy used in the document vector model. Each keyword can be given a score equal to the sum or the average of the corresponding keyword weights over all document vectors of the clusters; a predetermined number of terms achieving the highest scores can be ranked and assigned to the cluster.

When dimensional reduction is used, the original document vectors may no longer be available due to storage limitations. Nevertheless, meaningful keyword lists can still be extracted without the original vectors. The i-th keyword can be associated with a unit vector $z_i = (z_{i,1}, z_{i,1}, ..., z_{i,N})$ in the original document space, such that $z_{i,j} = 1$ if $i = j$, and $z_{i,j} = 0$ otherwise. Let $\phi(q)$ be the average of the document vectors belonging to a cluster q. Using this notation, the score for the i-th keyword can be expressed as $z_i \cdot \phi(q)$, which is the cosine of the angle between two vectors, z_i and $\phi(q)$, and indicates similarity of the two vectors.

$$\frac{z_i \cdot \phi(q)}{\|z_i\| \|\phi(q)\|} = cosangle(z_i, \phi(q)).$$

With dimensional reduction, the pairwise distance $cosangle(v, w)$ between vectors v and w of the original space is approximated by $cosangle(v', w')$, where v' and w' are the respective equivalents of v and w in the reduced dimensional space. Hence, we could approximate $cosangle(z_i, \phi(q))$ by $cosangle(z_i', \phi'(q))$, where z_i' and $\phi'(q)$ are the reduced-dimensional counterparts of vectors z_i and $\phi(q)$, respectively. The value $cosangle(z_i', \phi'(q))$ can in turn be approximated by $cosangle(z_i', \phi''(q))$, where $\phi''(q)$ is the average of the reduced-dimensional vectors of cluster q. Provided that the vectors z_i' have been precomputed for all $1 \le i \le n$, a ranked set of keywords can be efficiently generated by means of a nearest-neighbor search based on $\phi''(q)$ over the collection of reduced-dimensional keyword vectors $\{z_i' | 1 \le i \le n\}$.

3   Hyperlinks Analysis

Most of HTTP servers records access log, in which each record contains the IP address of the client, timestamp, the request line (a command in HTTP), and so on. From the information, we can see which HTML files are viewed in a visit. Figure 3 is an example of the access log of an HTTP server.

```
9.116.10.152 - yasi [1/Jun/2006:13:55:36 +0900] "GET A.htm HTTP/1.0" 200 312
9.116.10.152 - yasi [1/Jun/2006:13:55:38 +0900] "GET B.htm HTTP/1.0" 200 242
9.116.10.152 - yasi [1/Jun/2006:13:55:42 +0900] "GET C.htm HTTP/1.0" 200 98
133.41.55.120 - - [1/Jun/2006:14:50:00 +0900] "GET A.htm HTTP/1.0" 200 532
133.41.55.120 - - [1/Jun/2006:14:50:16 +0900] "GET C.htm HTTP/1.0" 200 98
202.181.97.53 - - [1/Jun/2006:14:51:30 +0900] "GET A.htm HTTP/1.0" 200 332
202.181.97.53 - - [1/Jun/2006:14:51:48 +0900] "GET D.htm HTTP/1.0" 200 212
```

Fig. 3. Access Log of HTTP Server

We can consider that one user from 9.116.10.152 viewed "A", "B", and "C" in a visit and another from 133.41.55.120 viewed "A" and "C" in a visit, though we could not identify users exactly. Based on the observation, we can estimate the probability of page transitions for each page. Let $p(A)$ be the probability of page views for page A in a visit and let $p(A + B)$ be the probability that both of A and B are viewed in a visit. We can compute *association rules* for each frequently accessed pair of pages of the form $L \Rightarrow_{conf} R$, where each of L and R is a page and *conf* is the conditional probability $p(L + R)/p(L)$ called *confidence* [11,12]. In this example, we can compute following rules:

$$A \Rightarrow_{0.33} B \qquad A \Rightarrow_{0.66} C \qquad A \Rightarrow_{0.33} D$$
$$B \Rightarrow_{1.00} A \qquad B \Rightarrow_{1.00} C$$
$$C \Rightarrow_{1.00} A \qquad C \Rightarrow_{0.50} B$$
$$D \Rightarrow_{1.00} A$$

Association rule mining function can efficiently compute all frequent rules that has more than the minimum support value, which is a user specified minimum probability of page view [12,13]. In general, a huge number of rules is produced if the log is large.

We consider the set of the association rules as a huge weighted directed graph $G = (V, E)$ where V is a set of web pages and E is a set of the rules between two pages. In the graph, there exists a directed edge from $L \in V$ to $R \in V$ ($L \neq R$) if there exist a rule $L \Rightarrow_{conf} R$ that satisfies the conditions.

Assume we have a set of association rules and its corresponding graph as shown in Figure 4. The rule "$V1 \Rightarrow V2, 0.5$", for example, implies that a surfer visited $V1$ tended to visit $V2$ with confidence 0.5.

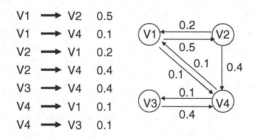

V1 → V2	0.5
V1 → V4	0.1
V2 → V1	0.2
V2 → V4	0.4
V3 → V4	0.4
V4 → V1	0.1
V4 → V3	0.1

Fig. 4. Weighted Directed Graph of Rules

We can consider the clustering problem on the weighted directed graph of web pages. We want to group pages into clusters so that pages in a cluster are close to each other but are not close to pages in other clusters. We use the weight to compute the closeness between two pages in the graph. For example, assume we are focusing on the closeness between $V1$ and $V4$. Though there exists a direct edge from $V1$ to $V4$ with 0.1 confidence, the sequence $V1 \Rightarrow V2 \Rightarrow V4$ has higher (transitive) confidence, $0.5 * 0.4 = 0.2$. We consider the closeness of $V1$ to $V4$ is 0.2 because of the existence of the (transitive confident) path $V1 \Rightarrow V2 \Rightarrow V4$.

If there is no direct hyperlink from a web page $V1$ to another page $V4$, confidence value of $V1 \Rightarrow V4$ may be very small or close to zero. However, if there is a popular transitive link from $V1$ to $V4$, we should consider the two pages is close. In order to evaluate the closeness, we should examine the existence of such transitive paths, which we call optimal path, for each pair of pages.

We solve the all-pairs optimal path problem in which we compute all effective, in other words not negligible, optimal path for all pairs of vertices in a weighted directed graph. We have to handle huge number of pages especially in a large enterprise. Moreover, thanks to the rapid growth of the internet, the size will be expected to become larger and larger. Though there are several efficient all-pairs optimal path algorithms [14], conventional algorithms can not handle such huge graph well. We developed a space efficient algorithm and its implemantation for computing all-pairs optimal path. We make clusters after we compute the closeness, which is defined by the optimal path, for each pair of web pages.

3.1 Path and Path Value

Let us consider a path p from a vertex to another along directed edges in the graph. A path is a sequence of different vertices of the form:

$$v_1 \Rightarrow_{val(v_1, v_2)} v_2 \Rightarrow_{val(v_2, v_3)} \cdots \Rightarrow_{val(v_{n-1}, v_n)} v_n,$$

where v_i $(i = 1, \cdots, n)$ is the i-th vertex in the sequence of n $(n \geq 2)$ vertices. We call v_1, the leftmost vertex, a cause and v_n, the rightmost vertex, an effect[1]. Transitive weight of the path, which we call *path value*, can be computed based on the confidence value of its constituent directed edges. The path value of $p = \langle v_1, v_2, \cdots, v_n \rangle$ is the product of the confidence of its constituent directed edges, $val(p) = \prod_{i=1}^{n-1} val(v_i, v_{i+1})$ where $val(v_i, v_{i+1})$ $(0 < val(v_i, v_{i+1}) \leq 1)$ is confidence of the directed edge from v_i to v_{i+1}. Note that since we assume $0 < val(v_i, v_{i+1}) \leq 1$, $val(p)$ will monotonically decrease if the number of edge along the path increases.

3.2 Effective Path and Optimal Path

A path from a cause vertex to an effect vertex is the optimal sequence of directed edges whose path value is the maximum among all possible sequences from the cause to the effect. We call a path *effective* if path value is is not worse than a user specified *threshold* value. Notice that we do not have to use paths whose path value are too small to be grouped into a cluster.

---

[1] In the literature of association rule mining, especially in sequential pattern mining, the left-hand side of a rule is considered as a cause (or an antecedent) event, while the right-hand side is considered as en effect (or a consequent). We inherit the terms, though we think the terms are not adequate in this analysis.

There may exist many effective paths from v_i to v_j. Let $P(v_i, v_j)$ be a set of all effective paths from v_i to v_j. We call it *optimal path* from v_i to v_j if its path value is the maximum among $P(v_i, v_j)$.

We define the maximum path value val^* from v_i to v_j by

$$val^*(v_i, v_j) = \begin{cases} max\{val(p)|p \in P(v_i, v_j)\} & \text{if } P(v_i, v_j) \neq \emptyset, \\ 0 & \text{otherwise.} \end{cases}$$

The optimal path from v_i to v_j is defined as any path $p \in P(v_i, v_j)$ with $val(p) = val^*(v_i, v_j)$. The optimal path must not have any cycles in its sequence because the path value decreases monotonically if the number of edges along the path increases.

The optimal path from the cause to the effect itself is valuable if we are interested in the cause or the effect. So, we also call it the optimized transitive association rule from the cause to the effect [15].

3.3   Optimal Subpath Property

If a path is the optimal effective path, any subpath of the optimal path is also the optimal path within the subpath.

Let $p = \langle v_1, v_2, \cdots, v_n \rangle$, be the optimal path from v_1 to v_n. For any i, j such that $1 \leq i \leq j \leq n$, let $p_{ij} = \langle v_i, \cdots, v_j \rangle$ be the subpath of the optimal path p.

Lemma 1. Any subpath p_{ij} of the optimal effective path p must be the optimal path from v_i to v_j.

Proof. Assume three paths $p_{1i} = \langle v_1, \cdots, v_i \rangle$, $p_{ij} = \langle v_i, \cdots, v_j \rangle$, and $p_{jn} = \langle v_j, \cdots, v_n \rangle$ are subpaths of the optimal path p from v_1 to v_n.

The path value of p is, by definition, $val(p) = val(p_{1i}) * val(p_{ij}) * val(p_{jn})$. Assume there is a path p'_{ij} from v_i to v_j, which is different from p_{ij}, with $val(p'_{ij}) > val(p_{ij})$. The consecutive path p_{1i}, p'_{ij}, and p_{jn} is a path from v_1 to v_n whose $val(p_{1i}) * val(p'_{ij}) * val(p_{jn})$ is greater than $val(p)$, which contradicts the assumption that p is the optimal path from v_1 to v_n.

The problem that we want to solve is to compute the optimal effective path from v_i to v_j for every pair of vertices v_i and v_j in a weighted directed graph $G = (V, E)$. And, this property can be used in dynamic programming well and effectively.

3.4   CE-Hash

The problem to find the optimal path from v_i to v_j for every pair of vertices v_i and v_j in a graph $G = (V, E)$ may require $O(|V|^2)$ space. Even though there are many web pages, the number of edges is relatively small in general. For example, the number of web pages that a surfer views in a visit is not so large even though the number of pages is huge. Therefore, we can assume $|E| \ll |V|^2$.

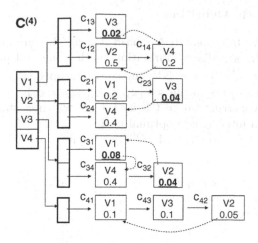

Fig. 5. Cause-Effect Hash Structure

In order to compute all-pairs optimal path, we maintain $O(|E|)$ path information. We use a hash structure as shown in Figure 5, which we call *Cause-Effect Hash* or CE-hash in short, for storing optimal paths. Let C be a CE-hash of $G = (V, E)$. Assume there are $|V|$ vertices in G and all vertices are numbered like $v_1, v_2, \cdots, v_{|V|}$. Let c_{ij} be an entry of C containing information of the optimal path from v_i to v_j, $p_{ij} = \langle v_i, \cdots, v_j \rangle$, and its path value, $val(p_{ij})$. Thanks to the Lemma 1, the elements of any prefix path of p_{ij} must be in C. Hence, c_{ij} contains a pointer to the prefix path instead of containing all sequence of the optimal path, which is space saving.

The outer hash with size $O(|V|)$ in C corresponds to the cause vertex of optimal paths. There exists an entry of a cause vertex in the outer hash if there exists an effective outgoing edge from the vertex. In each existing entry of the outer hash contains a secondary hash. If there is an entry for a cause vertex v_i, it contains optimal path information c_{ij} for all j such that there is an effective path to v_j from v_i. Size of a secondary hash is proportional to the best path value among all outgoing edges from the corresponding cause item. If v_i has an outgoing edge whose probability, for example, is large, the size of the secondary hash for the cause v_i is large.

Figure 5 is the CE-hash of the graph in Figure 4. In the figure, the entry labeled C_{32}, for example, contains the optimal path information from v_3 to v_2. The entry can be found by searching the outer hash entry for the cause v_3 and then search the secondary hash entry for v_2. C_{32} contains path value 0.04 of p_{32} and a pointer (dotted arrow in the figure) to the prefix path. If we want to know the vertices along the path, it can be computed by traversing the pointer. For example, the prefix path of C_{32} is C_{31}. Similarly, the prefix path of C_{31} is C_{34}, which has no pointer to the prefix path. Therefore, we can see the optimal path of C_{32} is $\langle v_3, v_4, v_1, v_2 \rangle$.

3.5   Optimal Path Algorithm

Let $C^{(k)}$ be the k-th intermediate CE-hash. Let $c_{ij}^{(k)}$ be an entry of $C^{(k)}$, which contains the k-th intermediate information (path value and prefix) of the path from v_i to v_j. Let $p_{ij}^{(k)}$ be the k-th intermediate optimal path for which all stopovers from v_i to v_j are in the set $\{v_1, v_2, \cdots, v_{k-1}, v_k\}$.

First of all, we construct initial CE-hash $C^{(0)}$, which contains edges in E in $G(V, E)$ as the 0-th interdediate optimal paths.

We can construct initial CE-hash, $C^{(0)}$, of Figure 4 as $C^{(0)}$ in Figure 6.

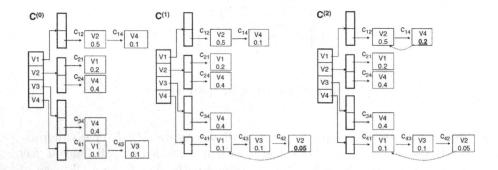

Fig. 6. Intermediate Cause-Effect Hash

In this example, we set the minimum confidence value to zero so that we can understand the algorithm easily. Since all elements of $C^{(0)}$ are edges of the original graph, the pointer for the prefix path for each element is null.

The Floyd-Warshall algorithm solves the all-pairs optimal path problem for a weighted directed graph [16,14]. The algorithm uses a dynamic programming that relies on the property that an optimal path between two vertices contains other optimal paths within it. It computes all-pairs optimal paths in $O(|V|^3)$ time.

Thanks to the property of Lemma 1, we can use a dynamic programming that maintains $C^{(k)}$ by applying the the Floyd-Warshall algorithm. Let *cutval* be the minimum confidence. Figure 7 shows the algorithm, which we call "On Memory Algorithm," for computing all-pairs optimal paths, which is suitable for cases where $|E| \ll |V|^2$.

The dynamic programming, shown in Figure 7, computes $C^{(k)}$ from $C^{(k-1)}$ for $k = 1, \cdots, |V|$. By definition, $p_{ij}^{(k)}$ is the k-th intermediate optimal path from v_i to v_j and all of the stopovers from v_i to v_j are vertices in the set $\{v_1, v_2, \cdots, v_k\}$. Consider the highest numbered vertex v_k in the set of stopovers. If v_k is not included in the stopover of $p_{ij}^{(k)}$, all of the stopovers are in the set $\{v_1, v_2, \cdots, v_{k-1}\}$. Thus, $p_{ij}^{(k)} = p_{ij}^{(k-1)}$. If v_k is included as a stopover in $p_{ij}^{(k)}$, we decompose $p_{ij}^{(k)}$ into a subpath from v_i to v_k and a subpath from v_k to v_j. By Lemma 1, the former path is the optimal path from v_i to v_k and all of the stopovers are in the set $\{v_1, v_2, \cdots, v_k\}$, which is $p_{ik}^{(k)}$. Since the optimal path must not have cycles, v_k is not included in the stopovers of $p_{ik}^{(k)}$. Thus, $p_{ik}^{(k)} = p_{ik}^{(k-1)}$. Similarly, $p_{kj}^{(k)} = p_{kj}^{(k-1)}$.

As for $c_{ij}^{(k)}$, $p_{ij}^{(k)} = p_{ij}^{(k-1)}$ if $val(p_{ij}^{(k-1)})$ is greater then $val(p_{ik}^{(k-1)}) * val(p_{kj}^{(k-1)})$. Otherwise, $p_{ij}^{(k)}$ becomes the consecutive path of $p_{ik}^{(k-1)}$ and $p_{kj}^{(k-1)}$, we use $p_{ik}^{(k-1)} + p_{kj}^{(k-1)}$ notation for such consecutive path. The path value of $p_{ij}^{(k)}$ is

$$val(p_{ij}^{(k)}) = max(\ val(p_{ij}^{(k-1)}),\ val(p_{ik}^{(k-1)}) * val(p_{kj}^{(k-1)})\ ).$$

```
0   Algorithm ONMEMORY(C^(0),cutval)
1      For each k = 1, ···, |V|
2         For each c_ik^(k-1) ∈ C^(k-1)
3            For each c_kj^(k-1) ∈ C^(k-1)
4               If val(p_ij^(k-1)) ≥ val(p_ik^(k-1)) * val(p_kj^(k-1))
5                  c_ij^(k) = (p_ij^(k-1), val(p_ij^(k-1)));
6               else
7                  If val(p_ik^(k-1)) * val(p_kj^(k-1)) > cutval
8                     c_ij^(k) = (p_ik^(k-1) + p_kj^(k-1),
9                               val(p_ik^(k-1)) * val(p_kj^(k-1)));
```

Fig. 7. On Memory Algorithm

Algorithm in Figure 7 computes $C^{(1)}$ from $C^{(0)}$ as in Figure 6. In this figure, $c_{42}^{(1)}$ has a pointer to $c_{41}^{(1)}$, as shown in the dotted arrow, which means the prefix path of $c_{41}^{(1)}$ is the path of $c_{42}^{(1)}$. By traversing the dotted arrow, the path from an effect to a cause can be computed. For example, $p_{42}^{(1)}$ of $c_{42}^{(1)}$ is $\langle v_4, v_1, v_2 \rangle$. Similarly, we can compute $C^{(2)}$ from $C^{(1)}$ as in the Figure 6. $C^{(3)}$ is the same as $C^{(2)}$ in this example. We, then, compute $C^{(4)}$ from $C^{(3)}$ as in Figure 5.

This algorithm incrementally updates $C^{(k)}$ and finally outputs $C^{(|V|)}$, which contains optimal effective path for each v_i and v_j $(i \neq j)$ in $G = (V, E)$. In the incremental update process, we can prune ineffective paths by examining path value of new candidate path $p_{ij}^{(k)} = p_{ik}^{(k-1)} + p_{kj}^{(k-1)}$ in step 7 of the algorithm.

On Memory Algorithm needs $O(|E|)$ space for maintaining $C^{(|V|)}$. Since the space for $C^{(|V|)}$ is much smaller than $O(|V|^2)$ and the CE-hash structure is space-efficient, we are allowed to store them on main memory if the graph is moderate size. However, in some large databases like log files, we can not store $C^{(|V|)}$ on main memory. In order to handle such large databases, we need to modify the algorithm.

We need to perform following preprocess before performing modified algorithm, which we call "Stepwise Algorithm." We sort the $|E|$ weighted directed edges according to their confidence values in descending order, which takes $O(|E| \log |E|)$ time. Note that a merge sort algorithm, which is a divide-and-conquer external sorting algorithm, can compute with small working space even though $|E|$ is too large to store on main memory [17,14]. After the sorting process, we divide the sorted edges into $|E|/m$ almost-equal-sized subsets, where m is the number of edges in a subset. Each subset S is numbered sequentially like $S_1, S_2, \cdots, S_{|E|/m}$ so that confidence value of any edge in S_x must be equal or larger than that of any edge in S_y such that $x > y$. When $|E|$ is so large that the $O(|E|\log|E|)$ time is not affordable, we can improve the efficiency of the preprocess by using a random sampling technique [18].

In Stepwise Algorithm, we perform On Memory Algorithm for each subset S_x from $x = 1$ to $|E|/m$ incrementally. We set adequate m so that we can perform On Memory Algorithm for every subsets within moderate working space. Now, we define tentative optimal paths that are calculated after the x-th iteration of On Memory Algorithm for $x < |E|/m$ as *candidate paths*.

Let $C'(x)$ be the CE-hash that contains candidate paths after the x-th iteration of On Memory Algorithm, and let $p'_{ij}(x)$ be a candidate path from v_i to v_j, which is stored in $C'(x)$. As an input of the Stepwise Algorithm, we prepare an empty hash $C'(0)$.

The algorithm in Figure 8 computes all candidate paths with smaller working space by iterating On Memory Algorithm and pruning candidate paths from the CE-hash. In the x-th iteration, the algorithm computes $C'(x)$ from $C'(x-1)$ as follows. In the Step 2-6, we add new edges into $C'(x-1)$ and prepare $C'(x)^{(0)}$. If there exists effective path from v_i to v_j in $C'(x-1)$ when we add a path $p_{ij} \in S_x$, we choose a path whose path value is larger. We use the $C'(x)^{(0)}$ in the x-th execution of On Memory Algorithm. The dynamic programming in the x-th execution computes $C'(x)^{(k)}$ from $C'(x)^{(k-1)}$ for each $k = 1, 2, 3, ..., |V|$, and then we have $C'(x)$.

```
0    Algorithm STEPWISE(C'(0), S_{1,⋯,|E|/m}, cutval)
1        For each x = 1, ⋯, |E|/m
2            For each p_{ij} ∈ S_x
3                If val(p_{ij}) > val(p'_{ij}(x − 1))
4                    c'_{ij}(x)^{(0)} = (p_{ij}, val(p_{ij}));
5                else
6                    c'_{ij}(x)^{(0)} = c'_{ij}(x − 1);
7            ONMEMORY(C'(x)^{(0)}, cutval);
8            For each c'_{ij}(x) ∈ C'(x)
9                If val(p'_{ij}(x)) * min_x < cutval
10                   Move c'_{ij}(x) from C'(x) to disk.
11       Move all elements of C'(x) to disk.
```

Fig. 8. Stepwise Algorithm

Candidate paths in $C'(x)$ are used for further iteration if $x < |E|/m$. In the further iteration, the confidence values of upcoming paths in $S_{x+1}, \cdots, S_{|E|/m}$ must be equal or smaller than the minimum path value in subset S_x. Let min_x be the minimum value in subset S_x, a candidate path, $p'(x)$, whose path value does not satisfy inequation $val(p'(x)) * min_x > cutval$ can not be a subpath of any candidate path in the further iteration. Thanks to the property, we can move such candidate paths that can not be a subpath of any candidate path in the further iteration from $C'(x)$ to the disk as in the Step 8-10 of the algorithm in Figure 8. Since some of such pruned candidate paths may be the final optimal path, we need to store in the disk.

After completing all the stepwise execution for S_x for $x = 1, \cdots, |E|/m$, all remaining elements of the CE-hash, i.e., $C'(x)$ are also moved into the disk. Since the output of Stepwise Algorithm contains paths that are not optimal, we have to remove such non-optimal paths.

In order to find optimal paths in the disk, we sort all the candidate paths to find optimal paths. This postprocess can work with small working space by using an external sorting algorithm like merge sort.

4 Spatial Knowledge Extraction

A geospatial association like Table 1 is a set of two dimensional point objects augmented with concepts. We have developed several spatial data mining functions that find relationship between such objects with location.

4.1 Co-Location Pattern

Each element of the Concepts list of geospatial association is considered to be a class label of the corresponding point. In our example, records of the business database could be classified into "profitable branches" and "unprofitable branches".

The *co-location patterns* are sets of classes whose objects are spatially close to one another [19][2], according to some minimum distance threshold. Assume that point objects from three classes — **circle**, **triangle**, and **square** — are distributed on a map as in Figure 9 (left). In this example, there are four occurrences of a **circle** point situated close to a **triangle** point. Similarly, there are three occurrences of a **circle** point lying near a **square** point, and two occurrences of a **triangle** point appearing next to a **square**. Moreover, there are two occurrences in which all three kinds of points lie close to one another. The {**circle**, **triangle**} is an example of a co-location pattern whose frequency is four (within the map). Our spatial mining function enumerates all such patterns whose frequency is no less than a user specified *minimum support* value.

The function may find a pattern, for example,

({"profitable branch", "education", "sports"}, 250),

which indicates there are 250 instances (triples) consist of a "profitable branch" object, an "education" object, and a "sports" object. From the high frequency of the pattern, we may deduce a relationship among education, sports, and the profitability of the product.

---

[2] The author called them "neighboring class sets" in the original paper.

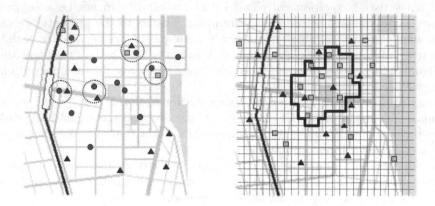

Fig. 9. Co-Location Pattern (left) and Optimal Region (right)

We performed this function for a real geospatial association for web pages of California's Silicon Valley area. We found that "non-commercial" and "font" refer to locations in close proximity to one another. Similarly, "software" and "telephone" form a co-location pattern.

4.2   Optimal Region

A geospatial association can also be utilized to compute an optimal connected pixel grid region with respect to some criterion. We first make a pixel grid of an area of interest as shown in Figure 9 (right). For each pixel, we can compute the density of a specific concept. For example, if we are interested in the concept "profitable branch," we could compute the density of point objects associated with "profitable branch." Minimization or maximization of density over a collection of grid regions can lead to significant spatial knowledge discoveries.

Although the problem of finding an optimal connected grid region is NP-hard, if we limit the grid region to be of *x-monotone* or *rectilinear* shape, we can use efficient algorithms, which is developed for computing optimized two-dimensional rules [20]. With this shape restriction, we can efficiently compute optimal x-monotone or rectilinear grid regions maximizing the density of a target compact.

In the experiment for web pages in Silicon Valley area. We found many dense regions of a concept related to "genome" and "software".

4.3   Hyperlink-Based Clustering

We computed association rules of web pages for access log files that are provided in "http://ita.ee.lbl.gov/html/contrib/NASA-HTTP.html". The files contain two month's worth of all HTTP requests to the NASA Kennedy Space Center WWW server in Florida. There are 3,404K records and 38K different pages. We compute optimal paths and its confidences, which imply distance, for all pairs of the web pages.

We made 10 clusters of web pages based on the closeness calculated by the path value of optimal path. The centroid of a dominant cluster is "news/sci.space.shuttle/archive/sci-space-shuttle-22-apr-1995-40.txt". Some members of the cluster are "shuttle/countdown/liftoff.html", "shuttle/missions/sts-73/mission-sts-73.html", and "shuttle/countdown/video/livevideo.jpeg". Since most of the web pages in those clusters are related to a liftoff of a shuttle, it makes sence that those pages are grouped into a cluster. Table 3 shows the statistics of the 10 clusters, which we computed from the access log.

Table 3. Cluster Statistics

ID	#pages	percent	ID	#pages	percent
0	1477	(15%)	5	1067	(11%)
1	1009	(10%)	6	705	(7%)
2	1660	(17%)	7	960	(10%)
3	484	(5%)	8	1052	(11%)
4	601	(6%)	9	688	(7%)

5   Conclusion

We presented a method for extracting spatial knowledge from collections of web pages that containing location information and access log of web pages.

In the geospatial association tables generated from web pages, false positive records are possible; that is, some pages may contain location information that are unrelated to the keyword concepts expressed. Such false positive records lead to incorrect spatial knowledge. We found many examples that seem to be false positive records in the experimental results so far. Further investigation is therefore needed to refine the spatial insights found from web pages.

The typical causes of false positive records are: (1), the many portal web pages containing large lists of addresses with miscellaneous contexts; and (2), the many pages quoting addresses that are not directly related to the main topic of the pages. Web pages are too numerous, too large, and too unstructured to allow the custom annotation of each page according to the use and relevance of the spatial information quoted.

One of our direction for future research is to use the hyperlink based clusters to eliminate outliers of both location and concept.

References

1. Manning, C., Schutze, H.: Foundations of Statistical Natural Language Processing. MIT Press, Cambridge (2000)
2. Buyukokkten, O., Cho, J., Garcia-Molina, H., Gravano, L., Shivakumar, N.: Exploiting geographical location information of web pages. In: Proc. of Workshop on Web Databases (WebDB) (1999)
3. McCurley, K.: Geospatial mapping and navigation of the web. In: Proc. of World Wide Web (WWW), pp. 221–229 (2001)
4. http://www.census.gov/geo/www/tiger/
5. Malassis, L., Kobayashi, M.: Statistical methods for search engines. Technical Report RT-413 (33 pages), IBM Tokyo Research Laboratory Research Report (2001)
6. Jarvis, R.A., Patrick, E.A.: Clustering using a similarity measure based on shared nearest neighbors. IEEE Transactions on Computers (11), 1025–1034 (1973)
7. Ertoz, L., Steinbach, M., Kumar, V.: Finding topics in collections of documents: A shared nearest neighbor approach. Technical Report Preprint 2001-040 (8 pages), University of Minnesota Army HPC Research Center (2001)
8. Guha, S., Rastogi, R., Shim, K.: Rock: A robust clustering algorithm for categorical attributes. Information Systems 25(5), 345–366 (2000)
9. Houle, M.E.: Navigating massive data sets via local clustering. In: Proceedings of the Ninth ACM SIGKDD International Conference on Knowledge Discovery and Data Mining, ACM, pp. 547–552. ACM Press, New York (2003)
10. Morimoto, Y., Aono, M., Houle, M.E., McCurley, K.S.: Extracting spatial knowledge from the web. In: SAINT, pp. 326–333. IEEE Computer Society Press, Los Alamitos (2003)
11. Agrawal, R., Imielinski, T., Swami, A.: Mining association rules between sets of items in large databases. In: Proc. of ACM SIGMOD Conference, pp. 207–216. ACM Press, New York (May 1993)
12. Agrawal, R., Srikant, R.: Fast algorithms for mining association rules. In: Proc. of VLDB Conference, pp. 487–499 (1994)
13. Han, J., Pei, J., Yin, Y.: Mining frequent patterns without candidate generation. In: Proceedings of the ACM SIGMOD Conference on Management of Data, pp. 1–12. ACM Press, New York (2000)
14. Cormen, T.H., Leiserson, C.E., Rivest, R.L., Stein, C.: Introduction to Algorithms. MIT Press, Cambridge (2001)
15. Morimoto, Y.: Optimized transitive association rule: Mining significant stopover between events. In: Proc. of the ACM Symposium on Applied Computing, pp. 547–548. ACM Press, New York (2005)
16. Floyd, R.W.: Shortest path. Communications of the ACM 5(6), 345 (1962)
17. Knuth, D.E.: Sorting and searching. The Art of Computer Programming 1(3) (1973)
18. Fukuda, T., Morimoto, Y., Morishita, S., Tokuyama, T.: Mining optimized association rules for numeric attributes. J. of Computer and System Sciences 58(1), 1–15 (1999)
19. Morimoto, Y.: Mining frequent neighboring class sets in spatial databases. In: Proc. of ACM SIGKDD Conference on Knowledge Discovery and Data mining (KDD), pp. 353–358. ACM Press, New York (2001)
20. Fukuda, T., Morimoto, Y., Morishita, S., Tokuyama, T.: Data mining with optimized two-dimensional association rules. ACM Trans. on Database Systems 26(2), 179–213 (2001)

A Semantic and Structural Similarity Approach to Personalize Location-Based Services

Arianna D'Ulizia, Fernando Ferri, and Patrizia Grifoni

IRPPS-CNR, via Nizza 128, 00198 Roma, Italy
{arianna.dulizia,fernando.ferri,patrizia.grifoni}@irpps.cnr.it

Abstract. In this paper the problem of refining user queries on a trader of location-based services (LBSs) is faced by using two kinds of similarity (structural and semantic). LBSs typically concentrate on providing users with information about points of interest or on the support for navigation and routing tasks. Personalization is a key feature of LBSs that improves their usability and involves a process of gathering user information during interaction with the user. This information can be used to deliver appropriate content and services, tailor made to the user's needs. The approach proposed in the paper can be applied to any kind of location-based services, it is independent from the specific domain, it takes into account spatial information and, finally, it allows the user profiles to be acquired explicitly and implicitly.

Keywords: Location-based services, structural similarity, semantic similarity, personalization.

1 Introduction

The increasingly widespread use of telecommunication technologies, including wireless communications, Internet, portable devices and geographic information systems (GIS), along with the emerging capability of terminals and/or mobile network infrastructure to pinpoint the location of personal mobile communications devices on the hearth, has lead location-based services (LBSs) to become the future trend of wireless and mobile services.

LBSs typically concentrate on providing users with information about points of interest or on the support for navigation and routing tasks. Generally, user requires LBSs by using a trader (or trading service or discovery service) that matches the service's properties desired by the user against the properties of the services registered in the trader's repository, returning to the user details for those services that match. In particular, the trader allows the user to find LBSs by formulating a query in which the user specify the features of the service he/she is looking for (e.g. "Find all the museums located in Rome"). An efficient way to improve the responsiveness of the trader and consequently the usability of LBSs is to adapt these services to the preferences, interests and device characteristics of the end user.

Personalization is thus a key aspect as it allows users' preferences to be integrated into the process of delivering any information-related content or outcome of service

S. Bhalla (Ed.): DNIS 2007, LNCS 4777, pp. 33–47, 2007.

computing. Generally, information about the end user (preferences, interests, goals, personal characteristics, etc) is referred to as *user profile*. The needed services will vary depending on the user profile. In several cases, the preferences, the interests or the goals expressed by the user involve concepts that are not directly known by the trader, that is they have no instances in the trader's repository. So the trader could not provide any result to the user request. It would therefore be useful to obtain as result also similar services by using a similarity model.

The aim of this paper is twofold and use two kinds of similarity (structural and semantic) to obtain the goal to refine user queries. Therefore, on the one hand, we focus on the problem of discovering LBSs personalized on the base of the user profile. In particular, we propose an approach based on structural similarity between user profile and service profile to filter the offers that the trader provides to the user and to make the user request more precise and selective. On the other hand, we study the problem of providing approximate answers in the case that the user request involves information about a service with no instances in the trader's repository. In this case, we propose the use of a semantic similarity model that provides the missing service to be replaced by a similar one that is present in the trader's repository.

In this paper we present a specific application on tourism, however, differently from other approaches in literature, our proposal and the implemented system can be applied to any kind of location-based services. In fact the approach is independent from the specific domain, it takes into account spatial information and the user profiles are explicitly and implicitly (during the interaction) acquired.

The paper is organized as follows. Section 2 describes the background on Location Based Services. Section 3 discusses the issue of the personalization of location-based services. In section 4 the theoretical foundations of the semantic and structural similarity models are given. Section 5 shows an application of the similarity approach both to discover services personalized on the base of the user profile and to provide approximate results for the user request. In section 6 the tests that we performed to validate the proposed approach are described. Finally, section 7 concludes the paper and discusses future work.

2 Background

The literature provides various definitions of LBSs. Some authors restrict these services to a mobile environment, as Virrantaus et al. [11] that define them as "services accessible with mobile devices through the mobile network and utilizing the ability to make use of the location of the terminals". Other authors focus on GIS functionalities at the base of LBS, like Koeppel [8] that considers LBS as "any service or application that extends spatial information processing, or GIS capabilities, to end users via the Internet and/or wireless network".

According to these two definitions, mapping/route guidance, mobile yellow pages, mobile advertisements and traffic alerts can be considered examples of LBS, as they represent services that use the location of the terminals to provide spatial information and GIS functionalities to end users through the mobile, Internet or wireless network.

To be more precise, the first two examples (mapping/route guidance, mobile yellow pages) belong to a kind of LBS, called "pull" services, in which the user makes explicitly a request of information to the service center. In this case the information is delivered from LBS to the user's device in a "pull" mode. In the other two examples (mobile advertisements and traffic alerts) the position of the user's device is utilized to estimate if he/she is a potential customer of the service and, if it is so, the information is automatically delivered to the user without need of his/her request. These services deliver information in a "push" mode and are consequently called "push" services. LBSs can be classified also in other ways, according to their functionalities [11].

Personalization is a key feature of LBSs that improves their usability. It involves a process of gathering user's information during interaction, which is then used to deliver appropriate content and services, tailor made to the user's needs [2]. Personalization can be explicit or implicit. The former implies a direct participation of the user in the adjustment of the service, that is the user clearly indicates the useful information and his/her profile. The latter does not require any user involvement but it uses learning strategies for acquiring knowledge about user's habits, interests and behaviors.

One of the research challenges for LBSs is to correctly understand the user request and consequently provide relevant results on the base of the user profile. This challenge has been faced by several authors in literature. Zein and Kermarrec [13] propose a metamodel that uses ontologies to describe the characteristics of a service. The problem of replacing a specific service, required by the user and that does not exist in the trader's repository, is solved by using the inheritance relationship that allows the user to navigate in the ontology and to obtain information from the concepts that inherit the concept of the requested service. The authors present also a model of user profile for adapting services to the needs of the user.

Ontologies are also used in [3] to capture the semantics of the user's query, of the services and of the contextual information that is relevant in the service discovery process. In particular, the authors propose an approach composed of two main phases. First of all, they match the user query with a set of available service descriptions obtaining as result a classification of the available services into three matching types (precise match, approximate match and mismatch). Secondly, they use 'concept lattices' for clustering services with similar attributes and consequently ordering these services by their relevancy for the user.

As our approach, both these two works use user and context information to personalize the provided services but they don't deal with the eventuality that the user query mismatches with the set of available services. In this case we propose a similarity model that allow to identify the most semantically and structurally similar concept, expressed in the user query but not having instances in the trader's repository.

Therefore, another research challenge for LBSs, and, more generally, in any other fields in which we need of providing answers to queries which involve unavailable information, is the evaluation of similarity between the concepts involved in the user request and concepts available in the database. Several authors addressed this objective in literature, with reference both to semantic similarity and structural similarity. In particular, Lin [9] faces the evaluation of semantic similarity between concepts in an Is-a taxonomy based on the notion of information content. More precisely, the author investigates an information-theoretical definition of similarity that is applicable as a probabilistic model and shows how this definition can be used to measure similarity in a number of different domains.

Regarding structural similarity, the ontology matching is the main field dealing with its evaluation. Formica and Missikoff [5] present a proposal for evaluating concept similarity in an ontology management system, called SymOntos. The authors consider four notions of similarity: *tentative similarity*, defined as a preliminary degree of similarity among concepts; *flat structural similarity*, that considers the concept's structure (attributes, values) and evaluates the similarity of each concept; *hierarchical structural similarity*, pertaining to hierarchically related pairs of concepts, and *concept similarity*, obtained by combining the structural and the tentative similarity.

Another paper dealing with the similarity assessment within an ontology is [1]. Like the previous work, this paper considers the instances of concepts and proposes to assess the semantic similarity among instances by taking into account both the structural comparison between two instances in terms of the classes that the instances belong to, and the instances comparison in term of their attributes and relations.

Our approach differs from the previous ones, as we start from a taxonomy of concepts built by using a lexical database (we have chosen WordNet [12] but any other lexical database for the English language can be used as well) and consequently the results are not based on a specific ontology. Moreover, this paper aims at combining the evaluation of semantic and structural similarity with the problem of discovering personalized LBSs on the base of the user profile.

3   User and Service Profiles

User profiling is the key to personalize LBSs and to provide tailored information. Generally, user profiling exploits both explicit and implicit personalization. In particular, in the initial phase the user profiling approach requires an explicit personalization as the user has to enter the necessary information (preferences, interests, goals, personal characteristics, etc) to allow the system to build the initial user profile. In the following phases of the user profiling approach an implicit personalization is performed. User and service profiles are defined in the database as set of couples attribute-values as represented in Table 1a and 1b. In user and service profiles the couples attribute-values include temporal and spatial (location) information.

Moreover, couples attribute-values in user profile are dynamically managed, in fact some couples attribute-values are specified explicitly by the user and other couples attribute-values are specified by the system depending from the history of the interaction of the user and queries formulated.

Table 1. Schema of the user and service profiles

Person	Instance of person
Attributes	Attribute 1: values Attribute 2: values .... Attribute n: values

Concept	Instance of service
Attributes	Attribute 1: values Attribute 2: values .... Attribute m: values

(a)   (b)

For example, instantiations of the user and service profiles' schema that we will also use in the following sections are shown in Table 2 and 3, respectively.

Table 2. An example of user profile

Person	Mario Rossi
Attributes	Residence: Florence Profession: researcher Age: 35 Language: {English, French} Sport: {football, basketball} Cuisine: {vegetarian, japanese} Hobby: {gardening, how-to, stamp collection}

Table 3. Service profile for a museum

Concept	Museum
Attributes	Location: Rome Ticket price: 5-10 Euro Style: {modernism, futurism, surrealism} Opening day: all except Thursday Opening hours: 08:00am – 06:00pm Tour time: 5 hours Guide language: {English, French, Spanish, German}

Moreover, the following concepts, corresponding to the available LBSs, and their attributes are stored in the trader's repository of our running example.

Museum (Location, Ticket price, Style, Opening day, Opening hour, Tour time, Guide language)
Art gallery (Location, Ticket cost, Opening day, Opening hour, Artistic style)
Art exhibition (Location, Ticket cost, Artist, Opening day, Opening hour)
Archeological site (Location, Ticket price, Historic period, Opening day, Opening hour)
Church (Name, Location, Foundation, Opening hour)

4 The Similarity Approach

The goal of a similarity model is to obtain better, more flexible matches between the information or the service expected by the user and the request received from the system. More precisely, a similarity model facilitates the identification of concepts that are close but not identical. Two main types of similarity models can be taken into account in the case of discovering personalized LBSs: semantic and structural similarity models.

4.1 Semantic Similarity

Semantic similarity is evaluated according to the *information content* approach [9]. The method's starting assumption is that concepts, which represent services available in the trader's repository, are organized according to a taxonomy. Furthermore, each concept is associated with a weight, standing for the *probability* that an instance belongs to the given concept. In the literature, probabilities are estimated according to the *frequencies* of concepts. Given a concept c, the *probability p(c)* is defined as follows:

$$p(c) = freq(c)/M$$

where *freq(c)* is the *frequency* of the concept c estimated using noun frequencies from large text corpora, such as the *Brown Corpus of American English* [6], and M is the total number of concepts. In this paper probabilities have been defined according to *SemCor* project [4], which labels subsections of Brown Corpus to senses in the *WordNet* lexicon [12], but any other lexical database for the English language can be used as well. According to *SemCor*, the total number of observed instances of nouns is *88,312*.

The association of probabilities with the concepts of the taxonomy allows us to introduce the notion of a weighted taxonomy (see Figure 1). The root of a weighted taxonomy is labeled by the most general concept, in this case Entity, whose probability is equal to 1.

In a weighted taxonomy, the information content of a concept c is defined as *-log p(c)*; that is, as the probability of a concept increases, its informativeness decreases, therefore the more abstract a concept, the lower its information content.

According to [9], the *information content similarity (ics)* between two concepts c_1 and c_2 is essentially defined as the maximum information content shared by the concepts divided by the information content of the compared concepts.

$$ics(c_1, c_2) = 2 * \frac{\log p(sharedconcept)}{\log p(c_1) + \log p(c_2)} \tag{1}$$

For instance, consider the concepts Science museum and Art museum. According to the probabilities shown in Figure 1, the following holds:

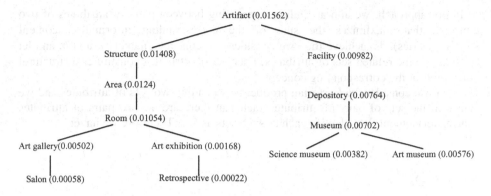

Fig. 1. Fragment of WordNet weighted taxonomy

*ics (Science museum,Art museum) = 2 log p(Museum)/(log p(Science museum) + log p(Art museum)) = 2 * 2.153 / (2.418 + 2.239) = 0.92*

By calculating the *ics* among all concepts of the weighted taxonomy, we can built the *information content similarity graph* in which nodes are labeled by concepts and edges are labeled by the *ics*. Figure 2 shows this graph for the fragment of weighted taxonomy in Figure 1.

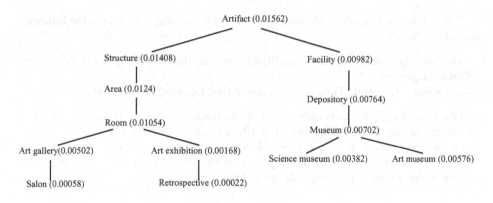

Fig. 2. Information content similarity graph

4.2 Structural Similarity

The theory from which we have started to develop the structural similarity approach is the *maximum weighted matching* problem in bipartite graphs [7]. This problem, starting from an undirected graph in which each edge has a weight associated with it, aims at finding a matching, that is a subset of the edges such that no two edges share a vertex, such that it has the maximum total weight.

In our approach, we aim at finding a matching between pairs of attributes of two concepts that maximizes the sum of the corresponding information content similarities (ics). To achieve that, we consider two concepts, namely a and b, and let c_a and c_b be the related sets of attributes. Each set of attributes constitutes a structural constraint for the corresponding concept.

Then, we consider the Cartesian product between the two sets of attributes and we select all the sets of pairs of attributes such that there are no two pairs of attributes sharing an element. We identify each of such sets as $S_{a,b}$. Therefore we have:

$$S_{a,b} = \{(a_1,b_1),....,(a_n,b_n) \in c_a \times c_b \ / a_h \neq a_k \wedge b_h \neq b_k, \forall 1 \leq h,k \leq n\}$$
$$S = \{S_{a,b}\}$$

Then, for each selected set $S_{a,b}$ of pairs we consider the sum of the information content similarity (ics) of the attributes of each pair of the set. Finally, the set of pairs of attributes that maximizes this sum is chosen, and the maximal sum corresponds to the *structural distance* between the two concepts. The maximal sum has to be divided for the maximal cardinality between c_a and c_b that we indicate as $|c_{max}|$.

Formally, the *structural distance* can be calculated in this way:

$$d = \frac{1}{|c_{max}|} \max_{S_{a,b} \in S} \sum_{\langle i,j \rangle \in S_{a,b}} ics(i,j) \tag{2}$$

For instance, we consider *Museum* and *Art Gallery* concepts that have the following sets of attributes:

Museum (Location, Ticket price, Style, Opening day, Opening hour, Tour time, Guide language)
Art gallery (Location, Ticket cost, Opening day, Opening hour, Artistic style)

For calculating the structural distance between these two concepts we must calculate the *ics* among all pairs of attributes that belong to the same weighted hierarchy. We have four pairs of identical attributes, one pair of synonyms (*Ticket_price* and *Ticket_cost*) and a pair of attributes (*Style* and *Artistic_style*) that belong to the weighted taxonomy shown in Figure 3.

Fig. 3. Fragment of WordNet weighted taxonomy for attributes

The set of pairs of attributes of *Museum* and *Art Gallery* that maximizes the sum of the related *ics* is the following:

{(*Location,Location*),(*Ticket_price,Ticket_cost*),(*Style,Artistic_style*),(*Opening day,Opening day*),(*Opening hour,Opening hour*)}

since we have:

ics(*Location,Location*) = 1
ics(*Opening day,Opening day*) = 1
ics(*Opening hour,Opening hour*) = 1
ics(*Ticket_price,Ticket_cost*) = 1
ics(*Style,Artistic_style*) = 0.96

Therefore, the structural distance between the *Museum* and *Art Gallery* concepts is:

d (*Museum,Art Gallery*) = 1/7 * [1+1+1+1+0.96] = 0.71.

At this point we can introduce the structural similarity graph that connects all concepts by the corresponding structural distances, calculated as in formula (2). In this graph each concept is depicted as a node and structural distances are the labels of the links among nodes.

For instance, if we consider the concepts and the corresponding attributes of our repository example, the structural similarity graph is depicted in Figure 4.

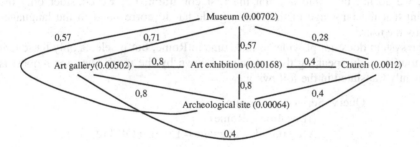

Fig. 4. The structural similarity graph

5   The Similarity Approach for Discovering Personalized LBSs

In the following, the problem of discovering services personalized on the base of user profile by using the semantic and structural similarity approaches is addressed. To achieve that, the starting assumption is that we dispose of the user and service profiles modeled as shown in Table 1.a and 1.b.

When the user requests a service, the request could be not specified by all the parameters that are needed to get a good result from the service. The user profile allows the request to be specified automatically to get better results. For example, if the user request is:

Query: Service = museum
{Location = Rome}

the query outcome could be improved by considering user's interests expressed in the user profile (see Table 2). The structural similarity model can be used in order to understand which attributes are meaningful in the user profile to better specify user request. To achieve that, we consider the set of attributes of the user profile and the set of attributes of the museum service profile given below (see also Table 2 and 3).

UserProfile (Residence, Profession, Age, Language, Sport, Cuisine, Hobby)
Museum (Location, Ticket price, Style, Opening day, Opening hour, Tour time, Guide language)

We calculate the *ics* among all pairs of attributes that belong to the same weighted hierarchy, as otherwise the ics is equal to zero. If two concepts are synonyms we assume that their *ics* is equal to 1. In our running example there is only a pair of attributes that belongs to the same weighted hierarchy (see Figure 5) and this is (*Residence, Location*). We have furthermore a pair of attributes, (*Language, Guide_language*), that contains the same concept and can be considered synonyms. So the attributes that can be matched between *UserProfile* and *Museum* are (*Residence, Location*) and (*Language, Guide_language*). This means that the user request can be improved by introducing two constraints that are *Location = Residence* and *Guide_language = Language*. As the user has already specified that "Location = Rome", we do not take into account the first constraint, but we consider only the constraint that the language of the tourist guide has to correspond to the languages known by the user.

So the system does not provide all museums in Rome, but it selects only these that have a tourist guide speaking the languages known by the user. The user request is consequently translated in the following:

Query: Service = museum
{(Location = Rome)
AND (Guide_language = English OR French)}

At this point the system can provide the user with an exact matching between user request and features of LBSs, by searching the service profiles that have "museum" as concept.

However, a large number of services, which are conceptually similar to the service requests by the user, are not considered. For example, services as art gallery or art exhibition does not match exactly the user request.

Therefore, a further improvement of the query's results can be obtained by using the semantic similarity model in order to approximate the matching between user's request and features of LBSs.

To achieve that, we consider the WordNet weighted taxonomy containing the concepts stored in our repository (see Figure 5) and we calculate the *ics* among the service required by the user (Museum) and these concepts (indicated by bold font). As shown in Figure 5, the *ics* between Museum and Art gallery, Art exhibition, Church and Archaeological site are 0.81, 0.73, 0.62 and 0.31, respectively. This means that the most semantically similar concept to "Museum" is "Art gallery".

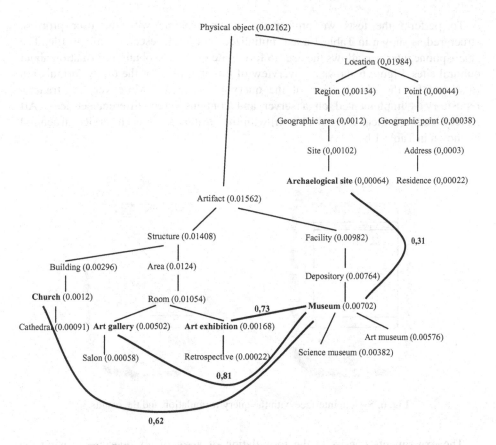

Fig. 5. Fragment of WordNet weighted taxonomy for the example

If "Museum" has no instances in the repository, it can be replaced by the concepts which are most semantically similar and do have instances in the repository. In the case that "Museum" has instances in the repository, the system can provide the user with both services that exactly match the user request and then similar services ordered according to the increasing *ics*.

The user request is consequently translated in the following:

> Query: Service = museum OR art gallery
> {(Location = Rome)
> AND (Guide_language = English OR French)}

6 Experiments and Evaluation

To validate the proposed approach, we have performed a set of experiments among the research staff of our Institute (CNR-IRPPS).

The total number of people involved in the experiments is twenty.

To perform the tests we predisposed a smart-phone with the user profiles, structured as shown in Table 1.a and initialized with each research staff profile. The smart-phone interface allows the user to formulate queries to obtain information about cultural sites. Figure 6 shows an overview of this interface for the query formulation (6.a) and for the visualization of the query results (6.b). Moreover, the trader's repository is implemented on a server and contains three different services (Art gallery, Archeological site and Church) with fifty instances for each service structured as shown in Table 1.b.

Fig. 6. System interface with the query formulation and the results

The experiments consist in the formulation of a set of six queries by using the smart-phone. Three queries concern services that haven't instances in the trader's repository (for example Museum and Art Exhibition) and consequently the semantic similarity between the concept expressed in the query and concepts in the repository is evaluated. The other three queries concern services that have instances in the trader's repository and the structural similarity between the user profile attributes and the service profile attributes is evaluated in order to understand the user attributes that can refine the query.

We adopted two different criteria for evaluating the functioning of our models. In both cases we used the concepts of recall and precision defined in [10] to evaluate the results of the model. In particular, the recall of a concept c can be defined as $|A \cap B|/|A|$ and the precision as $|A \cap B|/|B|$, where A is the set of similar concepts expected by the user and B is the set of similar concepts calculated by the model.

In the case of the semantic similarity model the evaluation of the results expected by the user is performed by asking the user to answer to a questionnaire in which we list all possible services similar to the requested services. We thus compare the set of similar services expected by the user with that produced by the model. For each query and research staff member we have calculated the recall and precision. Table 4 summarizes the results of this experiment. The three queries concerning services with no instances in the trader's repository are shown in the first column of Table 4, whereas services retrieved by the model, services expected by the user and precision and recall rates are listed in column 2, 3 and 4 respectively. Services retrieved by the model are ranked according to the *ics* to the query and only the services with *ics* greater than 0,5 are returned.

Table 4. Results of the semantic similarity model

Query	Results of the model	Expected results	Recall & Precision
Service= "Museum"	Art gallery (*ics*=0,81) Church (*ics*=0,62)	Art gallery	Recall: 100% Precision: 50%
Service= "Art exhibition"	Art gallery (*ics*=0,78) Church (*ics*=0,65) Archeological site (*ics*=0,56)	Art gallery	Recall: 100% Precision: 33%
Service= "Cathedral"	Church (*ics*=0,98) Art gallery (*ics*=0,69) Archeological site (*ics*=0,53)	Church	Recall: 100% Precision: 33%

The resultant average recall is equal to 100% while the average precision is equal to 39%. This means that all results expected by the user are returned by the trader and the trader provides also the 61% of results that the user doesn't expect.

Regarding the structural similarity model, the evaluation of results is performed by asking the user to improve the query by specifying his/her preferences about the requested service and then comparing the answers with the improvements performed by the trader. In particular, let C be the set of constraints expressed by the user and D the set of constraints found by applying the model. We can evaluate the recall and precision of the query results by using the formulas defined above. Table 5 summarizes the results of this experiment. The three queries concerning services with instances in the trader's repository are shown in the first column of Table 5. Instances of services retrieved by the model by applying the set of query constraints D are listed in column 2, whereas column 3 contains instances of services expected by the user according to the set C of query constraints. Finally, recall and precision rates are shown in column 4.

The resultant average recall is equal to 100% while the average precision is estimated equal to 70%. This demonstrates that the model is able to refine the query according to the preferences contained in the user profile even if the user expresses also other constraints that are not inferable from the user profile.

Table 5. Results of the structural similarity model

Query	Results of the model	Expected results	Recall & Precision
Service= "Art gallery"	Art gallery (instances=12) with(Location= Residence)	Art gallery (instances=6) with{(Location= Residence)AND(Artistic style=modern)}	Recall: 100% Precision: 50%
Service= "Church"	Church (instances=18) with(Location= Residence)	Church (instances=13) with{(Location= Residence)AND(Foundation <=1600 ac}	Recall: 100% Precision: 72%
Service= "Archeological site"	Archeological site (instances=14) with(Location= Residence)	Archeological site (instances=12) with{(Location= Residence)AND(Historic period<=1000 ac}	Recall: 100% Precision: 86%

7 Conclusions and Future Work

Combining location-awareness with user profiling opens the way for new possibilities in LBSs.

This paper proposes an approach based on semantic and structural similarity and devoted both to discover LBSs personalized on the base of the user profile and to provide approximate answers if a concept misses in the trader's repository.

To achieve the first goal a structural similarity model has been used to match attributes in the user profile with those of the service profiles in order to understand which attributes of the user profile are meaningful to better specify the user request. The second goal has been faced by using a semantic similarity model that allows the missing concept to be replaced by the most similar one that is present in the trader's repository.

The results of the tests we have performed demonstrate the effectiveness of the semantic similarity model. An improvement of the structural similarity model can be achieved by considering also similarities between instances of attributes and by refining the user profile structure.

References

1. Albertoni, R., De Martino, M.: Semantic Similarity of Ontology Instances Tailored on the Application Context. In: Meersman, R., Tari, Z. (eds.) OTM 2006. LNCS, vol. 4275, pp. 1020–1038. Springer, Heidelberg (2006)
2. Bonett M.: Personalization of Web Services: Opportunities and Challenges, ARIADNE, 28, (2001)

3. Broens, T., Pokraev, S., Sinderen, M.V., Koolwaaij, J., Costa, P.D.: Context-Aware, Ontology-Based Service Discovery. In: Markopoulos, P., Eggen, B., Aarts, E., Crowley, J.L. (eds.) EUSAI 2004. LNCS, vol. 3295, pp. 72–83. Springer, Heidelberg (2004)
4. Fellbaum, C.: A Semantic Network of English: the Mother of all WordNets. Computers and the Humanities 32, 209–220 (1998)
5. Formica, A., Missikoff, M.: Concept Similarity in SymOntos: an Enterprise Ontology Management Tool. The Computer Journal 45(6), 583–594 (2002)
6. Francis, W.N., Kucera, H.: Frequency Analysis of English Usage. Houghton Mifflin, Boston (1982)
7. Galil, Z.: Efficient algorithms for finding maximum matching in graphs. ACM Computing Surveys 18, 23–38 (1986)
8. Koeppel, I.: What are location services? From a GIS Perspective. ESRI white paper (2000)
9. Lin, D.: An Information-Theoretic Definition of Similarity. In: ICML 1998. Proceedings of the 15th International Conference on Machine Learning, Madison, WI, pp. 296–304 (1998)
10. Rodriguez, M.A., Egenhofer, M.J.: Determining semantic similarity among entity classes from different ontologies. IEEE Trans.Knowl.Data Eng. 15(2), 442–456 (2003)
11. Virrantaus, K., Markkula, J., Garmash, A., Terziyan, V., Veijalainen, J., Katanosov, A., Tirri, H.: Developing GIS-supported location-based services. In: Proceedings of the Second International Conference on Web Information Systems Engineering, Kyoto, Japan, pp. 66–75 (2001)
12. WordNet 2.1: A lexical database for the English language (2005), http://www.cogsci.princeton.edu/cgi-bin/webwn
13. Zein, O.K., Kermarrec, Y.: An approach for describing/discovering services and for adapting them to the needs of users in distributed systems. American Association for Artificial Intelligence (2004)

Exploring Structural and Dimensional Similarities Within—Lunar Nomenclature System Using Query Interfaces

Junya Terazono, Tomoko Izumita, Noriaki Asada, Hirohide Demura,
and Naru Hirata

The University of Aizu,
Tsuruga, Ikki-machi, Aizu-Wakamatsu,
Fukushima 965-8580, Japan
{terazono,izumita,asada,demura,naru}@u-aizu.ac.jp

Abstract. Scientists and researchers often search for lunar features from lunar names within geographic data from Moon. Currently, there are few facilities available for above search. Therefore, we propose starting construction of lunar nomenclature search system based on high-level query language interfaces, such as query by example (QBE) and Query-By-Object approach. Using this method, we can designate specific points of the moon using nearby name by lunar feature names. Currently, we are focusing on the region including Copernicus crater.

Keywords: lunar feature, query interface, GIS, moon.

1 Introduction

In recent years, lunar exploration has became an active area of study in the planetary science fields. The proposals of the exploration programs are proposed worldwide. Currently, many countries including United States, China, Russia, India and Japan have official lunar exploration programs. Especially, Japan has scheduled the launch of lunar explorer "KAGUYA" (SELENE) in August 2007. KAGUYA will orbit around the moon for approximately one year, and surveys composition of minerals, contents of elements of the moon, and tracks lunar gravity which is a key to clarify internal structure of the moon. There are some plans of exploration after KAGUYA in Japan, including landing, roving and sample return.

This worldwide rise of lunar exploration calls for the infrastructure of lunar exploration for the scientific domain. There are demands for GIS-like system for the moon among planetary scientists as scientific data size and its utilization are increasing. Web-based systems are particularly popular as similar systems are widely used in the terrestrial science.

There are several attempts for constructing Web-based planetary GIS. USGS (US Geological Survey) is constructing USGS Map-A-Planet [1] system which is targeted for the scientific use (Figure 1).

S. Bhalla (Ed.): DNIS 2007, LNCS 4777, pp. 48–53, 2007.
© Springer-Verlag Berlin Heidelberg 2007

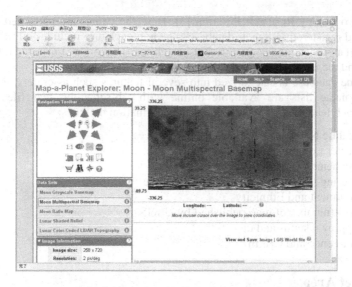

Fig. 1. The screenshot of the Map-a-Planet website [1]

USGS has also a planetary Web-GIS platform called PIGWAD [2]. This is Java-driven system. Search for several planetary bodies (Mars, Moon, Venus, Jovian satellites) are provided. Users can click lunar map on the web screen and locate nomenclature, geological maps, and other major features.

However, these planetary (or lunar) GIS have no function which can search from the lunar feature names. If one need to search from the name, he or she have to determine the point for search from the map, and then locate the feature. More sophisticated interface which can use names as a key is strongly demanded not only from the researcher but educational persons.

On the other hand, the adaptation or adding such function for existing lunar and planetary GIS needs some knowledge on computer language or database access language (SQL). As every scientist, particularly planetary scientist, are not the expert of SQL or database, it is very difficult requirement.

We propose to prepare a proper and easy-to-use interface to query. There are two approach to reduce users' query difficulties, QBE (Query-By-Example) [3] and QBO (Query-By-Object) [4]. QBE is used for table-based query and thus useful and user-friendly for query that the all conditions are prepared. However, it is inadequate to use for more complicated queries, such as "X near Y" and "X excluding Y". QBO approach is useful for the query of these types.

In our case, QBO is the most suitable approach because user will narrow down their query in several conditions. For example, "All craters near Rima Gay-Lussac" (Rima means valley) is very common query. QBO is the best selection for our objective.

2 Lunar Nomenclature

Every major feature in the planetary system has its own names. These feature names are acknowledged by IAU (International Astronomical Union) and managed by

USGS Astrogeology Research Program [5]. Here the names are classified into their features and registered into their database. The nomenclature includes all planets such as the Earth and Mars and some asteroids, satellites. The moon is also included in this database and the name list can be obtained via the web [5].

The name database is categorized into several features such as craters, valleys and seas. The name lists are available online with tab-separated text files [5]. For example, such feature such as the crater, the information includes:

- Feature Name
- Center Longitude and Latitude
- Diameter
- Starting / Ending Longitude and Latitude
- Continent and Ethnicity
- Map (number of map published by USGS)
- Approval status and date
- Reference and Origin

3   Target Area

We selected the target field for the area including Copernicus crater (Figure 2). The area is approximately from 0 to 14 N in latitude and from 16 to 27 W in latitude.

Copenicus crater is well-known for lunar observers and scientists for its prominent appearance. And it is also known that the crater is geologically young, created approximately one billion years ago. The lunar geological stratigraphic system has been determined first at here [6].

In this target area, there are six major (named) craters, 47 small craters associated with major craters, one Rima (valley) feature, one Montes (mountain) feature and one Mare (sea).

For example, Copernicus crater, the largest crater in this area, is located in 9.7 N and 20.1 W and has a 93 kilometers in diameter. As most craters can be approximated as the circle, we are attempting to register these values in the database. Database system is PostgreSQL (version 8.2) running on Solaris 8. No spatial extension such as PostGIS is used.

Currently, our registration is limited to the named crater feature as these are very simple and easy to start the implementation. We will extend the registration into other feature such as small (unnamed) craters and other features.

4   Database Structure

Currently, we are constructing basic database to evaluate performance and data structure. As lunar features has various type of characteristics, we are now starting from the construction using craters. As the crater is very simple compared to other structures such as seas and valleys, it is appropriate as a first step.

Crater is circular land features common in the planetary bodies including the moon. It is fundamentally characterized by its diameter and center coordinates (latitude and longitude). And, to search from the webpage, some corresponding information is required.

Current structure of table of the crater is shown in Table 1. The data structure is composed of two tables. One table is a name (Latin, English, and Japanese) and another table is the characteristics of a crater which are described in the USGS website [5].

Table 1. The database structure of the lunar crater

Name

varchar[n]	Latin Name
varchar[n]	English Name
varchar[n]	Japanese Name

Crater

varchar[n]	Latin Name
varchar[n]	Feature type
FLOAT4	Center Latitude
FLOAT4	Center Longitude
INT4	Diameter
FLOAT4	Starting Latitude
FLOAT4	Starting Longitude
FLOAT4	Ending Latitude
FLOAT4	Ending Longitude
varchar[n]	Continent
varchar[n]	Ethnicity
varchar[n]	Map
varchar[n]	Quadrangle
INT4	Approval Status
INT4	Approval Date
varchar[n] or TEXT	References
varchar[n] or TEXT	Origin

Here, all data structures were defined along the PostgreSQL-compliant data type for portability. The Latin Name will be the key for the query, as several blurring may exist in English name and Japanese name.

The keys Map and Quadrangle means corresponding geological map published by USGS. These values are important for geological research.

Basically, when we define the region of the crater using this table structure, the three values (diameter, center latitude, center longitude) is required.

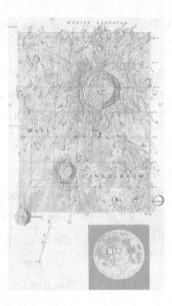

Fig. 2. The target area cited from reference [7]. The Copenicus crater is located in the center (the largest crater in this map). The relative location in the near side of the moon is shown in the bottom of the figure. The scale (full scale is 100 kilometers) is also shown in the left bottom.

5   Problems and Future Perspective

Registration of crater feature is relatively easy because they have only two parameters, the location and diameter basically. Other collateral information can be separated into another table in the database. This has another effect to make better performance for query as the large table makes query slower. As there are so many features on the lunar surface, database performance will be a issue on the system. We may need some re-designing to educe more performance by reclassification of database tables.

Other issue when we construct the database is the approximation of other features. One problem is a small crater. They are too small to appear in the planetary name system to find their exact diameter and location. We need to extract this information by hand from the lunar geological (or topological) maps.   Another is the approximation of other features. Most features such as mare and montes can be applied by polygon approximation. Polygon approximation is commonly used in GIS and it is not a problem. However, in lunar nomenclature, these features have no exact border. Therefore, we need some criteria to distinguish the feature. The altitude data can be used for these classifications, and geological data may be also useful if available. We are now estimating the possibility to classify these features and register into the database.

Moon is smaller than earth, the one fourth of the earth in diameter, the curvature of the surface will become the problem when we implement GIS system using this data structure. The coverage of crater is appropriate to measure the polar coordinate using (r, θ) from the center of the moon. We are now evaluating whether this approach is applicable not only the large craters but small craters.

6 Summary

As a basic step for future lunar GIS, we propose to construct lunar nomenclature data search system. This system is based on the query by any attribute of lunar feature and, using QBE and other high-level query language approaches such as QBO, user-oriented search. Currently (as of July 2007), implementation is undergoing for a test field (Copernicus crater and vicinity). It will be extended to include the implementation of global data in the future.

References

1. USGS Map-a-Planet. http://www.map-a-planet.org/
2. USGS PIGWAD. http://webgis.wr.usgs.gov/
3. Rahman, S.A., Bhalla, S.: Spatial QBE interface for Web GIS. Proc. 2005 Fifth International Conference on Computer and Information Technology (2005)
4. Rahman, S.A., Bhalla, S., Hashimoto, T.: Query-By-Object interface for Dynamic Access and Information Requirement Elicitation. In: Proc. Fourth International Conference on Mobile Business (2005)
5. Gazetteer of Planetary Nomenclature, USGS Astrogeology Research Group. http://planetarynames.wr.usgs.gov/
6. Shoemaker, E.M., Hackman, R.J.: Stratigraphic basis for a lunar time scale. In: Kopel, Z. (ed.) The Moon, pp. 289–300. Academic Press, London (1962)
7. Rükl, A.: Astronomy Atlas of The Moon, Kalmbach books (1990)

Semantic Tracking in Peer-to-Peer Topic Maps Management

Asanee Kawtrakul[1], Chaiyakorn Yingsaeree[1], and Frederic Andres[2]

[1] NAiST Research Laboratory, Kasetsart University, Bangkok, Thailand
asanee.kawtrakul@nectec.or.th, chaikorn@gmail.com
[2] National Institute of Informatics, Tokyo, Japan
andres@nii.ac.jp

Abstract. This paper presents a collaborative semantic tracking framework based on topic maps which aims to integrate and organize the data/information resources that spread throughout the Internet in the manner that makes them useful for tracking events such as natural disaster, and disease dispersion. We present the architecture we defined in order to support highly relevant semantic management and to provide adaptive services such as statistical information extraction technique for document summarization. In addition, this paper also carries out a case study on disease dispersion domain using the proposed framework.

1 Introduction

This paper gives an overview of a generic architecture we are currently building as part of the Semantic Tracking project in cooperation with the FAO AOS project [32]. Initiated by FAO and KU, the Semantic Tracking project aims at providing a wide area collaborative semantic tracking portal for monitoring important events related to agriculture and environment, such as disease dispersion, flooding, or dryness.

This implies to deal with any kind of multilingual internet news and other online articles (e.g. wiki-like knowledge and web logs); it describes the world around us rapidly by talking about the update events, states of affairs, knowledge, people and experts who participate in.

Therefore, the Semantic Tracking project targets to provide adaptive services to large group of users (e.g. operator, decision makers), depending on all the knowledge we have about the environment (users themselves, communities they are involved in, and device he's using). This vision requires defining an advanced model for the classification, the evaluation, and the distribution of multilingual semantic *resources*. Our approach fully relies on state of the art knowledge management strategies. We define a global collaborative architecture that allows us to handle resources from the gathering to the dissemination.

However, sources of these data are scattered across several locations and web sites with heterogeneous formats that offer a large volume of unstructured information. Moreover, the needed knowledge was too difficult to find since the traditional search

S. Bhalla (Ed.): DNIS 2007, LNCS 4777, pp. 54–69, 2007.
© Springer-Verlag Berlin Heidelberg 2007

engines return ranked retrieval lists that offer little or no information on semantic relationships among those scattered information, and, even if it was found, the located information often overload since there was no content digestion. Accordingly, the automatic extraction of information expressions, especially the spatial and temporal information of the events, in natural language text with question answering system has become more obvious as a system that strives for moving beyond information retrieval and simple database query.

However, one major problem that needs to be solved is the recognition of events which attempts to capture the richness of event-related information with their temporal and spatial information from unstructured text. Various advanced technologies including name entities recognition and related information extraction, which need natural language processing techniques, and other information technologies, such as geomedia processing, are utilized part of emerging methodologies for information extraction and aggregation with problem-solving solutions (e.g. "know-how" from livestock experts from countries with experiences in handling bird flu situation). Furthermore, ontological topic maps are used for organizing related knowledge.

In this paper, we present our proposal aiming to integrate and organize the data/information resources dispersed across web resources in a manner that makes them useful for tracking events such as natural disaster, and disease dispersion. The remainder of this paper is structured as follows: Section 2 describes the key issues in information tracking as nontrivial problems; In Section 3 we introduce the framework architecture and its related many-sorted algebra. Section 4 gives more details of the system process regarding the information extraction module. Section 5 discusses the personalized services (e.g. knowledge retrieval service and visualization service) provided for collaborative environments. Finally, in Section 6, we conclude and give some forthcoming issues.

2 Key Issues of Information Tracking

Collecting and extracting data from the Internet have two main nontrivial problems: overload and scattered information, and salient information and semantic extraction from unstructured text. Many experiences [20, 21, and 35] have been done regarding event tracking or special areas or areas related to events monitoring (e.g. the best practice for governments to handle bird flu situation), the collection of important events and their related information (e.g. virus transmission from one area to other locations and from livestock to humans).

Firstly, target data used for semantic extraction are organized and processed to convey understanding, experience, accumulated learning, and expertise. However, sources of these data are scattered across several locations and websites with heterogeneous formats. For example, the information about Bird Flu consisting of policy for controlling the events, disease infection management, and outbreak situation may appear in different websites as shown in Fig. 1.

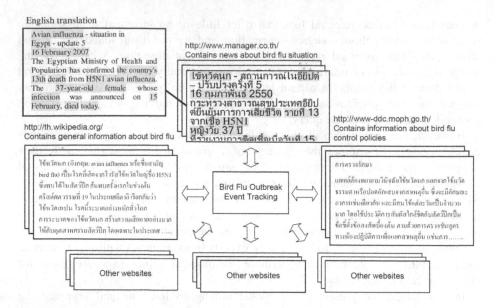

Fig. 1. Information required for tracking bird flu outbreak over Internet

Consequently, collecting required information from scattered resources is very difficult since the semantic relations among those resources are not directly stated. Although it is possible to gather those information, the collected information often overload since there is no content digestion. Accordingly, solving those problems manually is impossible. It will consume a lot of time and CPU power. The system that can collect, extract and organize those information according to contextual dimensions automatically, is our research goal for knowledge construction and organization.

Secondly, only salient information must be extracted to reduce time consumption for users to consume the information. In many case, most of salient information (e.g. time of the event, location that event occurred, the detail of the event) are left implicitly in the texts.

For example: in the text in Fig. 1, the time expression "15 February" mentioned only "date and month" of the bird flu event but did not mention the 'year'. The patient and her condition (i.e. '37-year-old female', and 'died') was caused by bird flu which is written in the text as 'Avian influenza' and 'H5N1 avian influenza'. Accordingly, the essential component of computational model for event information capturing is the recognition of interested entities including time expression, such as 'yesterday', 'last Monday', and 'two days before', which becomes an important part in the development of more robust intelligent information system for event tracking.

Information extraction in traditional way processes a set of related entities in the format of slot and filler, but the description of information in Thai text such as locations, patient's condition, and time expressions can not be limited to a set of related entities because of the problems of using zero anaphora [17]. Moreover, to activate the frame for filling the information, name entity classification must be robust as it has been shown in [5].

3 A P2P Framework of Information Extraction for Event Tracking

In this section, we give an overview of the modeling we are providing. Preliminary parts of our framework have been previously introduced to the Natural Language Processing and Database community [15]. In the following, we present our P2P framework and related many-sorted algebra modeling.

3.1 Our Framework's Semantic Tracking Algebra

Let us introduce our design approach of an ontological topic map for event semantic tracking. The ontological topic map [22] helps to establish a standardized, formally and coherently defined classification regarding event tracking.

One of our current focus and challenges has to develop a comprehensive ontology, which defines the terminology set, data structure and operations regarding semantic tracking and monitoring in the field of agriculture and environment.

The Semantic Tracking Algebra is a formal and executable instantiation of the resulting event tracking ontology. Our algebra has to achieve two tasks: (1) first, it serves as a knowledge layer between the users (e.g. agriculture experts) and the system administration (e.g. IT scientists and researchers).

Let us remind the notion of many sorted algebra [13]. Such algebra consists of several sets of values and a set of operations (functions) between these sets. Our Semantic Tracking Algebra is a domain-specific many-sorted algebra incorporating a type system for agriculture and environment data. It consists of two sets of symbols called sorts (e.g. topic, RSS postings) and operators (e.g. tm_transcribe, semantic_similarity); the function sections constitute the signature of the algebra. Its sorts, operators, sets, and functions are derived from our agriculture ontology. Second order signature [14] is based on two coupled many-sorted signatures where the top-level signature provides kinds (set of types) as sorts (e.g. DATA, RESOURCE, SEMANTIC_DATA) and type constructors as operators (e.g. set).

To illustrate the approach, we assume the following simplified many-sorted algebra:

Kinds DATA, RESOURCE, SEMANTIC_DATA, TOPIC_MAPS, SET
Type constructor

-> DATA	topic
-> RESOURCE	rss, htm // resource document type
-> SEMANTIC_DATA	lsi_sm, rss_sm, htm_sm // Semantic and metadata vectors
-> TM	tm(topic maps)
TM ->SET	set

Unary operations

$\forall$ resource in RESOURCE, resource $\rightarrow$ sm: SEMANTIC_DATA,tm **tm_transcribe**
$\forall$ sm in SEMANTIC_DATA sm $\rightarrow$ set(tm)   **semantic_similarity**

The notion sm:SEMANTIC_DATA is to be read as "some type sm in SEMANTIC_DATA," and means there is a typing mapping associated with the tm_transcribe operator. Each operator determines the result type within the kind of SEMANTIC_DATA, depending on the given operand resource types.

Binary operations

$\forall$ tm in TOPIC_MAPS, (tm)$^+$ $\rightarrow$ tm   **topicmaps_merging**

$\forall$ sm in SEMANTIC_DATA , $\forall$ tm in TOPIC_MAPS,
$\qquad$ sm,tm tm $\rightarrow$ tm   **semantic_merging**

$\forall$ topic in DATA, $\forall$ tm in TOPIC_MAPS,
$\quad$ set(tm) x (topic $\rightarrow$ bool) $\rightarrow$ set(tm) **select**

The semantic merging operation takes two or more operands that are all topic maps values. The select takes an operand type set (tm) and a predicate of type topic and returns a subset of the operand set fulfilling the predicate. From the implementation of view, the resource algebra is an extensible library package providing a collection of resource data types and operations for agriculture and environment resource computation. The major research challenge will be the formalization and the standardization of cultural resource data types and semantic operations through ISO standardization.

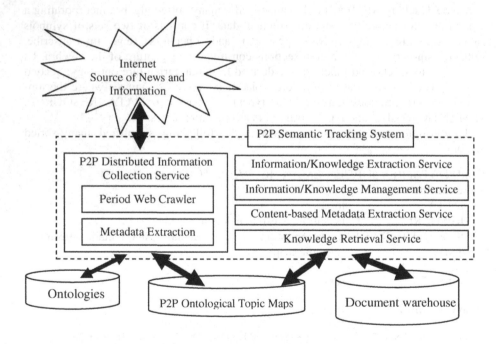

Fig 2. Architecture of our proposed P2P Semantic Tracking System

3.2   Framework of P2P Semantics Tracking System

As shown in Fig. 2, the proposed framework consists of six main services. The detail of each service is outlined as followed:

Information and Knowledge Extraction and Management Services: To generate useful knowledge from collected documents, two important modules, information extraction and knowledge extraction, are utilized. Ontological topic maps and domain-related ontologies defined in OWL [9] are used as a knowledge base to facilitate the knowledge construction and storage process as it has been shown in Garsho's review [11]. The standard ISO/IEC Topic Maps (ISO 13250) facilitates the knowledge interoperability and composition. The information extraction and integration module is responsible for summarizing the document into a predefined frame-like/structured database, such as <disease name, dispersion location and time, status of patient's condition>. The knowledge extraction and generalization is responsible for extracting useful knowledge (e.g. general symptom of disease) from collected document. Latent semantic analysis will be applied to find new knowledge or relationships that are not explicitly stored in the knowledge repository. Language engineering and knowledge engineering techniques are key methods to build the target platform. For language engineering, word segmentation [31], named entity recognition [6], shallow parsing [28], shallow anaphora resolution and discourse processing [6, 7, and 12] have been used. For knowledge engineering, ontological engineering, task-oriented ontology, ontology maintenance [16] and Topic Maps [5] model have been applied.

Distributed Information Collection Management Service: The information, both unstructured and semi-structured documents are gathered from many sources. Periodic web crawler and HTML Parser [33] are used to collect and organize related information. The domain specific parser [17] is used to extract and generate meta-data (e.g. title, author, and date) for interoperability between disparate and distributed information. The output of this stage is stored in the document warehouse.

Content-based Metadata Extraction Service: To organize the information scattered at several locations and websites, Textual Semantics Extraction [27] is used to create a semantic metadata for each document stored in the document warehouse. Guided by domain-based ontologies associated to reasoning processes [23] and Ontological topic map, the extraction process can be taught of as a process for assigning a topic to considered documents or extracting contextual metadata from documents following Xiao's approach [36].

Knowledge Retrieval Service: This module is responsible for creating response to users' query. The query processing based on TMQL-like requests is used to interact with the Knowledge management layer.

Knowledge Visualization: After obtaining all required information from the previous module, the last step is to provide the means to help users consume that information in an efficient way. To do this, many visualization functions is provided. For example, Spatial Visualization can be used to visualize the information extracted from the

Information Extraction module and Graph-based Visualization can be used to display hierarchal categorization in the topic maps in an interactive way [27].

Due to page limitation, this paper will focus in only Information Extraction module, Knowledge Retrieval Service module and Knowledge Visualization Service module.

4 Information Extractions

The proposed model for extracting information from unstructured documents consists of three main components, namely Entity Recognition, Relation Extraction, and Output Generation, as illustrate in Fig. 3. The Entity Recognition module is responsible for locating and classifying atomic elements in the text into predefined categories such as the names of diseases, locations, and expressions of times. The Relation Extraction module is responsible for recognizing the relations between entities recognized by the Entity Recognition module. The output of this step is a graph representing relations among entities where a node in the graph represents an entity and the link between nodes represents the relationship of two entities. The Output Generation module is responsible for generating the n-tuple representing extracted information from the relation graph. The details of each module are described as followed.

4.1 Entity Recognition

To recognize an entity in the text, the proposed system utilizes the work of H. Chanlekha and A. Kawtrakul [6] that extracts entity using maximum entropy [2], heuristic information and dictionary. The extraction process consists of three steps. Firstly, the candidates of entity boundary are generated by using heuristic rules, dictionary, and statistic of word co-occurrence. Secondly, each generated candidate is then tested against the probability distribution modeled by using maximum entropy. The features used to model the probability distribution can be classified into four categories: Word Features, Lexical Features, Dictionary Features, and Blank Features as described in [7]. Finally, the undiscovered entity is extracted by matching the extracted entity against the rest of the document. The experiment with 135,000 words corpus, 110,000 words for training and 25,000 words for testing, shown that the precision, recall and f-score of the proposed method are 87.60%, 87.80%, 87.70% respectively.

4.2 Relation Extraction

To extract the relation amongst the extracted entities, the proposed system formulates the relation extraction problem as a classification problem. Each pair of extracted entity is tested against the probability distribution modeled by using maximum entropy to determine whether they are related or not. If they are related, the system will create an edge between the nodes representing those entities. The features used to model the probability distribution are solely based on the surface form of the word surrounding the considered entities; specifically, we use the word n-gram and the location relative to considered entities as features. The surrounding context is classified into three disjointed zone: prefix, infix, and suffix. The infix is further segmented into smaller chunks by limiting the number of words in each chunk. For example, to recognize the relation between VICTIM and CONDITION in the sentence "The [VICTIM] whose [CONDITION] was announced on ....", the prefix, infix and suffix in this context is 'the', 'whose', and 'was announced on ....' respectively.

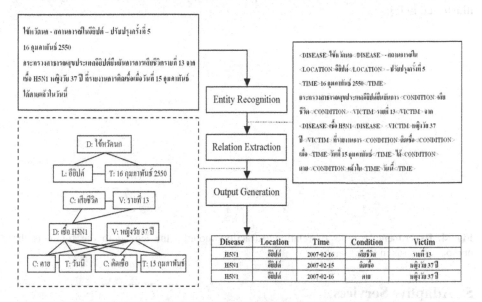

Fig. 3. Overview of the information extraction module

To determine and to assess the "best" n-gram parameter and number of words in each chunk of the system, we conduct the experiment with 257 documents, 232 documents for training and 25 documents for testing. We vary the n-gram parameter from 1 to 7 and set the number of words in each chunk as 3, 5, 7, 10, 13, and 15. The result is illustrated in Fig. 4. The evident shows that f-score is maximum when n-gram is 4 and number of words in each chunk is 7. The precision, recall and f-score at the maximum f-score are 58.59%, 32.68% and 41.96% respectively.

4.3 Output Generation

After obtaining a graph representing relations between extracted entities, the final step of information extraction is to transform the relation graph into the n-tuple representing extracted information. Heuristic information is employed to guide the transformation process. For example, to extract the information about disease outbreak (i.e. disease name, time, location, condition, and victim), the transformation process will starts by analyzing the entity of the type condition, since each n-tuple can contain only one piece of information about the condition. It then travels the graph to obtain all entities that are related to considered condition entity. After obtaining all related entities, the output n-tuple is generated by filtering all related entities using constrain imposed by the property of each slot. If the slot can contains only one entity, the entity that has the maximum probability will be chosen to fill the slot. In general, if the slot can contain up to n entities, the top-n entities will be selected. In addition, if there is no entity to fill the required slot, the mode (most frequent) of the entity of that slot will be used to fill instead. The time expression normalization using rule-based system and synonym resolution using ontology are also performed in this step to generalize the output n-tuple. The example of the input and output of the system are illustrated in Fig. 3.

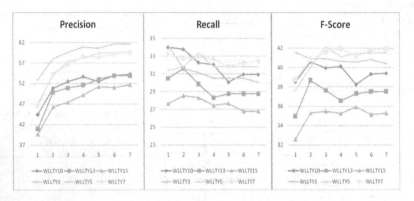

Fig. 4. Experimental results of relation extraction module (the legend in the figure is the number of words in each chunk)

5 Adaptive Services

Distributed adaptive and automated services require exploiting all the environmental knowledge stored in ontological topic maps that is available about the elements involved in the processes [24]. An important category of this knowledge is related to devices' states; indeed, knowing if a device is on, in sleep mode, off, if its battery still has autonomy of five minutes or four days, or if it has a wired or wireless connection, etc. helps adapting services that can be delivered to this device. For each device, we consider a state control that is part of the device's *profile*. Then, of course, we use the information contained in communities' and users' *profiles*. Personalized services rely on user-related contexts such as localization, birth date, languages abilities, professional activities, hobbies, communities' involvement, etc. that give clues to the system about users' expectations and abilities. In the remainder of this section, we present the two main adaptive services: the knowledge service and the knowledge visualization service based on our model.

5.1 Knowledge Retrieval Service

The Knowledge Retrieval Service module is responsible for interacting with the topic maps repository to generate answers to user's TMQL-like queries [33]. The framework currently supports three types of query. The detail of each query type is summarized in Table 1.

Table 1. Detail of four query types supported by the Knowledge Service module

Query type	Description
Query by Object	A mechanism employed when users know the object but want to acquire more information/knowledge about it. The query example is as following:
	SELECT qa_who,lblWho
	FROM {qa_who} ne:text {lblWho}
	WHERE (lblWho like "*เด็ก*")
	USING NAMESPACE ne = <http://naist.cpe.ku.ac.th/EventTracking#>
Query by Relationship	A mechanism employed when users know the relation label. For example, user can access knowledge repository such as "ne:atLocation". The query example is as following:
	SELECT disease,lblDisease,location,lblLocation
	FROM {disease} ne:text {lblDisease},
	* {disease} ne:atLocation {location},*
	* {location} ne:text {lblLocation}*
	WHERE (lblDisease like "*หวัดนก*") AND (lblLocation like "*เวียด*")
	USING NAMESPACE ne = <http://naist.cpe.ku.ac.th/EventTracking#>
Query by Instance	A mechanism employed when users know the instance or some parts of instance label that can access to knowledge repository such as "ne:Disease-3-10". The query example is as following:
	SELECT disease,lblDisease,location,lblLocation
	FROM {disease} ne:text {lblDisease},
	* {disease} ne:atLocation {location},*
	* {location} ne:text {lblLocation}*
	WHERE (lblDisease like "*หวัดนก*") AND (lblLocation like "*เวียด*")
	USING NAMESPACE ne = <http://naist.cpe.ku.ac.th/EventTracking#

5.2 Knowledge Visualization Service

The Knowledge Visualization Service is responsible for representing the extracted information and knowledge in an efficient way. Users require to access to concise organization of the knowledge. Schneiderman in [12] pointed that "the visual information-seeking mantra is overview first, zoom and filter, then details on-demand". In order to locate relevant information quickly and explore the semantic-related structure, our flexible approach regarding two ways of visualizations (spatial-based or graph-based visualization) is described in the following.

Spatial Visualization

The spatial-based visualization functions help users to visualize the extracted information (e.g. the bird flu outbreak situation extracted in Fig. 3.) using web-based geographical information system, such as Google Earth. This kind of visualization allows the users to click on the map to get the outbreak situation of the area according to their requests. In addition, by viewing the information in the map users can see the spatial relations amongst the outbreak situations easier than without the map. One usage example of Google Earth integrated system for visualizing the extracted information about bird flu situation is shown in Fig. 5.

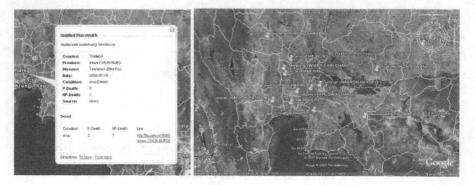

Fig. 5. Google Earth visualization for bird flu outbreak tracking

Graph-based Visualization

The graph-based visualization function is useful to show the global structure of topic maps and relations between different nodes in a 3D visual space. End-users can directly select the requested topics and related associations. The graph viewer provides a better global understanding of the content by exploring through graph nodes. The kind of intuitive visualization of topic maps allows browsing through all

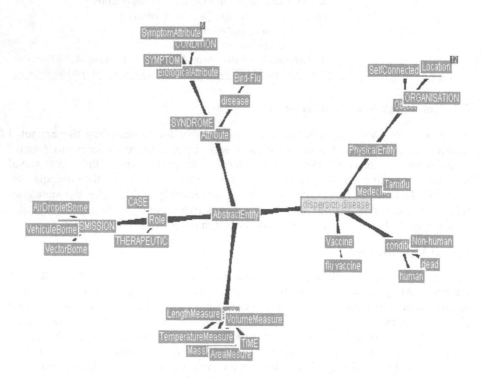

Fig. 6. Graph-based visualization of topic maps about dispersion disease

the topics and related relationships defined in the topic map as shown in Fig. 6. The graph can be moved and restructured along its topological view according to the user's need and selections.

6 Related Works

6.1 Distributed Services

We agree that distributed knowledge management has to assume two principles [4] related to the classification: (1) autonomy of classification for each knowledge management unit (such as community), and (2) coordination of these units in order to ensure a global consistency. Having a decentralized peer-to-peer knowledge management, the SWAP platform [25] is designed to enable knowledge sharing in a distributed environment. Pinto et al. provide interesting updates and changes support between peers. However, vocabularies in SWAP have to be harmonized; which implies to have some loss of knowledge consistency. But even if we share the approach of core knowledge structure that is expendable, the vocabulary, in our case, is common and fully shared by the community, so the knowledge evaluation and comparison can be more effective. Moreover, SWAP provides some kind of personalization (user interface mainly) but does not go as far as the Semantic Tracking does. From our point of view, SWAP definitely lacks environmental knowledge management that is required to perform advanced services; on the other hand, DBGlobe [26] is a service-oriented peer-to-peer system where mobile peers carrying data provide the base for services to be performed. Its knowledge structure is quite similar to our project as it is using metadata about devices, users and data within profiles; moreover, communities are also focused on one semantic concept. DBGlobe relies on AXML [3] in order to perform embedded calls to Web services within XML. Thus, it provides a very good support for performing services but does not focus on users and environments knowledge in order to offer optimized authoritarian adaptive services. Described as a P2P DBMS, AmbientDB [10] relies on the concept of Ambient Intelligence, which is very similar to our vision of adaptive services with automatic cooperation between devices and personalization. However, although AmbientDB is using the effective Chord Distributed Hash Table to index the metadata related to resources, it lacks the environmental knowledge management provided inside our project that is necessary to achieve adaptive collaborative distribution and personalized query optimization.

6.2 Semantic Tracking

The extraction framework described in this paper is closely related to ProMED-PLUS [37], a system for the automatic "fact" extraction from plain-text reports about outbreaks of infectious epidemics around the world to database, and MiTAP [8], a prototype SARS detecting, monitoring and analyzing system. The difference between our framework and those systems is that we also emphasize on generating the semantic relations among the collected resources and organizing those information by using topic map model.

The proposed information extraction model that formulates the relation extraction problem as a classification problem is motivated by the work of J.Suzuki et. al. [30]. This innovated work has proposed a HDAG kernel solving many problems in natural language processing. The use of classification methods in information extraction is not new. Intuitively, one can view the information extraction problem as a problem of classifying a fragment of text into a predefined category which results in a simple information extraction system such as a system for extracting information from job advertisements [38] and business cards [19]. However, those techniques require the assumption that there should be only one set of information in each document, while our model could support more than one set of information.

7 Conclusion and Forthcoming Issues

As communities generate increasing amounts of transactions and deal with fast growing data, it is very important to provide new strategies for their collaborative management of knowledge. In this paper, we presented and described our proposal regarding Information Modeling for Adaptive Semantic Management which aims at extracting information and knowledge from unstructured documents that spread throughout the Internet by emphasizing on information extraction technique, event tracking and knowledge organizing. We first motivated the need for such modeling in order to provide personalized services to users who are involved in semantic tracking communities. The motivation for this work is definitely to improve user's access to semantic information and to reach high satisfaction levels for decision making. Then, we gave an overview of our approach's algebra with its operators, focusing on update and consistency policies. We finally proposed and defined adaptive services that enable collaborative project to automatically dispatch semantic and to make the query results more relevant.

This challenging work needs more complicate natural language processing with deeply semantic relations interpretation.

Acknowledgement

The work described in this paper has been supported by the grant of National Electronics and Computer Technology Center (NECTEC) No. NT-B-22-14-12-46-06, under the project "A Development of Information and Knowledge Extraction from Unstructured Thai Document".

References

1. Kongwan, A., Kawtrakul, A.: Know-what: A Development of Object Property Extraction from Thai Texts and Query System. In: Proceedings of SNLP-2005, Bangkok, Thailand, pp. 157–162 (2005) ISBN: 974-9942-52-3
2. Berger, A.L., Pietra, S.-A.D., Pietra, V.-J.D.: A maximum entropy approach to natural language processing. Computational Linguistics 22, 39–71 (1996)

3. Biswas, D., Kim, I.: Atomicity for P2P based XML Repositories. In: ICDE 2007 (2007)
4. Bonifacio, M., Bouquet, P., Cuel, R.: The Role of Classification(s) in Distributed Knowledge Management. In: Proc. of KES, Podere d'Ombriano, Italy (September 16-18, 2002)
5. Biezunski, M., Bryan, M., Newcomb, S.: ISO/IEC JTC1/SC34. (May 22, 2002), Available at http://www1.y12.doe.gov/capabilities/sgml/sc34/document/0322.htm
6. Chanlekha, H., Kawtrakul, A.: Thai Named Entity Extraction by incorporating Maximum Entropy Model with Simple Heuristic Information. In: Su, K.-Y., Tsujii, J., Lee, J.-H., Kwong, O.Y. (eds.) IJCNLP 2004. LNCS (LNAI), vol. 3248, pp. 49–55. Springer, Heidelberg (2005)
7. Chareonsuk, J., Sukvakree, Y., Kawtrakul, A.: Elementary Discourse unit Segmentation for Thai using Discourse Cue and Syntactic Information. In: Proceedings of SNLP. Bangkok, Thailand, pp. 85–90 (2005) ISBN: 974-9942-52-3
8. Damianos, L., Bayer, S., Chisholm, M.A., Henderson, J., Hirschman, L., Morgan, W., Ubaldino, M., Zarrella, J.: MiTAP for SARS detection. In: Proceedings of the Conference on Human Language Technology. Boston, USA, pp. 241–244 (2004)
9. Dean, M., Connolly, D., van Harmelen, F., Hendler, J., Horrocks, I., McGuinness, D.L., Patel-Schneider, P.F., Stein, L.A.: OWL Web Ontology Language Reference. W3C Recommendation (February 10, 2004), Available at http://www.w3.org/TR/owl-ref/
10. Fontijn, W., Boncz, P.A.: AmbientDB: P2P Data Management Middleware for Ambient Intelligence. In: Proceedings. of PerCom Workshops, USA, pp. 203–207 (March 14-17, 2004)
11. Garshol, L.M.: Living with Topic Maps and RDF (May 2003), Available at
12. http://www.idealliance.org/papers/dx_xmle03/papers/02-03-06/02-03-06.html
13. Grosz, B., Joshi, A., Weinstein, S.: Centering: A Framework for Modeling the Local Coherence of Discourse. Computational Linguistics 21, 203–225 (1995)
14. Güting, R.H.: Gral: an extensible relational database system for geometric applications. In: Proceedings of the 15th international Conference on Very Large Data Bases. Very Large Data Bases, Amsterdam, The Netherlands, pp. 33–44. Morgan Kaufmann Publishers, San Francisco, CA (1989)
15. Güting, R.H.: Second-order signature: a tool for specifying data models, query processing, and optimization. In: Buneman, P., Jajodia, S. (eds.) SIGMOD 1993. Proceedings of the 1993 ACM SIGMOD international Conference on Management of Data, Washington, D.C., United States, May 25-28, 1993, pp. 277–286. ACM Press, New York (1993), http://doi.acm.org/10.1145/170035.170079
16. Kawtrakul, A., Permpool, T., Yingsaeree, C., Andres, F.: A Framework of NLP based Information Tracking and related Knowledge Organizing with Topic Maps. In: Kedad, Z., Lammari, N., Métais, E., Meziane, F., Rezgui, Y. (eds.) NLDB 2007. LNCS, vol. 4592, pp. 272–283. Springer, Heidelberg (2007)
17. Kawtrakul, A., Suktarachan, M., Imsombut, A.: Automatic Thai Ontology Construction and Maintenance System. In: Proceedings of Ontolex Workshop on LREC, pp. 68–74 (2004)
18. Kawtrakul, A., Yingsaeree, C.: A Unified Framework for Automatic Metadata Extraction from Electronic Document. In: Proceedings of The International Advanced Digital Library Conference. Nagoya, Japan (2005)
19. Kongwan, A., Kawtrakul, A.: Know-what: A Development of Object Property Extraction from Thai Texts and Query System. In: Proceedings of SNLP-2005. Bangkok, Thailand, pp. 157–162 (2005) ISBN: 974-9942-52-3

20. Kushmerick, N., Johnston, E., McGuinness, S.: Information extraction by text classification. In: Proceedings of IJCAI-2001 Workshop on Adaptive Text Extraction and Mining (2001)

21. Li, B., Li, W., Lu, Q., Wu, M.: Profile-based event tracking. In: SIGIR 2005 Proceedings of the 28th Annual international ACM SIGIR Conference on Research and Development in information Retrieval, Salvador, Brazil, August 15-19, 2005, pp. 631–632. ACM Press, New York (2005)

22. Chan, M.T., Hoogs, A., Sun, Z., Schmiederer, J., Bhotika, R., Doretto, G.: Event Recognition with Fragmented Object Tracks, icpr. In: Proceedings of 18th International Conference on Pattern Recognition (ICPR 2006), pp. 412–416 (2006)

23. Naito, M., Andres, F.: Application Framework Based on Topic Maps. In: Maicher, L., Park, J. (eds.) TMRA 2005. LNCS (LNAI), vol. 3873, pp. 42–52. Springer, Heidelberg (2006)

24. Pan, J.Z.: A Flexible Ontology Reasoning Architecture for the Semantic Web. IEEE Transactions on Knowledge and Data Engineering 19(2), 246–260 (2007)

25. Perich, F., Joshi, A., Finin, T.W., Yesha, Y.: On Data Management in Pervasive Computing Environments. Trans. Knowl. Data Eng. 16(5), 621–634 (2004)

26. Pinto, H.S., Staab, S., Sure, S., Tempich, C.: OntoEdit Empowering SWAP: a Case Study in Supporting DIstributed, Loosely-Controlled and evolvInG Engineering of oNTologies (DILIGENT). In: Bussler, C.J., Davies, J., Fensel, D., Studer, R. (eds.) ESWS 2004. LNCS, vol. 3053, pp. 10–12. Springer, Heidelberg (2004)

27. Pitoura, E., Abiteboul, S., Pfoser, D., Samaras, G., Vazirgiannis, M.: DBGlobe: a service-oriented P2P system for global computing. SIGMOD Rec. 32(3), 77–82 (2003)

28. Rajbhandari, S., Andres, F., Naito, M., Wuwongse, V.: Topic Management in Spatial-Temporal Multimedia Blog. In: the 1st IEEE International Conference on Digital Information Management (ICDIM 2006), Bangalore, India, pp. 81–88 (December 6-8, 2006) ISBN:1-4244-0682-X

29. Satayamas, V., Thumkanon, C., Kawtrakul, A.: Bootstrap Cleaning and Quality Control for Thai Tree Bank Construction. In: Proceedings of NCSEC-2005. Bangkok Thailand, pp. 849–860 (2005) ISBN: 974-677-541-3

30. Shneiderman, B.: The eyes have it: a task by data type taxonomy for information visualizations. In: Proceedings of 1996 IEEE Visual Languages, Boulder, CO, pp. 336–343 (1996)

31. Suzuki, J., Sasaki, Y., Maeda, E.: Kernels for structured natural language data. In: Proceeding of NIPS 2003 (2003)

32. Sudprasert, S., Kawtrakul, A.: Thai Word Segmentation based on Global and Local Unsupervised Learning. In: Proceedings of NCSEC-2003, Chonburi, Thailand (2003) ISBN: 974-3826-04-1

33. The AOS project, FAO, http://www.fao.org/aims/aos.jsp

34. Thamvijit, D., Chanlekha, H., Sirigayon, C., Permpool, T., Kawtrakul, A.: Know-who: Person Information from Web Mining. In: Proceedings of NCSEC. Bangkok, Thailand, pp. 849–860 (2005) ISBN: 974-677-541-3

35. Topic Map Query Language (TMQL), http://www.isotopicmaps.org/tmql/

36. Vargas-Vera, M., Celjuska, D.: Event Recognition on News Stories and Semi-Automatic Population of an Ontology, wi. In: IEEE/WIC/ACM International Conference on Web Intelligence (WI 2004), pp. 615–618 (2004)

37. Xiao, R., Tang, S., Li, L., Fang, L., Xu, Y., Chen, W., Xu, Y.: Using Categorial Context-SHOIQ(D) DL to Integrate Context-aware Web Ontology MetaData. In: Proceedings of Second International Conference on Semantics, Knowledge, and Grid (SKG 2006) (2006)

38. Yangarber, R., Jokipii, L., Rauramo, A., Huttunen, S.: Information Extraction from Epidemiological Reports. In: Proceedings of HLT/EMNLP-2005. Canada (2005)
39. Zavrel, J., Berck, P., Lavrijssen, W.: Information extraction by text classification: Corpus mining for features. In: Proceedings of the workshop Information Extraction meets Corpus Linguistics. Athens, Greece (2000)

Going Beyond Completeness in Information Retrieval

Masahito Hirakawa

Interdisciplinary Faculty of Science and Engineering, Shimane University
1060 Nishikawatsu, Matsue 690-8504, Japan
hirakawa@cis.shimane-u.ac.jp

Abstract. The importance of information (or content) is becoming more and more as computers penetrate everyday life. This has speeded up information management researches. We focus on information retrieval in this paper and present a classification of knowledge-based techniques which rely on semantic aspects of people and surroundings as well as data objects themselves. We also present the notion of cognition-based retrieval which counts our cognitive nature in retrieval processing as another promising approach.

Keywords: information retrieval, content-based retrieval, context-based retrieval, recommendation, the semantic web, cognition.

1 Introduction

Nowadays smart access to the information in computers is a primary concern of ours, though computers were first invented as machines to calculate numbers and then applied to edit documents and communicate with others.

Advances in computer technology have enabled us to have low-cost but powerful electronic devices such as digital cameras, camcorders, and camera phones. It is fun and easy to take pictures, videos, and sound data, resulting in increased volume of a data set. Besides, many people are enthusiastic in publishing documents on the web. It may be our nature that everybody wants to be recognized by others. This can also lead to information explosion. Sophisticated information management is a key to success in future computing. In this paper, we specially focus on information retrieval.

Here it is noted the information is interpreted under the existence of a viewer. In other words, the meaning of a data object is not determined by itself and varies depending on various factors which include his/her interest and current situation. This aspect is noticeable when we retrieve multimedia data. For example, if a certain portrait is given as a query, what should be returned? One may expect to get portraits of the same person. Another may want to have portraits which were drawn by the same painter. Obviously traditional simple keyword matching is not enough to this requirement.

In this paper, we will present a perspective of information retrieval techniques. We first summarize ideas which have been investigated so far. They are content-based retrieval, context-based retrieval, recommender systems, and the semantic web. Search tasks are carried out by referring to data objects and knowledge about them in

S. Bhalla (Ed.): DNIS 2007, LNCS 4777, pp. 70–80, 2007.
© Springer-Verlag Berlin Heidelberg 2007

a target domain. These types of knowledge-based retrieval help to make the resultant information trustworthy, compared to keyword-based retrieval.

Meanwhile, interests in those trials go mainly to how completely semantic aspects are modeled into a formal structure. But it can happen we don't catch the meaning of a data object as is presented in its expression because of our cognitive nature. The notion of cognition-based retrieval is presented in the paper as another interesting approach.

The rest of the paper is organized as follows. A classification of knowledge-based retrieval techniques is discussed extensively in Chapter 2. Chapter 3 presents cognition-based retrieval as a challenge to next-generation information retrieval. Finally, in Section 4, we will conclude the paper.

2 Knowledge-Based Retrieval Techniques

2.1 Overview

Information retrieval research has a long history. Wikipedia explains that information retrieval is interdisciplinary, based on computer science, mathematics, library science, information science, cognitive psychology, linguistics, statistics, and physics. Lau presented properties that information retrieval systems should support, as listed in the following [1]:

- Autonomous: systems should be able to autonomously select relevant information for their users with a minimum amount of human intervention.
- Proactive: systems should identify users' actual information retrieval requirements with reference to possible retrieval contexts.
- Adaptive: systems should be responsible to user's changing needs and adjust its information retrieval functions.
- Predictable: system's learning and retrieval behavior should be predictable and hopefully explanatory.
- Scalable: systems should be able to scale up to cope with the explosive growth of the amount of information.
- Balanced precision and recall: systems should strike for a better balance between recall-oriented and precision-oriented retrieval.

Our interests in this paper go to underlying techniques in finding valuable information. Traditionally, keyword-based retrieval was the common, where the database search is carried out by looking at a text string on its own. In multimedia databases, textual metadata is attached to every multimedia document as a clue. It is obvious the result of retrieval depends on its description which has been assigned by someone in advance. Keyword-based retrieval is simple but not satisfactory for retrieval tasks of today.

Retrieval based on certain knowledge is one of the challenges toward realization of advanced information retrieval facilities. It includes content-based retrieval, context-based retrieval, recommender systems, and the semantic web. Figure 1 shows a classification of those trials. It is noted, in knowledge-based retrieval, researchers try to specify knowledge as complete as possible.

Cognition-based retrieval which appears also in the figure is another approach to go beyond completeness, considering that we may not understand data objects as presented in the physical world.

In the subsequent sections we discuss knowledge-based approaches extensively. Cognition-based approach will be presented later in Chapter 3.

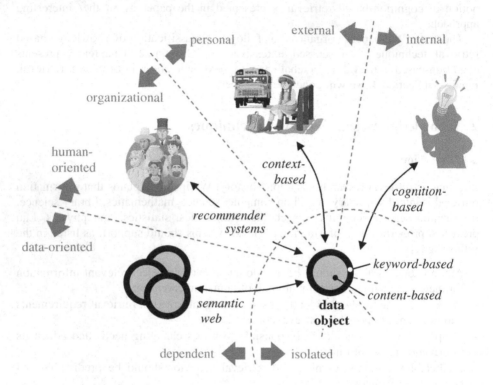

Fig. 1. Classification of information retrieval technologies. Starting with keyword-based retrieval, semantic relationships of every object with other objects, individuals, and environments are considered to have valuable information. Now it's time to take our cognitive aspects into account.

2.2 Content-Based Retrieval

It is not reasonable especially in a multimedia database environment to force the user to understand an object as is specified in metadata which has been assigned by someone in advance. Suppose you look at the painting "La Moulin de la Galette" by Renoir and want to retrieve the paintings whose subject is the same (that is, dancing people). However, if metadata do not contain the information which is necessary to respond to this inquiry, you must be frustrated.

Metadata is like sugar which coats tablets. It facilitates the use of data, but cannot be the replacement of data itself. Focus in retrieval should be applied to data objects stored in a database. Properties which are used for database search include color, texture, or shape for image databases. For videos, trajectory of moving objects and camera work are possible properties as well. This approach is the so called *content-based retrieval* [2], [3].

One of the key issues in content-based retrieval is user-friendliness in specifying queries. Unsurprisingly, in multimedia database retrieval, it is difficult in many cases to place a query in text. Example-based querying has been recognized as a powerful solution to this problem, where an example taking the form of an image or sound is given as a query. For example, Fig. 2 shows a query in which trajectory of a moving object is conditioned by a sequence of yellow rectangular boxes on a video shot having a certain camera motion [4], [5].

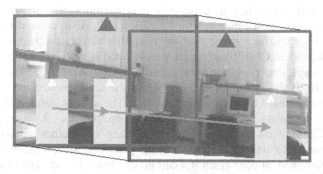

Fig. 2. A query to a video database is given by example. Trajectory of an object (yellow rectangular boxes) is specified as a condition on a mosaic picture which gives an overview of the video shot.

While most of the trials adopt low-level features for similarity evaluation, challenges to the content-based retrieval with reference to higher-level features have been active in years. The user is allowed to place queries such as "find images in which houses or buildings appear" or "find videos which make me happy". To make this possible, the system should have a substantial knowledge or concept about things in the world.

Let us explain our trial of movie retrieval system which adopts empirical rules so-called *grammar of films* [6]. Grammar of films is a collection of rules and techniques which have been identified by film directors and editors. It has been applied for movie production to emphasize certain meanings of the scenes. For example, a sequence of shots which are gradually shortened gives us the feeling of the increased tension. Based on such nature, the system finds video shots having a certain atmosphere (for example, calm/quiet, excitement, and increase/release of tension) as well as specific scenes (for example, conversation and chase scenes).

Meanwhile, content-based retrieval is not just for non-textual, i.e., multimedia documents. Full-text search is such example. It examines all the words in every document rather than any specially attached metadata.

2.3 Context-Based Retrieval

Content-based retrieval searches for properties subsisting in each data object. In other words, scope of property extraction and matching is limited within data object. When we consider the fact that every data object is identified and understood in relation to others, it is necessary for the system to consider another type of knowledge.

Based on this observation, an idea of *context-based retrieval* has been brought into a reality [7], [8]. If the system is capable of recognizing a context the user is situated and knowing the information he/she wants to get, the user can be freed from a burden of placing excessive query commands, even though there is a massive amount of information to be retrieved [9]. Suppose, for example, you are walking along a street. You receive an informative message from a nearby shop through your mobile phone or PDA without any special efforts.

Time and space are the most typical context information used in context-based retrieval. But it is not limited to those two. People/community, items, temperature, and weather are other possible context information [10]. Also, context-based retrieval is not a technology just for a single user, but is applicable for group activities.

Situation-dependent Chat Room we have developed is such an example (see Fig. 3) [11]. A premise is that people within a particular physical area, for example, an amusement park, stadium, and airport, would feel the same interest and objective, and they would like to share their ideas one another. The system helps a group of those people to communicate through their mobile equipments. Assume you are in Disneyland. Chatting on today's events or Mickey with other guests is fun and enjoyable. Meanwhile, when the user leaves the particular area, the connection link is closed, and reestablished later again if a certain condition is met. The system controls the communication channel by referring to his/her context or situation.

Now context-based retrieval systems are no longer at the laboratory level [12], [13]. AmbieSense, a European Union funded R&D project, developed localized information delivery applications for Seville city center and Oslo airport [12], and tested their practicality through experiments in which 238 travelers and tourists participated.

One promising research direction in context-based retrieval is life-log management. Rapid progress in computer and sensor technologies enables the system to record our daily life in detail. Here, as imagined, life-log data are broad and their amount is quite large. Since remembering the context of experiences is easier than remembering their details, context-based retrieval is expected to work effectively for life-log data search.

A context-based life-log retrieval system in [14] helps the user to navigate through a large collection of image, video and audio data which have been captured during his/her daily life. Possible triggers or clues for retrieval include time, location, and user's activity such as the operation of a computer and meeting with someone. MyLifeBits [15], DARPA's LifeLog, and SenseCam [16] are some other life-log research trials.

There are still debates about context, though. The objection which has been commonly made is the following [13]: Because context is tightly intertwined with users' internal and social – continuously changing – interpretations, it seems difficult to capture context in any general sense that would support design. Consequently, there have been doubts about whether the entire concept is of any use.

I would say context-based retrieval is promising, though a wide variety of studies are needed.

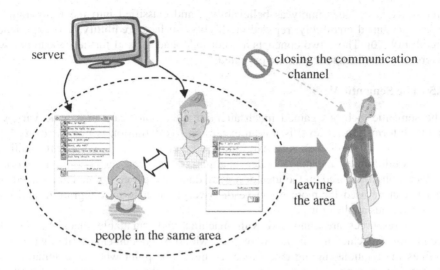

Fig. 3. People in the same area (context) can communicate one another. After they have left the area, the communication is terminated automatically.

2.4 Recommender Systems

Most of us receive more information than we can possibly handle in a pleasant fashion, due especially to the explosive growth of the web. All the information is not equally important to us. *Recommender systems,* or information filtering in general, are an effective and powerful tool to locate the right information at the right time [17].

There are several approaches for building recommender systems. One relies on content descriptions such as genre, director, and actors in the case of movie application. Content descriptions of every data object in a database are matched against the learned preferences of the user.

Our focus here is paid to another and most smart approach, called collaborative filtering (sometimes referred to as social filtering). It relies on the opinions of possible partners having a similar preference to help the user receive the information most likely to be interesting to him/her or relevant to his/her needs. Success in Amazon.com has shown the effectiveness of collaborative filtering technique.

Collaborative filtering techniques are vital, but the "wheat and tares" problem still remains. How can we get the really useful, high-quality information? Challenges to reduce a risk of choosing tares are still needed.

As a matter of fact, recommender systems may fail to build a reasonable recommendation model if there are not enough historical data for evaluating similarity between objects in a database. To cope with this problem, feature-based recommendation algorithms were proposed [18].

Another research challenge toward generation of high-quality recommendations is trustworthiness and credibility of information content. Barber defines trust as the expectation of technically competent role performance [19]. Specifically, trust is interpreted as an agent's expectation of another agent's competence in providing opinions to reduce its uncertainty in predicting new items' ratings. On the other hand, credibility is defined simply as believability, and classified into the following four types: presumed credibility, reputed credibility, surface credibility, and experienced credibility [20]. Those two concepts are not independent, and further researches with reference to the semantic web are expected.

2.5 The Semantic Web

The semantic web has gained much attention in recent years in a wide variety of fields. Information retrieval is not an exception. Its definition is presented in [21] as follows: The semantic web is an extension of the current web in which information is given well-defined meaning, better enabling computers and people to work in cooperation. It is based on the idea of having data on the web defined and linked such that it can be used for more effective discovery, automation, integration, and reuse across various applications.

Web resources are annotated with machine understandable metadata. The key technology in achieving this is an ontology which is a document or file that *formally* defines the relations among data objects. This is the point where the semantic web differs from recommender systems. In other words, the semantic web is based on a hard (or governed) knowledge toward achieving intelligent retrieval, while recommender systems adopt a soft (or federated) knowledge.

Some point out worries about its feasibility and cost of setting up ontologies as a semantic backbone for web resources. Because of the fact that web resources cannot be stable and alive as we are, a scheme of keeping live ontologies is compulsory. While formal and theoretical research based on the deep insight and understanding about entities is valuable, folksonomy may hold another key role to this tough problem [22].

Meanwhile, one interesting issue associated with context-based retrieval concerns the interpretation of data values sensed from the surroundings. For example, a GPS receiver generates values of longitude and latitude. They are good for the computer, but we are usually not interested in such physical values. Abstraction of a physical value to a symbol or concept is necessary – for example, "my office." What is difficult in its interpretation is that the same physical values can be associated with different level concepts such as "Building 3" or "Shimane University". It is needed to choose the appropriate abstraction level for a given physical value or a set of values.

3 A Challenge over Semantics

A central idea in the trials explained above is how we specify semantic knowledge as complete as possible. I have no doubt that completeness is worthwhile. However it is also true that we may not catch the meaning of an object as is presented in its expression. Let me give you some examples.

The first example concerns illusions which are known as a distortion of a sensory perception. Researchers have demonstrated different types of illusions. Figure 4 shows an example of optical (visual) illusions, in which a cubic frame is seen in front of the eight disks despite the fact there is no cubic frame at all. Otherwise, you may see a cubic frame behind the black wall through the eight circular windows. The brain

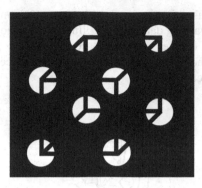

Fig. 4. An example of illusion. We may see a cubic frame in front of the black wall in spite of no actual drawing. Or a cubic frame may be seen through the black wall.

makes us see a thing as if it exists as long as there is no evidence for its non-existence [23]. As this example shows, we sometimes fail to acquire information correctly due to illusions.

Another example is Gal-Moji (characters of young girls) or Heta-Moji (awkward characters) which is used especially by young girls in Japan. They recognize Hiragana, the Japanese cursive syllabary, as figures and form a sentence by a combination of completely irrelevant characters. Figure 5 shows an example of Gal-Moji. Left-hand side is a regular Hiragana expression saying "How are you?". Right-hand side is its corresponding Gal-Moji expression. As you can see, for example, the second Hiragana character "ん" is expressed by a Greek character "ω". Gal-Moji users don't care about the language. What they are interested in is just the expressiveness of characters. Based on this observation, we proposed the notion of UniversalMedia in [24].

げんき？　→　（ナ"ω（ｷ？

Fig. 5. Gal-Moji saying "How are you?" is given on the right, while the left hand side is the Hiragana expression. Interestingly, people who enjoy using Gal-Moji treat the existing characters just as symbols or fragments, and express a sentence by concatenating those symbols so that it visually resembles its Hiragana expression.

One more example is non-photorealistic rendering and illustration-inspired visualization techniques. The former approach has advantages over realistic images by omitting extraneous detail, which focuses attention on relevant features, and by clarifying, simplifying, and disambiguating shape [25], [26]. Computer-generated caricatures are an example of this approach, where the difference between an average face and a particular face is computed and then exaggerated by a specified amount to augment the communication performance of human face images.

On the other hand, illustration-inspired techniques [27] are used to effectively convey motions in a time-varying dataset. Speedlines are the most typical example, as illustrated in Fig. 6. Of course, we can't see such lines at all in a physical environment, but we feel the motion without any inconvenience. It actually enhances our understanding of the meaning.

Fig. 6. Visualizing the motion with speedlines. Speedlines convey the meaningful information, even though they are actually nonsense.

Some of the notions in the above may be radical and exceptional. But, as they show, completeness of domain knowledge may not be the ultimate goal. It would be nice if information retrieval functions will be designed in consideration of the behavior of our cognitive system. We call this approach *cognition-based* retrieval.

One research issue is to investigate how media convey the information and affect people - the way of their understanding and thinking. There have been a variety of research trials on media in, for example, computer vision, speech understanding, and

Fig. 7. The input text is first decomposed into segments. Key characters are then extracted by referring to a pattern of their appearances in the segments. Finally an image is generated to each of the segments by arranging key characters on a stage and determining a camera work accordingly.

artificial intelligence. But those researchers focus on analysis of a certain media type (i.e., text, image, video, and sound). Less study on media themselves or interdependency among media entities has been done so far.

We have been investigating the mechanism of media understanding. As a very first step, we have proposed a scheme of generating a sequence of scenic pictures from a narrative text [28]. Figure 7 shows one example of the generated pictures.

Finally, we should work more with people in mathematics, library science, information science, cognitive psychology, linguistics, statistics, physics, and more, aiming at attaining *beyond completeness* features in information retrieval.

4 Conclusion

In this paper we classified information retrieval techniques from a user's perspective. Significant existent techniques discussed in the paper include content-based retrieval, context-based retrieval, recommender systems, and the semantic web. They focus on semantic aspects of objects, individuals, and surroundings which organize a social environment. Considering the fact that we may not catch the meaning of data objects as presented in their expression, the notion of cognition-based retrieval was presented as a new research direction. It considers our cognitive nature in searching for and presenting information, and would enable us to realize more powerful information retrieval systems having *beyond completeness* features.

References

1. Lau, R.Y.K.: Context-Sensitive Text Mining and Belief Revision for Intelligent Information Retrieval on the Web. Int. J. Web Intelligence and Agent Systems 1, 151–172 (2003)
2. Lwe, M.S., Sebe, N., Djeraba, C., Jain, R.: Content-Based Multimedia Information Retrieval: State of the Art and Challenges. ACM Trans. Multimedia Computing, Communications and Applications 2(1), 1–19 (2006)
3. Lewis, P., Dupplaw, D., Martinez, K.: Content-Based Multimedia Information Handling: Should we Stick to Metadata? Cultivate Interactive 6 (2002)
4. Assfalg, J., Del Bimbo, A., Hirakawa, M.: A Mosaic-based Query Language for Video Databases. In: Proc. IEEE Symp. Visual Languages, pp. 31–38. IEEE Computer Society Press, Los Alamitos (2000)
5. Hirakawa, M., Uchida, K., Yoshitaka, A.: Content-Based Video Retrieval using Mosaic Images. In: Proc. Symp. Cyber Worlds: Theories and Practices, pp. 161–167 (2002)
6. Yoshitaka, A., Ishii, T., Hirakawa, M., Ichikawa, T.: Content-Based Retrieval of Video Data by the Grammar of the Film. In: Proc. IEEE Symp. Visual Languages, pp. 310–317. IEEE Computer Society Press, Los Alamitos (1997)
7. Bierig, R., Goker, A.: Time, Location and Interest: An Empirical and User-Centred Study. In: Proc. ACM Conf. Information Interaction in Context, pp. 79–87. ACM Press, New York (2006)
8. Brown, P.J., Jones, G.J.F.: Context-Aware Retrieval: Exploring a New Environment for Information Retrieval and Information Filtering. J. Personal and Ubiquitous Computing 5, 253–263 (2001)

9. Hewagamage, K.P., Hirakawa, M.: Augmented Album: Situation-Dependent System for a Personal Digital Video/Image Collection. In: Proc. IEEE Int. Conf. Multimedia and Expo, pp. 323–326. IEEE Computer Society Press, Los Alamitos (2000)

10. Toyama, D., Kakimoto, M., Yoshitaka, A., Hirakawa, M.: A Community-Based Web Browsing System. In: Proc. IEEE Symp. Visual/Multimedia Approaches to Programming and Software Engineering, pp. 320–327. IEEE Computer Society Press, Los Alamitos (2001)

11. Hirakawa, M., Hewagamage, K.P.: Situated Computing: A Paradigm for the Mobile User Interaction with Multimedia Sources. Annals of Software Engineering 12(1), 213–239 (2001)

12. Lech, T.C., Wienhofen, L.W.M.: AmbieAgents: A Scalable Infrastructure for Mobile and Context-Aware Information Services. In: Proc. ACM Conf. Autonomous Agents and Multiagent Systems, pp. 625–631. ACM Press, New York (2005)

13. Tamminen, S., Oulasvirta, A., Toiskallio, K.: Understanding Mobile Contexts. J. Personal Ubiquitous Computing 8, 135–143 (2004)

14. Tancharoen, D., Yamasaki, T., Aizawa, K.: Practical Experience Recording and Indexing of Life Log Video. In: Proc. ACM Workshop Continuous Archival and Retrieval of Personal Experiences, pp. 61–66. ACM Press, New York (2005)

15. Gemmell, J., Bell, G., Lueder, R.: MyLifeBits: A Personal Database for Everything. Comm. ACM 49(1), 88–95 (2006)

16. Sellen, A., Fogg, A., Aitken, M., Hodges, S., Rother, C., Wood, K.: Do Life-Logging Technologies Support Memory for the Past? An Experimental Study Using SenseCam. In: Proc. ACM CHI, pp. 81–90. ACM Press, New York (2007)

17. Konstan, J.A.: Introduction to Recommender Systems: Algorithms and Evaluation. ACM Trans. Information Systems 22(1), 1–4 (2004)

18. Han, E.H., Karypis, G.: Feature-Based Recommendation System. In: Proc. ACM Conf. Information and Knowledge Management, pp. 446–452. ACM Press, New York (2005)

19. Barber, B.: The Logic and Limits of Trust. Rutgers University Press (1983)

20. Fogg, B.J., Tseng, H.: The Elements of Computer Credibility. In: Proc. ACM Conf. CHI, pp. 80–87. ACM Press, New York (1999)

21. Hendler, J., Berners-Lee, T., Miller, E.: Integrating Applications on the Semantic Web. J. Institute of Electrical Engineers of Japan 122(10), 676–680 (2002)

22. Wu, H., Zubair, M., Maly, K.: Harvesting Social Knowledge from Folksonomies. In: Proc. ACM Conf. Hypertext and Hypermedia, pp. 111–114. ACM Press, New York (2006)

23. http://www.brl.ntt.co.jp/IllusionForum/menu-e.html

24. Hirakawa, M.: From MultiMedia to UniversalMedia. Int. J. Computational Science and Engineering (2007) (in print)

25. Gooch, B., Reinhard, E., Gooch, A.: Human Facial Illustrations: Creation and Psychological Evaluation. ACM Trans. Graphics 23(1), 27–44 (2004)

26. Chen, H., Liu, Z., Rose, C., Xu, Y., Shum, H.Y., Salesin, D.: Example-Based Composite Sketching of Human Portraits. In: Proc. Symp. Non-Photorealistic Animation and Rendering, pp. 95–102 (2004)

27. Joshi, A., Rheingans, P.: Illustration-Inspired Techniques for Visualizing Time-Varying Data. In: Proc. IEEE Visualization, pp. 679–686. IEEE Computer Society Press, Los Alamitos (2005)

28. Yamamoto, M., Hikino, K., Kijima, S., Hirakawa, M.: Towards Understanding of Multimedia Documents: A Trial of Picture Book Analysis and Generation. In: Proc. IEEE Int. Symp. Multimedia, pp. 29–36. IEEE Computer Society Press, Los Alamitos (2005)

Query and Update Through XML Views

Gao Cong

Microsoft Research Asia
gaocong@microsoft.com

Abstract. XML has become a standard medium for data exchange, and XML views are frequently used as an interface to relational database and XML data. There have been a considerable number of studies on building and querying XML views, while updating related topics for XML views have not receive much attention. In this paper, we outline our work on building XML views defined in terms of update syntax and updating XML views of relations, and discuss some related work.

1 Introduction

XML was developed as a markup-language for arbitrary document structure, as opposed to HTML, which is a markup language for a specific kind of hypertext documents. An XML document consists of a properly nested set of open and close tags, where each tag can have a number of attribute-value pairs. The set of tags and their allowed combinations represent semantics of the enclosed text. The set of tags is not fixed, but can be defined to conform with XML DTD or XML schema. An example of XML application is to represent ontology in semantic Web.

Emerging from a document markup language, XML gains its popularity in data exchange in database community. XML has become a widely accepted standard for the representation and exchange of data on the Internet. XML also plays important roles in Web services and semantic Web, for instance, Web services that use XML-based descriptions in WSDL and exchange XML messages with SOAP. The data can be originally in the form of relation or XML. In both cases, XML views are required to exchange data on the internet. (1) If the data is originally in traditional relational databases or object-relational databases, there is a need to export the data into XML form in order to exchange them on the Internet. XML views come into place as intermediate schema that are extracted from relational data before they are mapped into XML data. In this case, XML views are always presented to users so that users can retrieve the underlying data through the XML views. Publishing languages are proposed to define the mapping between relational data and XML data. (2)If the data is originally in XML format, it is still required to present XML views on XML data for data exchange since users may want to exchange part of the data, possibly with some transformation.

There are several advantages for XML views. First, XML views provide application specific views of source data, and thus user can navigate and even update

S. Bhalla (Ed.): DNIS 2007, LNCS 4777, pp. 81–95, 2007.
© Springer-Verlag Berlin Heidelberg 2007

source data through their XML views. Second, XML views provide the ability of data access control by hiding sensitive data that users are not allowed to see. Third, XML views provide for a basis for further data integration. Finally, XML views enable us to exploit the potential of XML as the standard of data exchange.

It is clear that XML views are needed. There is a host of research on publishing relational data into XML data, and commercial database systems also support XML views of relations. An excellent survey of work along this line can be found in [27]. Moreover, XML views can also be defined on top of XML data, which are stored in traditional relations, object-relations, text files or native XML database.

With the increasing popularity of XML applications, publishing XML views and querying XML views with XQuery or XPath have received extensive attention. We next discuss two research topics related to both updates and XML views. First, building XML views defined in terms of query expressed with update syntax; second, updating though XML views.

XML view defined in terms of update syntax. Similar to updating in SQL for relations, it is also important to have its counterpart, i.e. an extension of declarative languages (e.g. XQuery and XPath), for XML, as XML applications grow in complexity. A number of update languages are proposed for XML to update selected nodes returned by embedded query in update expression. These update languages can be used to make persistent changes. Some XML applications may call for non-destructive update to the original XML document. Instead of updating the original XML file, a new copy of the original document is generated and updates are conducted on the new copy. So far, such a requirement has been met in two languages, namely XQuery Update[9] and XQuery![21]. Note that queries can be posed on the XML view defined in terms of non-destructive update syntax. In this paper, we will discuss the problems addressed in [18], the first research paper on efficient evaluation of such non-destructive update queries and effective composition of user queries with non-destructive update queries.

Updating though XML views. As discussed above, it is common to have XML views published from relations and XML data. For all the reasons that updating data through its relational views are needed, it is also important to update relational databases through their XML views. In this paper, we discuss previous work on updating through XML views and compare them in a uniform framework.

In the rest of this paper, we will focus on the above two topics. We first present some background on XML views, XML storage and XML query processing in Section 2. We outline our previous work on on building XML views defined in terms of update syntax and updating XML views of relations in Section 3. In section 4 we conclude this paper and discuss some open problems.

2   Preliminary

In this section, we briefly survey XML view, XML storage and XML query processing.

2.1 XML Views

We next discuss work on exporting XML views from relations and XML files. SilkRoute[19,20] adopts two declarative languages RXL and XML-QL to define and query views over relational data, respectively. XPERANTO[7,8] uses a canonical mapping to create a default XML view from relational data, and more views can be defined on top of the default view. Instead of adopting XQuery, ROLEX[31] composes XSLT stylesheet with defined XML views to produce a new XML view definition. Moreover, an algorithm is presented in [24] to translate XSLT scripts over XML views into efficient SQL queries. A new operator is introduced in [10] to support relation-valued variables in relational engines so that it can be enhanced for efficient XML publishing. Schema-directed XML publishing is studied in [3] to ensure that an XML view conforms to a predefined schema. In addition, major commercial database systems have provided the ability to export relational data to materialized XML views. In Oracle XML DB[36], XML views are defined by using SQL/XML, which is an extension to SQL. Oracle XML DB can only support XPath queries on XML views, which will be translated into an equivalent SQL query. Microsoft SQL Server 2000 [39] defines an XML view with an annotated XSD XML schema and supports XPath queries over the annotated XML Schema. IBM DB2 XML Extender [22] uses a Document Access Definition (DAD) file to define an XML view. However, it does not support any XML query languages over the XML view.

There has also been work on building XML views of XML data. For example, the task in [15,16] is to design efficient algorithm to translate an XML view query into SQL, when the XML data is stored in relational database. Xyleme [1] presents XML view for XML data stored as text files or in native XML storage. Xyleme [1] defines an XML view by connecting one abstract DTD to a large collection of concrete DTDs with an extension of OQL as the query language.

2.2 XML Storage and Query

With XML views used as data exchange also comes the need for storing and querying XML data (view). Indeed, XML is also employed for data repository although it is not that popular as vehicles for data exchange. Regardless of whether XML is used for data repository or data exchange, one needs to consider how to physically store XML data. The problem of XML data storage remains largely to be open and there is no uniform solution to this problem. One common practice is to store XML data into traditional databases, and the other is to store XML data as text files or in native XML databases. We think it is a primary problem that should be addressed for XML database. Although most of work on XML in database community focuses on optimization of query engine, such as indexing (e.g.[33]), the effectiveness of the proposed query optimization techniques largely depends on the underlying storage solution.

Indeed it may not be easy to find a ubiquitous solution due to the varieties of XML documents. XML documents are divided into two groups based on their contents: data-centric and document-centric in [25]. The former contains highly structured data while the latter refers to unstructured text documents. XML query languages, xQuery and XPath, are usually effective in expressing structural queries on data-centric documents, favored by database community. However for document-centric data, the text fields call for text retrieval techniques to find relevant field while document structure usually contains additional semantic information being helpful to improve the precision of text retrieval. Querying and retrieving document-centric documents also receive attention from database community, such as [2,41]. It is important to provide concrete evidence that XML technology has all the potential to be well supported, and actually to scale to very large sizes, regardless of how XML data is stored.

There are many algorithms designed to process exact query. Approximate answers, however, are sufficient sometimes. Moreover, estimating the result sizes of XML queries will be useful for query optimization by providing a quick feedback about the queries. Most of existing studies have focused on the selectivity estimation of XML queries without order-based axes. Existing XML selectivity estimators (e.g. [34,37,46]) are designed specifically for XML queries without order axes, and the order information is typically not captured. A framework, together with methods within the framework, is developed in [32] to estimate the result sizes of XPath expressions with (or without) order-based axes by capturing and summarizing the path and order information of XML elements in compact data structures. These selectivity estimators are mainly designed for data-centric data and they usually model the structural part of the data.

In addition, it is also advocated that that XML storage technology will naturally be distributed, possibly with data fragmentation and replication because it is widely used in the representation of web service. To accommodate the distributed storage, distributed query processing comes into place. XML documents can be fragmented both horizontally and vertically over a number of sites. Due to the hierarchical structures of XML documents, the evaluation on one fragment depends on the results of other fragment, which makes parallel processing difficult. Algorithms based on a novel idea of partial evaluation are introduced for distributed evaluation of boolean query[6] and data selecting query[12] on distributed XML fragments, irrespective of horizontally and vertically over a number of sites. We assume that XML document is physically stored as files. The key idea is twofold 1) to decouple the dependency relationship among fragments by introducing variables to represent the evaluation results of related fragments, and 2) to decompose XPath query into multiple sub-queries so that we can evaluate them individually by making use of the introduced variables.

3   Query and Update XML Views

In this section, we will outline our previous work on querying XML document through update syntax [18] and updating relational data via their XML views [11].

3.1   Query Via Update Syntax

In addition to the queries expressed in XQuery and XPath, XML applications may also request to produce a document substantially similar to, but not the same as, an existing document. As an example, consider an XML document T_0 depicted in Fig. 2, which contains a list of *parts*. Each *part* has a *pname* (part name), a list of *suppliers* and a *subpart hierarchy*; a *supplier* in turn has a *sname* (supplier name), a *price* (offered by the supplier), and a *country* (where the supplier is based). In this example, to track imports we might want to write a query that finds all the information in T_0 *except price*; in other words, the query is to return an XML tree that is almost identical to T_0 except that it does not have the *price* elements. Such a query cannot be easily expressed using xQuery without complicated user-defined recursive functions. Fortunately, the xQuery Update working draft of W3C [9] provides transform statement to express such queries. For example, using the syntax of xQuery Update [9], we can write the above query as

  transform copy $a := doc($"foo"$) **modify do delete** $a//price$ **return** $a

which would *return* the tree formed by applying "**delete** $a//price$ " to the input document doc("foo"). It should be noted that such an expression is non-updating on the underlying data. It is in fact a *query*, and would *not* modify the value of the underlying data doc("foo"). Similarly, such queries can also be readily expressed with the update language proposed by [21].

  We call such a query as transform queries by following the terms used in the xQuery Update working draft of W3C [9]. We first briefly introduce some representative application of transform query as discussed in [18].

Hypothetical queries. Suppose that a user is concerned that a planned tariff will cause a 15% increase in the price of parts imported from a number of countries, and wants to find out the new costs of those parts affected by the changes. However, (s)he cannot update T_0 in place before the new tariff policy takes effect. This is a hypothetical query for XML. One approach to achieving this is by creating a separate copy of T_0, updating the copy and then computing the costs by posing queries on the updated copy. An attractive alternative is to use transform query to build an XML view. The user can learn the effects of the update on the query without updating the original T_0 by posing query on the result of transform query.

Security views. In an organization, a number of user groups with access to T_0 may be subject to different access-control policies. Such a policy might, for example, prevent disclosure of *price* information from suppliers of countries for which the group of users was not responsible. To enforce the access control policy, each group is provided with a security view [17] that returns a document containing all the data from T_0 *not* about the sensitive *price* information. Since each user group has a slightly different view, it is not in general reasonable to materialize and maintain each of the provided security views; thus the views should be *virtual*. Unfortunately, as mentioned above, such views are far from trivial to write by hand in XQuery: the *price* information may appear at arbitrary depths in T_0. In contrast, it is conceptually straightforward to "delete" the

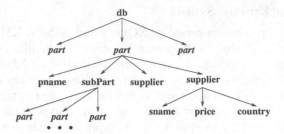

Fig. 1. An example XML document

price data. Using the syntax in [9], we can write such a view as conceptually straightforward manner as a transform query. Note that the intention is *not* to delete this data *in the source*; instead, it is merely to define the security view of a client with the update syntax. The view prevents the disclosure of price information of those suppliers of countries for which the group of users is not responsible.

"Updating" virtual views. Consider allowing users to pose an update on T_0, while T_0 is actually a *virtual* document defined as an XQuery view. In this case, there may be *no sensible* notion of performing an update on the virtual data; but one could still query a new document that would result from such an update on the document. This can be done by (a) writing a transform query Q in terms of the desired update, (b) composing Q with the view definition, and (c) composing user query with the composition of Q and the view definition. Again *no materialization* of the view is actually performed while one can query *an "updated" virtual view*.

Message transformation. Another application of transform queries is XML message transformation where one often wants to create a modified version of the original XML message without destroying it.

Formally, transform queries [9] are of the form:

transform copy $a := doc("T_0")$ **modify do** $u(\$a)$ **return** $a

where $u(\$a)$ is an *embedded update* expression. Here we study primitive updates supported by most proposals for XML update languages [9,21,28,30,38] which are of one of the following four forms:

insert	e **into** a/p	**delete**	a/p
replace	a/p **with** e	**rename**	a/p **as** l

where p is an XPath expression, e is a constant XML element (subtree), and l is a label.

To illustrate the semantics of these operations, consider an XML tree T with root r. The **insert** operation finds all the *elements* reachable from r via p in T, and adds the new element e given by *const-expr* as the last child of each of those elements. More specifically, (1) it computes $r[\![p]\!]$; (2) for each element v in $r[\![p]\!]$, it adds e as the rightmost child of v. Similarly, **delete** first computes $r[\![p]\!]$

and then removes all the nodes in $r[\![p]\!]$ (along with their subtrees) from T. The operation **replace** computes $r[\![p]\!]$ and then replaces each v in $r[\![p]\!]$ with e defined by *const-expr*. Finally, **rename** computes $r[\![p]\!]$ and for each v in $r[\![p]\!]$, changes the label of v to s.

Example 1. Referring to the XML tree T_0 of Fig. 2, let e be a *supplier* element with name HP. Then, one can apply the following update operations to T_0:

(1) **insert** e **into** p_1, where p_1 is $\mathcal{X}$ expression *//part[pname = 'keyboard']* *//part[¬supplier/sname = 'HP' ∧ ¬ supplier/price < 15]*; this is to first find every *keyboard* in T_0, and then for each of its subparts that is supplied neither by HP nor at a price lower than \$15 by any supplier, add e as a supplier;

(2) **delete** p_2, where p_2 is *//part[pname = 'keyboard']/subpart//supplier* *[¬sname = 'HP' ∧ ¬price < 15]*; this is to remove from T_0 the suppliers of all subparts of any *keyboard* except for supplier *HP* and those suppliers selling at a price lower than \$15;

(3) **replace** p_3 **with** e, where p_3 is *//part[pname = 'keyboard']/supplier [sname = 'Compaq']*; this is to substitute e for the supplier *Compaq* of any *keyboard*;

(4) **rename** *//country* **as** *address* changes the label *country* to *address* for every *country* in T_0.

Each operation may incur multiple changes at an arbitrary depth of T_0, since the same *part* element may occur at different places of T_0, due to the subpart hierarchy.

We next give three user cases in details to explain the transform query and its composition with user query.

Example 2. Composing a query with an update. Recall the XML document shown in Fig. 2. The goal of the query below is to return a list of suppliers from the *Bahama* with all information in the supplier subtree other than price. The query is a *composition* of two queries, namely, a user query and a transform query.

```
<result> {
    let $n := transform copy $a := doc("foo") modify do delete //price return $ a
    for $x in $n//supplier
    where $x/country = 'Bahama'
    return $x
} </result>
```

Example 3. Security views. We show how to enforce security for XML by using transform queries. Suppose that we want to enforce an access control policy such that a user group cannot see suppliers from country 'A'. We provide the user group with a security view without releasing information about suppliers from country 'A'. The user can then pose queries in XQuery on the security view. Below we give the transform query to define the virtual security view, as well as a user query on the view that is to find suppliers for *keyboard*. The composition of the two can be written as follows.

```
<result> {
    let $n := transform copy $a := doc("foo") modify do delete $a//price return $ a
    for $x in $n/part[pname = 'keyboard']/supplier
    return $x
} </result>
```

Example 4. Hypothetical queries. Suppose that DELL will start a promotional campaign in half a year: for all competitive parts from HP, DELL will sell them at 80% discount. A customer wants to check the lowest price for all subparts of keyboard if the promotion is in place. The first part can be expressed with a transform query, which is to replace the price offered by DELL with 20% of its original price if a part is also supplied by HP. The second part can be expressed with a user query posed on the results of the transform query.

```
<result>{
    let $n := transform copy $a := doc(T0) modify do
            replace $a //part[supplier/sname = 'HP']/supplier[name= 'DELL']/price
            with 20%*price return $a
    for $x in $n//part[pname = 'keyboard']//part/pname
    let $s := $x/supplier
    let $p := for $m in $s return if ($m/price = min($s/price)) then $m
    return <minprice pname = "$x"> $p </minprice>
} </result>
```

To our knowledge, [18] is the first work that seriously considers the evaluation of transform queries and their composition with user query. One obvious solution of evaluating transform queries seems to first make a copy T of the original XML T_0 and then update T to obtain T' using XML updates supported by an XQuery engine. Such a solution needs to maintain a separate version of the transformed (updated) documents. This is clearly not efficient for hypothetical queries that do not request the whole updated document but aim to query the updated document. This solution is not feasible for updating virtual views that have no materialized XML views. Worse still, very few publicly available XQuery engines support updates and they usually employ DOM tree, which is very memory intensive and makes them difficult to cope with large documents.

Two solutions are proposed in [18] to evaluate transform query. 1)The first solution consists of three *algorithms implemented on top of* XQuery *engine*. The three algorithms are completely realized with XQuery , and thus can be evaluated on existing XQuery engines. In other words, they remain side-effect free and thus can benefit from the existing and upcoming optimizations utilizing side effect free property in XQuery engines. Moreover, their implementations can be readily migrated to any XQuery engine conforming to XQuery 1.0. The first algorithm is based on a Naive *query rewriting* technique to translate transform queries (and their embedded updates) into standard XQuery . It has quadratic-time data complexity in the worst case. To remedy the high complexity, two algorithms with *linear-time data complexity* are developed using automaton techniques. 2) The second solution consists of an algorithm implemented within XQuery engines. This algorithm can cope with large XML documents while the first solution on top of existing XQuery engines cannot since most existing XQuery engines represent XML documents as memory intensive DOM trees, and do not handle large XML documents very well. In contrast to the first solution, this algorithm needs to extend existing XQuery engines.

In addition, efficient algorithms are proposed to compose user and transform queries into a single query in standard XQuery such that it is not required to separate the evaluations of user query and transform query. The algorithm significantly outperforms the conceptual strategy that sequentially evaluates transform query and user query one by one, since it shares the common computation of transform query and user query, and avoid unnecessary update to answer the user query in many cases.

Experimental study in [18] shows that the proposed algorithms are efficient and scale well, both *when* implemented in XQuery on top of query processor and as part of the query processor implementation; the second solution can handle very large XML documents while the memory overhead is very small; and the composition technique is effective.

Our final note is that the algorithms for evaluating transform query yield a convenient approach to supporting XML update functionality when update support is not available on a particular platform. For XML data stored as a file in a file system, the *lower bound* of time required to update a document is linear in the size of the data (for uploading the data from and re-serializing out to the file system), which is comparable with the efficiency of transform queries produced by our algorithms. Furthermore, translating updates to queries allows us to use a uniform optimizer for both queries and updates.

3.2   Update Via XML Views

In the section, we discuss updating relations though their XML views. The view update for relations is a class technical problem and has attracted extensive attention (see, e.g. [13,14,26,29]). Given an update on relational view, the problem of updating relational data through relational views is often defined as translating the view update into a side-effect free update on the underlying relational data, where the update on base-table can ensure that the view is successfully updated as request, and will not result in side-effect.

It is increasingly common to find XML views published from relational databases [22,36,39], often as an interface to the databases. Similar to their relational counterpart, the need of updating relational databases through their XML views arises. Given an XML view of a relational database, we want to propagate updates of the XML view to the original relational tables, without compromising the integrity of the XML view and the underlying relational data.

Example 5. Consider a *materials* database I_0, which maintains *part* data, *supply* records, *supplier* data and a relation *supply*. It is specified by the relational schema R_0 (with keys underlined):

part(<u>pno</u>, pname),   supplier(<u>sno</u>, name, price, country),
supply(<u>pno, sno</u>),   component(<u>pno1, pno2</u>)

where a tuple *(pno1, pno2)* in *component* indicates that *pno2* is a subpart of *pno1*. That is, *component* gives the component hierarchy of parts.

From the relational database an XML view T_0, depicted in Fig. 2 (the dotted lines will be illustrated shortly), is published for the material database by extracting part-supplier data from I_0. The view is required to conform to the DTD D_0 below (the definition of elements whose type is PCDATA is omitted):

```
<!ELEMENT  db        (part*)>
<!ELEMENT  part      (pno, pname, subpart, supplyBy)>
<!ELEMENT  subpart   (part*)>
<!ELEMENT  supplyBy  (supplier*)>
<!ELEMENT  supplier  (sno, name, price, country)>
```

As depicted in Fig. 2, the XML view contains a set of *parts*. Each *part* has a *pname* (part name), a set of *suppliers* and a *subpart hierarchy*; a *supplier* in turn has a *name* (supplier name), a *price* (offered by the supplier), and a *country* (where the supplier is based and is ignored in the figure for clarity).

Note that the view is defined recursively since the DTD D_0 is recursive (*part* is defined indirectly in terms of itself via *subpart*).

We next illustrate the main problems that should be considered with an example. Now consider an XML update Δ_X = insert T' into P_0 posed on the XML view T_0, where P_0 is a (recursive) XPath query *part[pno=P650]//part[pno= P320]/subpart*, and T' is the subtree representing the part P200. To carry out Δ_X, we need to find updates Δ_R on the underlying database I_0 such that such updates will not cause side-effects on XML view T_0.

Given the above example, the first problem that should be considered is how the XML view T_0 is materialized and stored. As in [11], it is compressed into a DAG by sharing, e.g. *part* subtrees that are *subparts* of multiple *parts*, as indicated by the dashed lines in the figure and the DAG is stored in relations. One motivation for the compression is that it is often significantly smaller than the original tree and may even lead to exponential saving in space. In contrast, XML

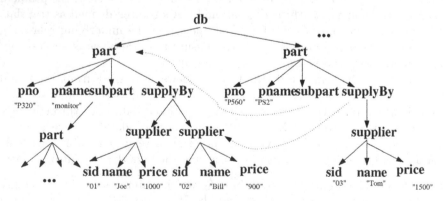

Fig. 2. Example XML view

view T_0 can also be materialized without sharing common subtree instances, which may need a different solution from that in [11].

The second problem is how to locate the nodes to be updated. The evaluation algorithm is dependent on the storage of XML views, considered in the first problem. As observed by [27], translation from (recursive) XPath queries (resp. updates) over recursive XML views (stored in relations) to SQL queries (resp. updates) is nontrivial. Add to the complexity is that one also needs to check side-effect of XML view in the process of computing nodes to be updated, the next problem to be considered.

The third problem is to check whether the update may cause XML view side effects. To explain side-effect of XML views, referring to the example above, Δ_X asks for inserting P200 as a *subpart* of *only* those P320 nodes below P650, whereas P320 has a unique *subpart* hierarchy (published from the same relational records) and thus the insertion will result in side effects if the P_0 do not cover every paths reaching node P320. The reason for XML view side effects is that the same relation objects are published to multiple instances in XML views. Hence if we allow this to occur, we need to cope with XML view side effect. Obviously the side-effect detection should be effectively coupled together with the evaluation algorithm in the second problem to achieve better performance.

The fourth problem that should be considered is the update semantics. Referring to the example above, if we follows the update semantics used in XML update [9,40], it is often the case that the update will result in side effects. In order to be consistent with the semantics of the XML view, [11] resolves the side effect, by revising the insert semantics such that the insertion will be performed at *every* P320 node. The DAG storage in [11] makes it easier to accommodate such a semantics; otherwise one needs to find all occurrences of P320 node to perform such a update. Note that the effect of side effects on delete semantics is even more subtle and calls for a new semantics (see [11]).

As a summary, in order to map updates over the XML view to updates over the corresponding relational views, the following sub-questions should be considered: 1) how are XML views materialized and stored? 2) how is the update semantics defined for XML view update? 3) how to locate the nodes that need to be updated? and 4) how to check whether the update will generate side-effect on XML views? What adds the complexity is that these sub-questions should not be dealt with in isolation.

While these are new issues beyond what we have encountered in relational view updates, automated processing of relational view updates to translate updates over relational views to updates on underlying base table is already intricate, even under various restrictions on the views [13,14,26]. In fact even the updatability problem, i.e. the problem of determining whether a relational view is updatable w.r.t. given updates, is mostly unsolved and few complexity results are known about it [5,13]. This tells us that it is unrealistic to reduce the XML view update problem to its relational counterpart and then rely on the commercial DBMSs to do the rest. A *key-preservation* condition on select-project-join views is identified, which is less restrictive than the conditions imposed by previous work [13,14,26], and some techniques are proposed to handle relational view update under key-preservation condition.

In addition to [11], the other work [4,?,42] on updating XML views also approaches the problem in two steps: 1) map from updates on XML views to updates on relational views, and 2) map from updates on relational views to updates on base-table. We next see how the other work on updating XML views fits with the two step framework by investigating how do they deal with the above outstanding issues. In [4], XML views are defined as query trees and are mapped to relational views; XML view updates are translated to relations only if XML views are well-nested (i.e., key-foreign key joins), and if the query tree is restricted to avoid duplication. Since it does not allows duplication in XML views and it assumes that the set of nodes to be updated are already specified in update statement, the update will not cause side-effect on XML views and it does not need to compute the set of nodes to be updated. Moreover, without worry about XML side effect, it can follow the update semantics in XML update work. The work in [42] requires a *round-trip* mapping that shreds XML data into relations in order to ensure that XML views are always updatable. More general XML views where duplication is allowed is considered in [43]. A detailed analysis on deciding whether or not an update on XML views is translatable to relational updates, along with detection algorithms, are provided in [45]. A framework for [45] is presented in [44]. XML view side-effects are considered in [45]. However it considers neither detecting side-effects in the process of computing the set of nodes to be updates, nor modifying update semantics to accommodate XML view side-effects.

In contrast with the work in [11], other work on XML view updates [4,42,44,45] cannot handle recursively defined XML views although DTDs (and thus XML view definitions) found in practice are often recursive. Nor do they support XML updates defined with recursive XPath queries, as opposed to proposals, e.g. [28,9], for XML update languages. Another limitation of the previous approaches is that they focus on translating XML view updates to relational view updates, and assume that the underlying relational DBMS handles the rest; however, most commercial DBMSs only have limited view-update capability [23,35,39].

Commercial database systems [23,35,39] provide support for defining XML views of relations and restricted view updates. IBM DB2 XML Extender [23] supports only propagation of updates from relations to XML but not vice-versa. Oracle XML DB [35] does not allow updates on XML (XMLType) views. It provides XMLType views to wrap relational tables in XML views using SQL statement. In SQL Server [39], users are allowed to specify the "before" and "after" XML views using *updategram*, a data structure for users to express changes in XML data, instead of update statements; the system then computes the difference and generates SQL update statements. The views supported are very restricted: only key-foreign key joins are allowed; neither recursive views nor updates defined in terms of recursive XPath expressions are supported.

4   Discussions and Conclusions

XML views are becoming popular for both traditional relational data and XML data. There are a number of studies on the exporting and querying XML views. We present our work on building XML views using update syntax, and updating though XML views. There are a couple of open problems for XML views that are worth being investigated, and we next list three specific subjects.

1) XML storage techniques. XML storage techniques are extreme important for XML applications. Debates on possible storage techniques are still ongoing in database community, and unfortunately there is still no widely accepted solution. Different XML storage techniques will introduce new problems for both building XML view defined in terms of update syntax and updating though XML views. For example, our work [18] on building XML views with update syntax assumes that XML documents are stored in files or native XML database. However, if the XML data are fragmented and distributed over a number of sites, or are stored in relations, how to efficiently build such XML views with update queries. As another example, if the source data are XML data distributed over a number of sites, how to update them though their XML views?

2) Document-centric XML documents. Compared with data-centric XML documents, document-centric XML documents received less attention in database community. However, many XML data are document-centric data, such as bioinformatics data and some Web data, and it might be more promising to employ XML repository for document-centric data than data-centric data. Document-centric XML data often require IR and data mining queries. How to efficiently evaluate such queries? How to generate XML views for document-centric data using update syntax and queries of IR types?

3) Concurrency control for XML updates. The concurrency control should be considered for both problems discussed in this paper. This would open up some new research topics. In addition, our work only considers a subset of XPath language. How to extend it to more general query?

Acknowledgement. I would like to thank my co-authors, Philip Bohannon, Byron Choi, Wenfei Fan, and Stratis Viglas. This summary paper is mainly based on our published work.

References

1. Aguilera, V., Cluet, S., Milo, T., Veltri, P., Dan Vodislav: Views in a large-scale xml repository. The VLDB Journal 11(3), 238–255 (2002)
2. Amer-Yahia, S., Curtmola, E., Deutsch, A.: Flexible and efficient xml search with complex full-text predicates. In: SIGMOD Conference, pp. 575–586 (2006)
3. Benedikt, M., Chan, C.Y., Fan, W., Rastogi, R., Zheng, S., Zhou, A.: Dtd-directed publishing with attribute translation grammars. In: Bressan, S., Chaudhri, A.B., Lee, M.L., Yu, J.X., Lacroix, Z. (eds.) CAiSE 2002 and VLDB 2002. LNCS, vol. 2590, p. 838. Springer, Heidelberg (2003)
4. Braganholo, V.P., Davidson, S.B., Heuser, C.A: From XML view updates to relational view updates: old solutions to a new problem. In: VLDB (2004)

5. Buneman, P., Khanna, S., Tan, W.: On propagation of deletions and annotations through views. In: PODS (2002)
6. Buneman, P., Cong, G., Fan, W., Kementsietsidis, A.: Using partial evaluation in distributed query evaluation. In: VLDB (2006)
7. Carey, M.J., Florescu, D., Ives, Z.G., Lu, Y., Shanmugasundaram, J., Shekita, E.J., Subramanian, S.N.: XPERANTO: Publishing object-relational data as XML. In: Suciu, D., Vossen, G. (eds.) WebDB 2000. LNCS, vol. 1997, Springer, Heidelberg (2001)
8. Carey, M.J., Kiernan, J., Shanmugasundaram, J., Shekita, E.J., Subramanian, S.N.: XPERANTO: Middleware for publishing object- relational data as XML documents. In: VLDB (2000)
9. Chamberlin, D., Florescu, D., Robie, J.: XQuery Update. W3C working draft, http://www.w3.org/TR/xqupdate/
10. Chaudhuri, S., Kaushik, R., Naughton, J.F.: On relational support for XML publishing: Beyond sorting and tagging. In: SIGMOD (2003)
11. Choi, B., Cong, G., Fan, W., Viglas, S.: Updating recursive xml views of relations. In: ICDE, pp. 766–775 (2007)
12. Cong, G., Fan, W., Kementsietsidis, A.: Distributed query evaluation with performance guarantees. In: SIGMOD Conference, pp. 509–520 (2007)
13. Cosmadakis, S.S., Papadimitriou, C.H.: Updates of relational views. In: PODS (1983)
14. Dayal, U., Bernstein, P.A.: On the correct translation of update operations on relational views. In: TODS 7(3) (1982)
15. DeHaan, D., Toman, D., Consens, M.P., Özsu, M.T.: A comprehensive xquery to sql translation using dynamic interval encoding. In: SIGMOD Conference, pp. 623–634. ACM Press, New York (2003)
16. Deutsch, A., Fernandez, M., Suciu, D.: Storing semistructured data with stored. In: SIGMOD Conference, pp. 431–442. ACM Press, New York (1999)
17. Fan, W., Chan, C.Y., Garofalakis, M.: Secure XML querying with security views. In: SIGMOD (2004)
18. Fan, W., Cong, G., Bohannon, P.: Querying xml with update syntax. In: SIGMOD Conference, pp. 293–304 (2007)
19. Fernandez, M.F., Kadiyska, Y., Suciu, D., Morishima, A., Tan, W.C.: SilkRoute: A framework for publishing relational data in XML. TODS 27(4), 438–493 (2002)
20. Fernandez, M.F., Morishima, A., Dan Suciu: Efficient evaluation of XML middleware queries. In: SIGMOD (2001)
21. Ghelli, G., Christopher, R., Simon, J.: XQuery!: an XML query language with side effects. In: Ioannidis, Y., Scholl, M.H., Schmidt, J.W., Matthes, F., Hatzopoulos, M., Boehm, K., Kemper, A., Grust, T., Boehm, C. (eds.) EDBT 2006. LNCS, vol. 3896, Springer, Heidelberg (2006)
22. IBM: IBM DB2 Extender Reference, ftp://ftp.software.ibm.com/ps/ products/db2/info/vr82/pdf/enUS/ db2sxe81.pdf
23. IBM: IBM DB2 Universal Database SQL Reference
24. Jain, S., Mahajan, R., Suciu, D.: Translating xslt programs to efficient sql queries. In: WWW 2002 (2002)
25. Kamps, J., Marx, M., de Rijke, M., Sigurbjörnsson, B.: Structured queries in xml retrieval. In: CIKM, pp. 4–11. ACM Press, New York (2005)
26. Keller, A.: Algorithms for translating view updates to database updates for views involving selections, projections, and joins. In: PODS (1985)

27. Krishnamurthy, R., Kaushik, R., Naughton, J.: XML-SQL query translation literature: The state of the art and open problems. In: Bellahsène, Z., Chaudhri, A.B., Rahm, E., Rys, M., Unland, R. (eds.) Database and XML Technologies. LNCS, vol. 2824, Springer, Heidelberg (2003)
28. Laux, A., Martin, L.: XUpdate - XML Update Language (2000), http://www.xmldb.org/xupdate/xupdate-wd.html
29. Lechtenborger, J., Vossen, G.: On the computation of relational view complements. TODS 28(2), 175–208 (2003)
30. Lehti, P.: Design and implementation of a data manipulation processor for an XML query processor. Technical report, Technical University of Darmstadt, Diplomarbeit (2001)
31. Li, C., Bohannon, P., Narayan, P.P.S.: Composing xsl transformations with xml publishing views. In: SIGMOD 2003: Proceedings of the 2003 ACM SIGMOD international conference on Management of data, pp. 515–526. ACM Press, New York (2003)
32. Li, H., Lee, M.L., Hsu, W., Cong, G.: An estimation system for xpath expressions. In: ICDE conference, p. 54. IEEE Computer Society, Washington, DC (2006)
33. Li, Q., Moon, B.: Indexing and Querying XML Data for Regular Path Expressions. In: 27th International Conference on Very Large Data Bases (2001)
34. McHugh, J., Widom, J.: Query Optimization for XML. In: 25th International Conference on Very Large Data Bases (1999)
35. Oracle. SQL Reference
36. Oracle XML DB: Oracle Database 10g R1 XML DB Technical Whitepaper. http://download-uk.oracle.com/technology/tech/xml/xmldb/current/twp.pdf
37. Polyzotis, N., Garofalakis, M.N.: Structure and Value Synopses for XML Data Graphs. In: 28th International Conference on Very Large Data Bases (2002)
38. Rys, M.: Proposal for an XML data modification language. Microsoft (2002)
39. SQL server: MSDN Library
40. Tatarinov, I., Ives, Z.G., Halevy, A.Y., Weld, D.S.: Updating XML. In: SIGMOD (2001)
41. Theobald, M., Schenkel, R., Weikum, G.: An efficient and versatile query engine for topx search. In: VLDB, pp. 625–636 (2005)
42. Wang, L., Mulchandani, M., Rundensteiner, E.: Updating XQuery Views Published over Relational Data: A Round-trip Case Study. In: Xsym (2003)
43. Wang, L., Rundensteiner, E.A.: Updating XML views published over relational databases: Towards the existence of a correct update mapping. Technical Report, Worcester Polytechnic Institute (2004)
44. Wang, L., Rundensteiner, E.A., Mani, M.: Ufilter: A lightweight XML view update checker. In: ICDE (2006)
45. Wang, L., Rundensteiner, E.A., Mani, M.: Updating XML views published over relational databases: Towards the existence of a correct update mapping. DKE (to appear)
46. Wu, Y., Patel, J.M., Jagadish, H.V.: Estimating Answer Sizes for XML Queries. In: 8th International Conference on Extending Database Technology (2002)

A Formalism for Navigating and Editing
XML Document Structure

Frithjof Dau and Mark Sifer

School of Information Systems and Technology
University of Wollongong, Australia
{dau,msifer}@uow.edu.au

Abstract. The use of XML has become pervasive. It is used in a range
of data storage and data exchange applications. In many cases such XML
data is captured from users via forms or transformed automatically from
databases. However, there are still many situations where users must read
and possibly write their own XML documents. There are a variety of both
commercial and free XML editors that address this need. A limitation of
most editors is that they require users to be familar with the grammar
of the XML document they are creating. A better approach is to provide
users with a view of a document's grammar that is integrated in some
way to aid the user. In this paper, we formalise and extend the design
of such an editor, Xeena for Schema. It uses a grammar tree view to
explicitly guide user navigation and editing. We identify a key property
that such an editor should have, stable reversable navigation, then via
our formal treatment extend the Xeena for Schema design to satisfy it.

1 Introduction

The eXtensible Markup Language (XML) [16] has become a standard foundation
for both data exchange and electronic documents. Many XML languages for
display and data exchange have been defined: XHTML [19] for web browsers,
SVG and X3D for 2D and 3D graphics, MathML for mathematical equations,
and DocBook for written documents such as technical reports. Documents in
these languages can be displayed through suitably configured web browser or
dedicated viewers. XML is already adopted by OpenOffice 2.0, and Microsoft
is adopting XML for it's office applications as well, as they define languages
for files created by their word processor and spreadsheet applications. Universal
business language (UBL) is one of many languages developed for the exchange
of business data. These are only a small selection of the many document and
data oriented XML languages that exist.

Dedicated tools usually create XML documents. Web authoring tools provide
a WYSIWYG interface that allows users to drag and drop web page elements
from a toolbar and save the generated web pages as XHTML documents. Similar tools support other document-oriented languages such as MathML and Doc-
Book. Specific business tools and applications convert data from a variety of
sources into XML languages for exchange and processing. Another approach for

S. Bhalla (Ed.): DNIS 2007, LNCS 4777, pp. 96–114, 2007.

the entry of data documents is web based interactive forms generated from a DTD or schema by systems such as XForms [18] and Forms-XML [7].

There are XML languages for which dedicated editors are either not available or not freely available. Because an XML document is a text file, it can be created and edited with any text editor. An XML enabled web browser such as Microsoft's Internet Explorer can check it for well-formedness and validity. However, this requires authors to observe XML syntax and grammar constraints as they create and edit. Authors that rarely use XML must learn how to write with XML tags and attributes, while authors that are familiar with XML must still learn the document's grammar. If an author does not use a language regularly or the language itself is very large or complex this can be a large learning task. A better choice is an XML editor.

XML editors integrate knowledge of XML syntax and grammar constraints when given a DTD or Schema. They support the creation of arbitrary XML documents, providing automatic assistance to keep generated documents well formed and valid. The significant number of commercial XML editors [15, 20, 21] and free XML editors [2, 17] that are currently available, confirm user demand for such editors. These editors provide a grammar view that shows one level of potential elements.

The Xeena for Schema XML editor [12, 13] integrates a view of a document's grammar to aid the user. It uses a more powerful deeper grammar view that explicitly assists navigation and guides editing. This paper extends earlier work [12, 13] in two respects. Firstly, we present here a *formal* treatment of Xeena for Schema. We identify and formally prove nice key properties of the original Xeena for Schema design. Our formalism describes a grammar view tree, a document view tree and the structure preserving mappings between them that facilitate navigation and editing. Secondly, though a navigational limitation of the original design is identified, we show how this limitation can be overcome.

2   Related Work

When text documents need to conform to a grammar, authors need to know the grammar or receive some interactive assistance, in order to create them manually. Such documents include: computer source codes that must conform to a particular programming language, technical documents for which allowable structures have been defined, and more recently XML documents. Many editors that provide such interactive assistance with some kind of grammar directed editing have been created. The earliest were syntax directed program source code editors [3, 14, 22] which have had limited success, as evidenced by the fact that existing commercial program editors do not use grammar directed editing, but typically support syntax highlighting or keyword completion only. Many editors for structured documents, usually large technical document have also been developed. These have been more successful. In this section we review these document editors and the reasons for their relative success. We also review recent XML navigation tools and editors.

2.1 Structured Document Editors

Grammar based editors called structured document editors within the technical document community have been successful. This is evidenced by the range of commercial and free structured document editors such as SGML editors currently available. A key driver in the design of document preparation systems was the need to separate content from presentation, so that the same content could be re-used in a variety of settings. This required content to be structured in some way, so document portions could be referred to or located in a systematic way for presentation processing. Content was represented as a tree of portions or elements whose arrangement satisfied a grammatical structure.

Unlike computer programs, structured documents contain markup; start tags and end tags that delimit an element and explicitly type the enclosed content. In many document languages, each grammar rule corresponds to an element in a document. This means that regular grammars are often sufficient for many structured document languages rather than the richer context free grammars required for programming languages. For most computer programming languages, the non-terminal grammar rules have no corresponding visible artifact in a program. To contrast, a structured documents nested element tree can reveal its grammar while a program sources lack such an explicit tree structure. This suggests users will have greater success manipulating a structured document's more visible element structure via its grammar.

Grif [6] is an early structured document editor. It provides several document views. A plain text global view for content entry, an outline or table of contents view, a presentation view that shows final presentation and a specialised view for editing mathematical formula. During editing in the global view a popup menu shows the elements that can be validly added or inserted at the current document location, while the various views are coordinated around this current location. Grif does not provide a raw text view that includes markup tags so its documents are always well formed, and because only valid edit operations are offered the element structure is always consistent with the grammar. Grif has also been proposed as a HTML editor for web documents [10].

Another early structured document editor is Rita [5]. Its authors noted "since various types of documents require tags with different placement, the creator of a document must learn and retain a large amount of knowledge". That is, because many documents grammars are large, it can be demanding for users to be familiar enough with a document's collection of tags to manually create one. To aid users, the Rita interface provided a presentation view, an element structure view, and a dynamic menu of element tags. A user could select an element in the structure view and be presented with the choice of elements that could be validly inserted. Users could also transform elements via a menu of valid transformations. To support greater editing flexibility invalid portions of a document could be marked with a special tag so they were treated as text, while a patch area allows arbitrary cut and pasting. Both Grif and Rita guide user editing with one visible grammar level.

2.2   Querying and Mining XML Structure

An integrated grammar view can assist user navigation of large structured documents. It can assist navigation from one instance of a grammar rule such as a section header to the next or previous instance. BBQ [8] is such a system for navigating and querying XML documents. It presents tree views of both a document and its grammar. The grammar view recursively presents each grammar rule as a node in an indented tree. The grammar tree view provides an overview of the grammar a user can explore to an arbitrary depth while also supporting navigation. BBQ also supports join-based queries that a user builds via multiple grammar tree views. Its focus is the application of a grammar tree view to the querying of large XML documents, while the focus of our work is the application of a grammar tree view to XML document editing.

Large grammars are also difficult to work with. Many DTDs and XML Schema are so large that most document instances only use a small subset of the defined elements. In such cases, presenting an author with all elements that can be added may be overwhelming. The difficulty is that most grammars are designed to organise content. They are not designed explicitly for visual presentation to a user. An alternative approach is to study the patterns of element use in a document, and initially only propose those elements or tree patterns the user is likely to use next. Because this approach works by studying existing document content it can be applied to XML documents have no DTD or XML Schema defined. This is the approach Chidlovskii's structural advisor for XML document authoring [4] takes. This approach could be integrated with BBQ style grammar tree view, by restricting the displayed grammar tree to nodes for common elements only. This would be a useful extension of our work.

2.3   XML Editors

A wide variety of Commercial and free XML editors that provide grammar assistance for editing are available. Some are element structure-oriented while others are content-oriented. Xeena [17] is a structure-oriented editor. Editing is done within a tree view of document elements. Users are guided with a DTD determined list of possible elements, shown in a separate panel. Users can select one of these potential elements and add or insert around the current document position. The Topologi document editor [15] is a content oriented editor for progressively convert arbitrary text documents into valid XML documents. It allows users to rapidly apply markup to large documents. Its primary editing view is a raw text view. It also offers grammar (SGML and XML) guidance when adding elements, providing users with a choice of valid elements.

Editors can also provide multiple editable views. XMetal [20] provides several editable views: a structure view, formatted view and formatted view with exposed tags. Like Xeena it offers a grammar sensitive list of valid elements to guide editing. XMLSpy [21] provides two editable document views: an indented tree view and a nested tree element view. It also offers a grammar sensitive list of valid elements to guide editing. The Amaya editor [2, 11] provides formatted, structure, source text and specialised views for markup languages such as

XHTML, SVG and MathML. It also provides a table of contents view to aid rapid navigation. All its views are editable. Again, elements can be added by choosing from a list of valid elements. Elements can also be added manually, in which case, the system tries to create a series of in-termediate elements down to the chosen type based on the grammar.

These editors use a document's grammar to determine the element list presented in a menu or panel that can be added into a document. Users make valid edits by inserting them. However, such a list provides a narrow view of the grammar that is only one level deep. It also does not indicate the grammar role of each potential element, whether it required, optional, repeatable or a choice. Our editor design is structure-oriented and uses a deeper BBQ like multi-level grammar tree view annotated with grammar roles to better support structure based navigation and editing. A design that we refine and formalise in this paper.

3   Building and Navigating

This section demonstrates Xeena for Schema editing and navigation to provide a context for the formal treatment.

3.1   Starting

On startup Xeena for Schema presents a document and its grammar. Figure 1 shows a new report document that contains a title and empty body in the right panel and its grammar tree in the left panel. The indented grammar tree has been manually unfolded to show several levels of the report grammar. It provides an overview that guides user editing. The grammar view uses the visual syntax shown in figure 2. In XML grammars, each element has a prefix called a *cardinality operator* that indicates how many times it may occur in a document according to the grammar. An '?' indicates that it is optional, i.e. it can occur zero times or once; an '*' indicates that is can occur arbitrary often, including zero times; and an an '+' indicates that is can occur arbitrary often, but it has to occur at least once. If no cardinality operator is given, the new element has to occur exactly one. In our ongoing formalization, this will be denoted by '1'. Besides the cardinality, each child element participates in either a sequence or a choice. In Xeena for Schema, round parentheses indicate the former and square parentheses indicate the latter. A grammar may be recursive. For example if the List node in figure 1 were unfolded it would contain another List node, and so on without limit. The grammar view also shows which elements have been instantiated in the document. Instantiated nodes have coloured icons while other nodes icons are grey. In figure 1 the document only contains a report, title and empty body element. Only their corresponding nodes in the grammar view are shown red. The grammar tree view provides both a comprehensive overview of potential structure and an indication of what has been instantiated so far.

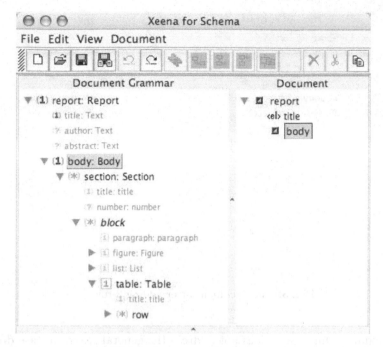

Fig. 1. A new report document

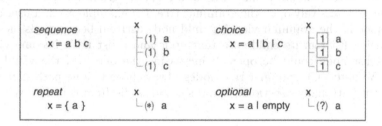

Fig. 2. Visual syntax of our grammar view

3.2 Building

A document is built by adding elements. The grammar tree view shows which elements can be added. Any grammar node that can occur multiple times or any node that can occur once but has not been instantiated (shown as grey) can be added. In figure 1 the report grammar was unfolded five levels deep to show the row elements. A user adds three row elements by just selecting it and clicking the add toolbar icon three times. Figure 3 shows the result. Three row elements, all required intermediate elements and all required compulsory elements are added by the editor.

An XML grammar may be recursive. Our grammar view supports this. For example in a report document, a list contains listItems which contain text and

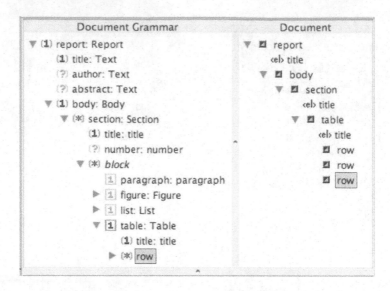

Fig. 3. The document after adding 3 rows

an optional list. Our design does not address the general recursive case directly because a user will never unfold a recursive grammar tree to an infinite depth, but it does address the finite unfoldings of a recursive grammar that a user does when they progressively open the grammar tree. For example when a user opens the list node in the grammar view its child node listItem becomes visible, then after opening listItem its children a text node and a list node become visible, then this list node could be opened; increasing the depth of the visible tree further. We note that grammar tree nodes always have unique path names, for example the path names section.list and section.list.listItem.list are distinct.

3.3   Unstable and Stable Navigation

The grammar view always shows the surrounding context for the current document cursor element. After adding the three row elements in figure 3, if the user selects the table node in the grammar view, the document cursor will move to the corresponding table element. Figure 4 shows this. To reverse this, the user can select the third row element in the document view again. The result would be figure 3 again. However, with Xeena for Schema if the user attempts to do this via the grammar view, i.e., if he clicks in the last step not on the third row element in the document view, but on the row element in the grammar view instead, the behaviour is different. The user would expect to return to the third row after this up-down navigation via the grammar view, but instead the document cursor has lost it's initial position (third row) and ended up on the first row. This is shown in figure 5. This is a limitation of the Xeena for Schema design that our formal treatment will characterise and address.

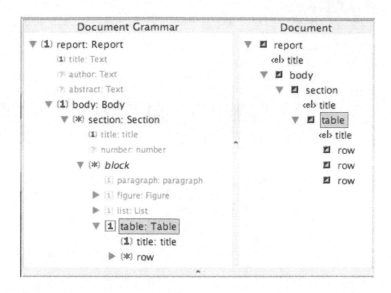

Fig. 4. The user selects the table node in the grammar view movings the document cursor to the corresponding table element

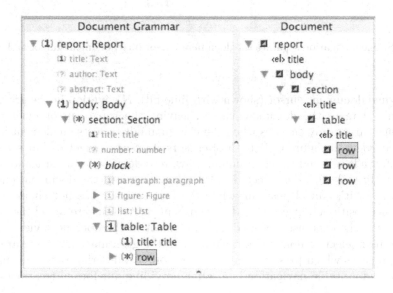

Fig. 5. The user selects the row node again in the grammar view moving the document cursor to the first row element

Figure 6 outlines a stable navigation design that addresses this. The document cursor is separated into two parts: (i) the element the user has clicked (shown with a black rectangle) which established the range of interest and (ii)

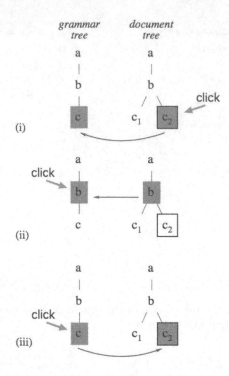

Fig. 6. Stable navigation separate the document cursor into document cursor and range cursor

the original document cursor (shown with blue fill). Navigation in the grammar view only changes the document cursor, leaving the range cursor unchanged. The range cursor only changes when the user manually selects an element in the document view. In figure 6 (i) both of cursors start as element c_2 but separate in (ii) after b is selected in the grammar view, into document cursor b and range cursor c_2. In (iii) after c is selected in the grammar view, the document cursors returns to c_2 it's initial position, while the range cursor has not changed and so remains positioned at c_2. This downwards navigation has retained the initial context, set when the user selected the c_2 element in the document view.

If the user selects a non-comaprable node in the grammar view there are several approaches which preserve stable navigation. We describe these alternative approaches for stable navigation later after providing our formal treatment.

4   Labelled Ordered Trees

In this section, XML grammars and XML documents are formalized as labelled ordered trees, according to the notions of [1] or [9]. We assume that the reader is familiar with trees. To clarify matters, trees are in our approach formalized as posets $(P, \leq)$, where the smallest element of P is the root of the tree. For

$p \in P$, we set $\downarrow p := \{q \in P \mid q \leq p\}$ and $\uparrow p := \{q \in P \mid q \geq p\}$. Technical terms like Upper/lower neigbours, minimal/maximal elements, childs, siblings and leaves etc. are defined as usual. The infimum of two elements p and p is denoted by $p \wedge q$. For $p, q \in P$ with $q < p$, let q^p denote the uniquely given upper neighbour of q with $q < q^p \leq p$. In the Hasse-diagram representation of trees, will draw them 'upside down', i.e., the bottom element of the tree is at the top of the representation.

Each element $p \in P$ gives rise to the path $\downarrow p$. In the following scrutiny on XML-trees, we are more interested in the paths than the elements of P, but as the elements stand in one-to-one to the paths, we do not provide a notation for the paths on its own. Note we have $p \leq q \Longleftrightarrow \downarrow p \subseteq \downarrow q$.

Labelled ordered trees will be defined to be trees, where the nodes in the tree correspond to XML elements, their labels denote the type names of the elements. Given a node, its children are totally ordered. For the ongoing formal treatment, it is convenient to extend this additional order to all elements of the tree.

Definition 1 (Additional Order on Trees). *Let $(P, \leq)$ be a tree. A partial order $\sqsubseteq$ on P is called order on siblings, iff[1] whenever two distinct elements p, q are comparable w.r.t. $\sqsubseteq$, they are siblings. If moreover siblings are always comparable w.r.t. $\sqsubseteq$, we call $\sqsubseteq$ a total order on siblings, and $(P, \leq, \sqsubseteq)$ is called ordered tree.*

To each order on siblings $\sqsubseteq$, we assign an extension $\circledast$ of $\sqsubseteq$ as follows: For two incomparable (w.r.t. $\leq$) elements $p_1, p_2 \in P$, let $q := p_1 \wedge p_2$, and we set $p_1 \circledast p_2 :\Leftrightarrow q^{p_1} \sqsubseteq q^{p_2}$. Let $\circledast$ be the reflexive closure of $\circledast$.

In the Hasse-diagram representation of trees, given two siblings p and q, we will draw p to the left of q iff $p \sqsubseteq q$.

Of course, $\circledast$ is not simply another order on P, but it has to respect some additional restrictions. For example, if a path A is left to a path B, then each subpath of A should be left to B as well. We first fix this idea.

Definition 2 (Respecting $\leq$). *A relation $R \subseteq P \times P$ respects $\leq$, if we have $R \cap \leq = \Delta_P \ (:= \{(p, p) \mid p \in P\})$, and for all p, q, p', q' with $p \neq q$, $p \leq p'$ and $q \leq q'$, we have $pRq \Leftrightarrow p'Rq'$.*

Now, given a tree $(P, \leq)$, the orders R on P which respect $\leq$ are exactly the orders obtained by extending orders on siblings. Moreover, such an R is obtained from extending a *total* order on the siblings iff R is a total order on the leaves of P. This is captured by the following lemma.

Lemma 1 (Extending Sibling Orders). *Let $(P, \leq)$ be a tree, let $\sqsubseteq$ be an order on siblings of P. Then $\circledast$ is an order extending $\sqsubseteq$ which respects $\leq$. Moreover, $\sqsubseteq$ is a total sibling order iff $\circledast$ is a total order on the leaves.*

Vice versa, let $(P, \leq)$ be a tree and R be an order on P which respects $\leq$. Then there exists a order $\sqsubseteq$ on the siblings with $R = \circledast$.

---

[1] iff: if and only if.

Proof: We start with the first propostion. We first show that $⊞$ is a strict order. As $⊞$ is defined only on pairs of incomparable (w.r.t. $\leq$) elements, it is obviously irreflexive. Now let $p_1 ⊞ p_2 ⊞ p_3$; we have to show $p_1 ⊞ p_2$. Let $q_1 := p_1 \wedge p_2$, $q_2 := p_2 \wedge p_3$, and $q := p_1 \wedge p_3$. We have $q_1^{p_1} \sqsubset q_1^{p_2}$ and $q_2^{p_2} \sqsubset q_2^{p_3}$. Moreover, as $q_1, q_2 \leq p_2$, we see that q_1 and q_2 are comparable. We do a case distinction. For $q_1 = q_2$, we have $q = q_1 = q_2$, thus $q^{p_1} \sqsubset q^{p_2} \sqsubset q^{p_3}$, hence $q^{p_1} \sqsubset q^{p_3}$, thus $p_1 \sqsubset p_3$. For $q_1 < q_2$, we have $q = q_1$, thus $q^{p_1} \sqsubset q^{p_2} = q^{p_3}$, hence $q^{p_1} \sqsubset q^{p_3}$, thus $p_1 \sqsubset p_3$. The case $q_1 > q_2$ is done analogously to the last case. So $⊞$ is a strict order, thus $⊞$ is an order.

Obviously, for siblings p_1, p_2 and $q := p_1 \wedge p_2$, we have $p_1 = q^{p_1}$ and $p_2 = q^{p_2}$, thus $⊞$ is indeed an extension of $⊞$. Moreover, $⊞$ respects $\leq$ by definition.

Next, let $\sqsubseteq$ be a total sibling order, and let p_1, p_2 be two distinct leaves of P. Then p_1 and p_2 are incomparable w.r.t. $\leq$. Let $q := p_1 \wedge p_2$, thus $q < p_1, p_2$. As q^{p_1}, q^{p_2} are siblings, they are comparable with respect to $\sqsubset$, thus, due to the definition of $⊞$, p_1 and p_2 are comparable with respect to $⊞$. Thus $⊞$ is a total order on the leaves.

Vice versa, let $⊞$ is a total order on the leaves, and let p_1, p_2 be two siblings. Then there are leaves p_1', p_2' with $p_1 \leq p_1'$ and $p_2 \leq p_2'$. Now p_1' and p_2' are comparable with respect to $⊞$, and $⊞$ respects $\leq$, so p_1 and p_2 are comparable with respect to $\sqsubseteq$.

As now the first proposition of the lemma is proven, we continue with the second. Let R be an order on P which respects $\leq$. Let $\sqsubseteq$ be the restriction of R to siblings. Let p_1, p_2 distinct elements of P. If p_1, p_2 are incomparable (w.r.t. $\leq$), for $q := p_1 \wedge p_2$ we have

$$p_1 \, R \, p_2 \overset{R \text{ resp. } \leq}{\Longleftrightarrow} q^{p_1} R \, q^{p_2} \overset{Def. \sqsubseteq}{\Longleftrightarrow} q^{p_1} \sqsubseteq q^{p_2} \overset{Def. ⊞}{\Longleftrightarrow} p_1 \sqsubseteq p_2$$

If p_1, p_2 are comparable (w.r.t. $\leq$), then $(p_1, p_2) \notin R$ by Def. 2 and $(p_1, p_2) \notin ⊞$ by Def. 1. We conclude $R = ⊞$. □

So a labeled ordered tree could be alternatively be defined to be a tree with an additional order which respects $\leq$ and which is a total order on all leaves of the tree. To ease the notation, if an ordered tree $(P, \leq, \sqsubseteq)$ is given, we will from now on identify the total order $\sqsubseteq$ on the siblings with its extension $⊞$. Note that for two distinct nodes p, q, they are incomparable w.r.t. $\leq$ iff they are comparable w.r.t. $\sqsubseteq$.

Definition 3 (Labeled Ordered Tree). *Let $\mathcal{L}$ be a set of labels, denoting the type names of the elements. An ordered, labeled tree (loto) is a structure $(P, \leq, \sqsubseteq, l)$ where $(P, \leq)$ is an ordered tree and $l : P \to \mathcal{L}$ is a labeling function.*

Let $(P, \leq^P, \sqsubseteq^P, l^P)$ and $(Q, \leq^Q, \sqsubseteq^Q, l^Q)$ be ordered, labeled trees. We say that $p \in P$ and $q \in Q$ are congruent and write $p \cong q$ iff the total orders $\downarrow p$ and $\downarrow q$ are 'the same sequence of labels', i.e., there exists a order-isomorphism $i :\downarrow p \to \downarrow q$ with $l^Q(i(s)) = i(l^P(s))$ for all $s \leq p$. An ordered, labeled tree $(P, \leq, \sqsubseteq, l)$ is called purified, iff there are no $p, p' \in P$ with $p \neq p'$ and $p \cong p'$. Finally, for two distinct siblings p, q, we set $p \sqsubseteq_l q :\Longleftrightarrow p \sqsubseteq q$ and $p \cong q$.

The acronym 'loto' is adopted from [9] and stands for 'labeled ordered tree object'. Note that, similarly to ⊴, the order $\sqsubseteq_l$ respects $\leq$.

Now we are finally prepared to define XML grammars and documents by means of labeled ordered trees.

Definition 4 (XML-Trees). *A document tree is a nonempty labeled ordered tree. A grammar tree is a structure $(P, \leq, \sqsubseteq, num, nk)$, where $(P, \leq, \sqsubseteq)$ is a nonempty, purified, labeled ordered tree, $num : P \rightarrow \{*, +, ?, 1\}$ is a mapping such that nk maps the bottom of P to 1, and $nk : P \rightarrow \{s, c\}$ is a mapping.*

The mapping num formalizes the cardinality operator of XML grammars, and the mapping nk (nodekind) indicates whether we have to make a choice among the siblings or not. Note that we consider only non-ambigous XML-grammars, that is, the corresponding XML-trees are purified.

5 Mapping Documents to Grammars

As described in section 3, in each state the user has selected an element in the XML grammar and an element in the XML document. Changing the selected element in the XML grammar affects the selected element in the XML document, and vice versa, changing the selected element in the XML document affects the selected element in the XML grammar. To describe the interdependence between the elements in the XML grammar and in the XML document formally, in this and the following section, we define mappings between the grammar tree and the document tree.

In this section, we start with the easier direction, which is mapping nodes in the document tree to nodes in the grammar tree. In order to do so, we have to define when a document tree conforms to a given grammar as well. This is done by the following definition.

Definition 5 ((Partial) Valid Docs). *Let $\underline{G} := (G, \leq^G, \sqsubseteq^G, l^G, num^G, nk^G)$ bea grammar-tree, let $\underline{D} := (D, \leq^D, \sqsubseteq^D, l^D)$ be document-tree. We say that $\underline{D}$ is a partial valid document w.r.t. $\underline{G}$, iff*

1. *For each element $d \in D$, there exists exactly[2] one $g \in G$ with $d \cong g$. This element will be denoted $\psi(d)$.*
2. *For all $d_1, d_2 \in D$ we have*

$$d_1 \sqsubseteq^D d_2 \implies \psi(d_1) \sqsubseteq^G \psi(d_2)$$

   That is., ψ respects the left-right-order $\sqsubseteq$.
3. *There do not exist a $d \in D$ with distinct children $d_1, d_2 D$ with $nk(d) = c$ and $d_1 \not\cong d_2$ (i.e., ψ respect choices).*

Let $d \in D$. We say that d is completed, iff we have:

---

[2] As $\underline{G}$ is purified, we could replace 'exactly' by 'at least'.

1. *If $nk(\psi(d)) = $ s, then:*
   (a) *For each child g' of $\psi(d)$ with $nk(g') = +$, there exists at least one child d' of d with $\psi(d') = g'$.*
   (b) *For each child g' of $\psi(d)$ with $nk(g') = $?, there exists at most one child d' of d with $\psi(d') = g'$.*
   (c) *For each child g' of $\psi(d)$ with $nk(g') = 1$, there exists exactly one child d' of d with $\psi(d') = g'$.*
2. *If $nk(\psi(d)) = $ c, and if there is a child g' of g with $num(g') \in \{+, 1\}$, then d has at least one child.*

We say that a partial valid document D is a complete valid document W.R.T. G iff all $d \in D$ are completed.

The mapping ψ is the mapping from grammar trees to document trees, which is used to describe the navigation formally. A first nice property in terms of *stable navigation* is that ψ respects $\leq$, and on purified downsets, it is even an order-embedding.

Lemma 2 (Properties of ψ). *Let $\underline{G} := (G, \leq^G, \sqsubseteq^G, l^G)$ be a grammar tree and $\underline{D} := (D, \leq^D, \sqsubseteq^D, l^D)$ be partial valid document tree w.r.t. $\underline{G}$. Then ψ respects $\leq^D$, i.e., for all $d_1, d_2 \in D$ we have $d_1 \leq^D d_2 \Rightarrow \psi(d_1) \leq^G \psi(d_2)$, and $\psi[D]$ is a downset of G (with $\psi[D] := \{\psi(d) \mid d \in D\}$).*

If $B \subseteq D$ is a purified downset, then ψ restricted to B is even an order embedding, i.e., for all $b_1, b_2 \in B$ we have $b_1 \leq^D b_2 \Leftrightarrow \psi(b_1) \leq^G \psi(b_2)$.

Proof: Let $d_1, d_2 \in D$ with $d_1 \leq^D d_2$. Then there is a $g' \in G$ with $g' \leq \psi(d_2)$ and $g' \cong \psi(d_1)$, thus $g' = \psi(d_1)$ due to the definition of ψ. Hence we get $\psi(d_1) \leq \psi(d_2)$. It can be similarly argued that $\psi[D]$ is a downset.

Now let B be a purified downset and $b_1, b_2 \in B$ with $\psi(b_1) \leq^G \psi(b_2)$. Then there exists $d \in D$ with $d \leq b_2$ and $d \cong \psi(b_1)$, hence $d \cong b_1$. As B is a downset, we have $d \in B$, and as B does not contain distinct and congruent elements, we have $d = b_1$. Hence we obtain $b_1 \leq b_2$.   □

6   Mapping Grammars to Documents

In the last section, we defined the mapping ψ from document trees to grammar trees. The definition of ψ is straight forward. In this section, we consider the other direction, that is, mapping nodes in the grammar tree to nodes in the document tree.

Note first of all, as the document tree will usually be only partial valid, but not completed, we cannot expect that we find for each grammar node a congruent document node. On the other hand, it might happen that for a given node in the grammar tree, we can have a number of different congruent nodes in the document tree.

The basic idea of our approach is that we have a selected node, called range cursor, in the document tree. This range cursor will be used to identify a set of nodes in the document tree which can be considered to be a *range of interest* fixed by the range cursor. It is the introduction of this range cursor into our formalism that makes stable navigation possible. The mapping from the grammar tree to the document tree will map each grammar node to a document node in this range of interest. We will use the letter φ to denote the mapping from grammar trees to document trees, and as φ As φ depends on on a range cursor r, it will be indexed with r (i.e., we write φ_r).

Assume we have an XML-document, and a range cursor, which is a node rc in the corresponding document tree. This node corresponds to a path. To which other paths can we move? We can move to subpaths of the given path $\downarrow r$, but not to paths which are congruent to a subpath of the given path $\downarrow r$ without already being a subpath. Or to put it another way: If we have a path which deviates from the given path, then the node where it deviates must have a different label. This idea will be fixed by a target area $TA(r)$. Now, the target area may still contain congruent, but different nodes. If we have such a choice, we choose the left-most path. The left-most elements in the target area will be called the range of r. We first have to fix these notions formally.

Definition 6 (Target Area, Range). *Let* $\underline{D} := (D, \leq, \sqsubseteq, l)$ *be a labeled, ordered tree, let* $r \in D$ *be an element. Let*

$$TA(r) := \{d \in D \mid \forall q \leq d \; \forall c \leq r : q \cong c \Rightarrow q = c\} \quad .$$

Now let $Rg(r)$ *be the set of all minimal elements of* $TA(r)$ *with respect to* $\sqsubseteq_l$. *We call* $Rg(r)$ *the range of* r.

Due to the informal description, it should be clear that the target area of a given document cursor is a down-set. This carries over to the range. Moreover, the range is purified. These claims are proven in the following lemma.

Lemma 3 (Properties of Range). *Let* $\underline{D} := (D, \leq, \sqsubseteq, l)$ *be a labeled, ordered tree, let* $r \in D$. *Then* $Rg(r)$ *is a downset w.r.t.* $\leq$, *and it is purified.*

Proof: We first prove that $Rg(r)$ is a downset. Let $d \in Rg(r)$ and $e < d$. We obviously have $Rg(r) \subseteq TA(r)$, thus $d \in TA(r)$, and $TA(r)$ is a downset w.r.t. $\leq$, thus $e \in TA(r)$ as well. Assume that e is not minimal w.r.t. $\sqsubseteq_l$, i.e., there exists $f \in TA(r)$ with $f \sqsubset_l e$. Then, as $\sqsubseteq_l$ respects $\leq$ due to Lem. 1, we have $f \sqsubset_l d$ as well, in contradiction to the minimality of d w.r.t. $\sqsubseteq_l$ in $TA(r)$. So we obtain $e \in Rg(r)$.

Assume $Rg(r)$ is not purified. So there are $d, e \in Rg(r)$ with $d \neq e$ and $d \cong e$. Then d, e are incomparable w.r.t. $\leq$. Let $q := d \wedge e$, thus $q < d, e$. From $q^d \leq d$, $q^e \leq e$, and $d \cong e$ we conclude $q^d \cong q^e$. Thus q^d, q^e are comparable w.r.t. $\sqsubseteq_l$, thus d, e are comparable w.r.t. $\sqsubseteq_l$. So d or e is not minimal w.r.t. $\sqsubseteq_l$, which is a contradiction. $\qquad\qquad\qquad\qquad\Box$

Now we can define the desired mapping φ_r and prove that it is, similar to ψ, order-preserving, and that it to some extent the inverse mapping to ψ.

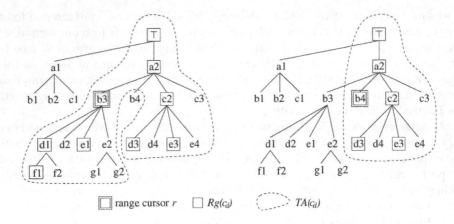

range cursor r $Rg(c_d)$ $TA(c_d)$

Fig. 7. Target areas and ranges

Lemma 4 (Mapping Induced by a Document Cursor). *Let a grammar-tree* $\underline{G} := (G, \leq^G, \sqsubseteq^G, l^G)$ *be given and let* $\underline{D} := (D, \leq^D, \sqsubseteq^D, l^D)$ *be partial valid document tree w.r.t.* $\underline{G}$, *let* r *be a document cursor. Let* $\varphi_r : G \to D$ *be defined as follows:*

$$\varphi_r(g) = \max_{\leq}\{d \in Rg(r) \mid \psi(d) \leq g\}$$

Then φ_r *well-defined and order-preserving (w.r.t.* $\leq$). *For each* $b \in Rg(r)$, *we have* $\varphi_r(\psi(b)) = b$. *Finally, we have* $\psi(\varphi_r(g)) \leq g$ *for each* $g \in G$.

Proof: Let $\Phi(g) := \{d \in Rg(r) \mid (\psi(d) \leq g\}$ for each $g \in G$. The root of D is an element of $Rg(r)$, thus it is an element of $\Phi(g)$, so $\Phi(g) \neq \emptyset$. Let $d_1, d_2 \in \Phi(g)$. Then as $\psi(d_1) \leq g$ and $\psi(d_2) \leq g$, $\psi(d_1)$ and $\psi(d_2)$ are comparable w.r.t. $\leq$. As $Rg(r)$ is a purified downset due to Lem. 3, from Lem. 2 we conclude that d_1 and d_2 are comparable w.r.t. $\leq$ as well. So $\Phi(g)$ is a nonempty total order, thus it has a maximal element, i.e., φ_r is well-defined. Moreover, for $g_1 \leq g_2$, we have $\Phi(g_1) \subseteq \Phi(g_2)$, thus $\varphi_r(g_1) = \max_{\leq}\Phi(g_1) \leq \max_{\leq}\Phi(g_2) = \varphi_r(g_2)$, so φ_r is order-preserving.

We have $\Phi(\psi(b)) = \{d \in Rg(r) \mid (\psi(d) \leq \psi(b)\} = \{d \in Rg(r) \mid d \leq b\}$ for each $b \in Rg(r)$ due to Lem. 2, which yields $\varphi_r(\psi(b)) = b$.

Finally, $\psi(\varphi_r(g)) \leq g$ follows immediately from the definition of φ_r. $\square$

For illustrating φ_r, we come back to the second example of Fig. 7. The corresponding mapping φ_r is depicted in Fig. 8.

7   Stable Navigation Approaches

Our formal description of the grammar to document mapping in the previous section introduced a range cursor that was distinct from the target area. That is, the document may contain two cursors from a users perspective. A cursor that captures where the user last clicked and a cursor that shows where the target of

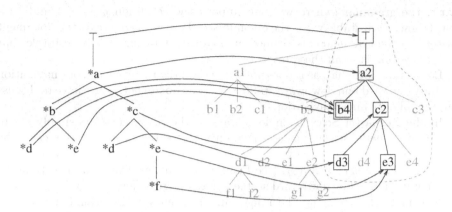

Fig. 8. Example for φ_r

the grammar cursor is. Figure 6 introduced a visual syntax for these cursors. In this section we describe how this separation provides the foundation for several stable navigation designs.

When using Xeena for Schema, in each state an element in the XML grammar, the *grammar cursor*, and an element in the XML document, the *document cursor*, is selected. Changing the selected element in the grammar view induces a new selected element in the document, and vice versa, changing the selected element in the document view induces a new selected element in the grammar. This changes have been formally captured by the mappings ψ and φ_r. The mapping φ_r relies on a range cursor r. For our formalization of navigation, let a state be a triple (c_g, c_d, r) with a grammar cursor $c_g \in G$, a document cursor $c_d \in D$, and an range cursor $r \in D$. Now the navigation can be captured as follows:

1. If a user clicks in a state (c_g, c_d, r) on a grammar node g, then let $(g, \varphi_r(g), r')$ be the new state.
2. If a user clicks in a state (c_g, c_d, r) on a document node d, then let $(\psi(d), d, r')$ be the new state.

We leave it for a moment open how r' is obtained.

We have shown that both ψ and φ_r are order preserving. That is, for example, if a user clicks in a given state on a grammar element below the grammar cursor, the document cursor changes to a new document cursor below the old one as well. So, the navigation is to some extent well behaved.

Now assume a user selects a grammar cursor g_1, then selecting a grammar cursor $g_2 \geq g_1$, and then he selects g_1 again. This yields a series of states (g_1, c_1, r_1), (g_2, c_2, r_2), (g_1, c_3, r_3). The question is: Does at the end of this 'hopping up and back' procedure on the grammar side, we are back on the same document cursor as well? If this is true, i.e. if we necessarily have $c_3 = c_1$, we call our navigation upward stable w.r.t. the grammar. The notions of downward stable w.r.t. the grammar (where we then impose the condition $g_2 \leq g_1$) and upward/downward stable w.r.t. the document (where we navigate on the document instead of the grammar) are defined analogously. Obviously, it is desirable that the navigation is in all these respects stable.

Let us first note that, as ψ does not depend on the range cursor, our navigation is always upward and downward stable w.r.t. the document. So we have to discuss only stable grammar navigation.

In the Xeena for Schema implementation, the range cursor is simply set to be the document cursor (i.e., we have only states (c_g, c_d, c_d). In this case, the navigation is upward stable w.r.t. the grammar. This can easily be seen, as we have $c_2 \geq c_1$, thus $c_1 \in Rg(c_2)$ (see Lemma 3). But, as discussed in section 3.3, this navigation is not downward stable w.r.t. the grammar.

Automatically changing the range cursor whenever the grammar cursor is changed is not appropriate. Certainly not when a user changes from a given grammar cursor to a new grammar cursor below, the range cursor should remain. Thus a different approach for changing the range cursor is needed. There are basically two different approaches:

1. The range cursor remains if the a user changes from a given grammar cursor to a new comparable grammar cursor (below or above). If the user changes to a new, incomparable grammar cursor, the range cursor is set to the new document cursor.
2. The range cursor is never changed when the user changes the grammar cursor. Instead, the user has to explicitly set the range cursor.

These approaches yield stable navigation. It is unlikely that a pure theoretical investigation can determine which of the given behaviours is best suited for users. The different approaches need to be implemented then evaluated in future work. But regardless of the choice, as they all yield stable navigation they are a significant improvement on the Xeena for Schema implementation.

8   Conclusion

We have presented a formal description that extends the grammar based navigation and editing of XML document trees implemented in Xeena for Schema. Our formal description abstracts beyond that implementation by dealing with generic XML grammar and document trees so that the design can be applied more broadly.

A key part of our formalism was the inclusion of a range cursor that captures user selection in the document view. This extra state allowed our revised design to support stable navigation, unlike Xeena for Schema which does not. This process of finding limitations and solving them is a major benefit of formalising designs in general, but is particularly so in the case of designing visual notations and languages and the interactive systems that use them.

Acknowledgment

The second author led the development of the Xeena for Schema XML editor while he was with the Knowledge Management group at IBM's Haifa Research Laboratory in Israel.

References

[1] Abiteboul, S.: Semistructured data: from practice to theory. In: Proc. of the IEEE Symposium on Logic in Computer Science, IEEE Computer Society Press, pp. 379–386. IEEE Computer Society Press, Los Alamitos (2001)

[2] Amaya editor/browser (accessed 2007), http://www.w3.org/Amaya/Overview.html

[3] Balance, R.A., Graham, S.L., Van De Vanter, M.L.: The pan language-based editing system. ACM Trans. on Software Engineering and Methodology 1(1), 95–127 (1992)

[4] Chidlovskii, B.: A structural Advisor for the XML document authoring. In: Proc. ACM Document Engineering, ACM Press, New York (2003)

[5] Cowan, D.D., Mackie, E.W., Pianosi, G.M., Smit, G.V.: Rita an editor and user interface for manipulating structured documents. Electronic Publishing 4(3), 125–150 (1991)

[6] Furuta, R., Quint, V., Andre, J.: Interactively editing structured documents. Electronic Publishing 1(1), 19–44 (1988)

[7] Kuo, Y.S., Shih, N.C., Tseng, L., Hu, H.C.: Generating form-based user interfaces for XML vocabularies. In: Proc. ACM Document Engineering, pp. 58–60. ACM, New York (2005)

[8] Munroe, K.D., Papakonstantinou, Y.: BBQ: A visual interface for integrated browsing and querying of XML. In: Proc. of IFIP Visual Database Systems, pp. 277–296. Kluwer Academic Publishers, Dordrecht (2000)

[9] Papakonstantinou, Y., Vianu, V.: DTD inference for views of xml data. In: Proc. of ACM PODS, pp. 35–46. ACM Press, New York (2000)

[10] Quint, V., Roisin, C., Vatton, I.: A structured Authoring Environment for the World-Wide Web. In: Proc. of the World Wide Web Conference (1995)

[11] Quint, V., Vatton, I.: Techniques for authoring complex XML documents. In: Proc. ACM Document Engineering, pp. 115–123. ACM Press, New York (2004)

[12] Sifer, M., Peres, Y., Maarek, Y.: Xeena for schema: creating xml data with an interactive editor. In: Bhalla, S. (ed.) DNIS 2002. LNCS, vol. 2544, pp. 133–146. Springer, Heidelberg (2002)

[13] Sifer, M., Peres, Y., Maarek, Y.: Browsing and editing xml schema documents with an interactive editor. In: Bianchi-Berthouze, N. (ed.) DNIS 2003. LNCS, vol. 2822, pp. 97–111. Springer, Heidelberg (2003)

[14] Teitelbaum, T., Reps, T.: The Cornell Program Synthesizer: A syntax-directed programming environment. Communications of the ACM 24(9), 563–573 (1981)

[15] Topologi markup editor (accessed 2007), www.topologi.com/products/tpro

[16] XML The extensible markup language 1.0 (3rd edn.), W3C recommendation (2004), www.w3.org/TR/2004/REC-xml-20040204

[17] Xeena at alphaworks (accessed 2007), www.alphaworks.ibm.com/tech/xeena

[18] Xforms 1.0 (second edition), W3C recommendation (2006), www.w3.org/TR/xforms

[19] XHTML 1.0 The extensible hypertext markup language (2nd edn.), W3C recommendation (January 26, 2000) (revised August 1, 2002), www.w3.org/TR/xhtml1

[20] XMetal Author(accessed 2007), na.justsystems.com/content.php?page=xmetal

[21] Xml Spy (accessed 2007), www.altova.com/manual2007/XMLSpy/SpyEnterprise/

[22] Zelkowits, M.: A small contribution to editing with a syntax directed editor. In: Proc. of the ACM Software Engineering Symposium on Practical Software Development Environments, ACM Press, New York (1984)

SSRS—An XML Information Retrieval System

Zhongming Han, Wenzheng Li, and Mingyi Mao

College of Computer Science and Technology
Beijing Technology and Business University
{hanzm,liwz,maomy}@th.btbu.edu.cn

Abstract. In this paper, a novel XML information retrieval system is proposed. It supports our extended IR language, which integrates our notion of relevance scoring. We develop the inverted index structure, which can efficiently term and phrase search. A novel scoring method is also presented to filter the irrelevant nodes. As our system supports two kinds of formats for the returning fragments, we take different measures to realize. Experiments show that the system can efficiently and effectively handle XMLIR style queries.

1 Introduction

XML data is now available in different forms ranging from persistent repositories to streaming data. The Initiative for the Evaluation of XML retrieval (INEX) [20], for example, was established in April 2002 and has prompted researchers worldwide to promote the evaluation of effective XML retrieval, which has attracted a lot of research attention. Consequently, some XML retrieval systems have been proposed.

In traditional information retrieval system, the retrieved results will be the whole documents and which elements should be considered is unknown to users. But in the XML IR system, it is not the same situation. Returning the whole XML document to users is not the desired form of results what users really expect. If the whole XML document is returned, users have to go through the whole document themselves to make sure which part of the documents they really care about. It's of great importance to filter the relatively irrelevant XML nodes. But deciding which XML nodes are relevant and which XML nodes are irrelevant to the corresponding XML query turns out to be a challenge for XML IR system. This challenge deals with a brand new and important problem: relevance scoring. As in the full-text query, the element could not be defined when one retrieval query is proposed. That means some query term cannot be explicitly assigned to any node. During the query process, according to the relevance score of nodes and the threshold value, some irrelevant nodes will be filtered from the retrieve results. Consequently, it is of great importance to utilize the document structural information in computing the relevance score. For traditional information system, relevance score is based on term frequency or vector space model. Currently, these technologies have been widely used to the query term explicitly assigned to a node. But to the query term not explicitly assigned to any node, the traditional technologies are not adequate and the structural information must be made good use of in order to more efficiently support the vagueness of XML IR style.

S. Bhalla (Ed.): DNIS 2007, LNCS 4777, pp. 115–131, 2007.

For different XML retrieval systems, another important problem is IR style query language. XPath and XQuery [18, 19], which provide powerful structured queries over XML documents, have only limited rudimentary capabilities for querying the unstructured data in XML documents using full-text search. These capabilities are primarily based on the Contains function, the expressiveness of which is limited. In addition, the Contains function cannot score query results which is necessary to compute the relevance of query answers when querying textual content in documents. Up till now, extensions of these languages have been proposed in order to cover document-centric processing as well. Our previous work proposes a new relevance scoring method. In order to efficiently integrate these relevance scoring methods into our system, we must also extend these languages.

Our main contributions are as follows:

1. EXIR language is presented, which is a full-text extension of XQuery and integrates our notion of structural and semantic relevance scoring mechanism.
2. Based on HiD index structure, we implement the three-level inverted list index and the algorithm for term and phrase searching.
3. A novel relevance scoring approach is proposed. Two methods to support the formats of the returning XML fragments are also presented.

The rest of the paper is structured as follows. In Section 2, the system architecture is introduced. The implementation of the system is detailed in Section 3. Comprehensive experiments are conducted to evaluate SSRS system in Section 4. In Section 5 we review the related work. Finally, we make some concluding remarks in Section 6.

2 System Architecture

Our system is called structural and semantic relevance scoring (SSRS) XML information retrieval system. Fig.1 depicts the architecture of the system implementation. At the core of our implementation, is the IR module that interacts with a set of other modules. During the preprocessing stage, the HiD index, which is created in indexing module by using the XML document from the XML document collection, is stored in the HiD index.

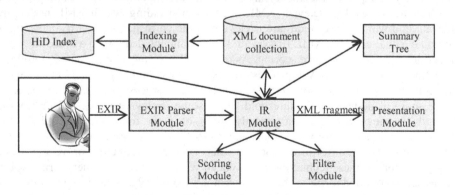

Fig. 1. System Architecture

Then it comes to retrieval stage. When an EXIR query is entered, the system parses the query and first identifies and evaluates any nested XQuery expressions in the full-text query. The EXIR query evaluation proceeds as follows. First, for all the predicates in the query, namely tag constraints, value-based constraints and node constraints, a scored pattern tree based on those constraints is constructed in EXIR parser module. When the scored pattern tree is built, it will be efficiently processed in IR module by using our HiD index information and our novel tree pattern query algorithms. After the query processing is finished, query results, together with term frequency *TFs*, inverted document frequency *IDFs*, and data path numbers *DPNums* of the nodes will be returned. These returned index information will be used to compute the relevance score of nodes in the scoring module with the notion of structural relevance in contrast to the traditional way of computing the relevance score. By doing so, the query results are turned into the initial scored data trees. Based on the relevance score and a threshold value, the irrelevant *IR-nodes* in the initial scored data trees will be filtered in the filter module and filtered data trees are created. As our system support two kinds of formats for the returning XML fragments, namely, unified and disunified, which different measures are taken to realize. For unified formats, the summary tree will be reconstructed to summon up the structure of the filtered data trees without the filtered *IR-nodes* in the summary tree module. Then the filtered data trees can be reconstructed in accordance to the reconstructed summary tree. In contrast, for disunified formats, no actions will be taken to the filtered data trees. Finally, in the scoring module, result data trees are produced by computing the relevance scores for the root nodes of the trees in unified or disunified format respectively. The semantic weights for relevance scoring are specified in the EXIR. The higher the score is, more relevant is the result data tree to the corresponding IR query. The scores of the root nodes can be used to rank the result data trees.

3 System Implementation

3.1 Extending XML Query Language with IR Functionality

We now illustrate how our relevance scoring features are incorporated into XQuery. Because our relevance scoring method has been discussed in detail in our previous work [22], here the language realizing the scoring mechanism is presented. The syntax of our Extended Information Retrieval Language (EXIR) expression FTAbout is introduced as follows:

FTAbout::= Expr("ftabout" FTConditions)? FTFormat?
FTConditions ::= FTOr | FTOr FTRelevanceWeight
FTRelevanceWeight ::= ("content" AlphaExpr) | ("structure" AlphaExpr) | ("content" AlphaExpr, "structure" AlphaExpr)
FTOr ::= FTAnd ('||' FTAnd)*
FTAnd ::= FTNot ('&&' FTNot)*
FTNot ::= FTWords | '!' FTWords
FTWords ::= String
FTFormat ::= "Unified"|"Disunified"
AlphaExpr ::= (". "Digits) | (Digits". "[0-9]*)

Expr is any XQuery expression that specifies the search context, which is the sequence of XML nodes over which the full-text search is to be performed. *FTConditions* specifies the full-text search condition with the added notion of *FTRelevance-Weight*. *AlphaExpr* is used to distribute the weight over semantic relevance and structural relevance when computing the relevance score. Two *AlphaExprs* will be assigned to semantic relevance and structural relevance respectively. If the user does not specify *AlphaExpr*, then a default value for *AlphaExpr* will be used. It is worth to mention the reason why different *AlphaExpr* should be used for different XML documents. Because better relevance can be obtained under different optimize *AlphaExpr* value for different XML document owing to their different characteristics. If the XML document has less semantic information, a small *AlphaExpr* for semantic weight could be better for effectiveness. In order to fully make use of the characteristic of the document from both semantic aspect and structure aspect, *AlphaExpr* will be used in the scoring method when computing the relevance score of the query results.

The *FTConditions* is basically a Boolean query plus context for the terms. From the definition of *FTConditions*, it is clear that the search text of the full text search can be single text as well as a text string connected by different Boolean operators. The IR engine can interpret these Boolean operators the way it would normally interpret any Boolean operators. And the conventional precedence of the Boolean operators is also followed.

The option *Format* denotes the format of the returning query results. Two formats, unified and disunified, are supported. The default format is unified. For the nodes in the search context, a higher value of the relevance score should imply a higher degree of relevance to *FTConditions*.

At last, an example is taken to explain how user's full text query can be expressed by using EXIR language.

Query 1: Find all document components in DBLP.xml that are part of an article written by an author with the last name 'John' and are about 'XML' Relevance to 'Information Retrieval' and the structural relevance weight and semantic relevance weight of is 0.8 and 0.2 respectively. Query 1 can be expressed by the following expression.
//book[author/lastname='John'] ftabout 'XML' && 'Information Retrieval'[structure 0.8, content 0.2]

3.2 Indexing Stage

In our previous work [21] an efficient and novel index structure has been proposed. The specific inverted list index structure is developed here to support EXIR.

This inverted index is a three-level structure as showed in Fig.2. The upper level is a hash table for vocabulary or lexicon of all the index terms for XML documents. We call this hash table term hash table (THT), the entry of which has the form *(Term $_n$, P)*, where *Term $_n$* stands for all the terms occur in the element and *P* is a pointer. What *P* points to is a node list *(NL)*, which consists of the middle level, records all the elements the *Term$_n$* has occurred in. In this node list, each node occupies an entry which has the form *(NodeID, IX, TF)*, where *NodeID* is the element identifier containing *Term $_n$*, and *IX* is value index, *TF* is frequency of *Term $_n$* in the element. As *IX* in the middle level is the value index, which is the virtue identifier of XML nodes in the value index, the value index and structure index can so be efficiently linked together.

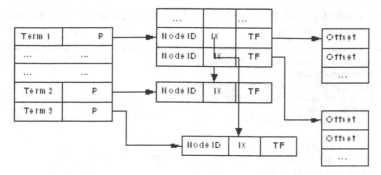

Fig. 2. The structural framework of the inverted index

In fact, the node list here not only includes text information but also structure index information. The third level is posting lists *(PL)*, which has all the actual occurrences positions of *Term_n* in the text of each node. Each posting is an offset value, which counts from the beginning of the node.

This inverted index is different from the traditional two-level inverted index in many aspects. The traditional inverted index, which has just two levels, corresponds to some extent to the first and second level of our inverted index in terms of index structure. The reasons why the traditional two-level inverted index should be improved into our three-level inverted index can be three-folded.

Firstly, the text value of nodes will be searched during the processing of XML query. So it is important to efficiently find all the nodes that have the text value of the corresponding term. The second level of our inverted index is designed to this goal. When the term in the term hash table is found in the first level, the pointer in the corresponding entry is pointed to a node list, in which all the elements include this term.

Secondly, as the text string value of a XML node can include a lot of different terms. All the position information of these term instances is important in the query processing stage. Consequently, the third level of our inverted index is so constructed as a posting list of all the occurrence offsets of terms in the same XML element.

Thirdly, phrase search method can be efficiently supported by the index structure. As showed in the value index structure, the individual term search can already be efficiently supported. Sometimes a lot of information retrieval queries deal with phrases instead of individual terms, such as "information retrieval" instead of "information" and "retrieval". In order to support phrase search method, one easy way is to indexing all the possible phrase in the term hash table, but it is obvious that indexing all the possible phrases will lead to the explosion of the hash table and great degradation of search performance. So it is not feasible for our hash table in the first level of value index to include all the possible phrases. Meanwhile, another possible way is to utilize the individual term information and get the needed information to support phrase search method, which is adopted here. The implementation follows the following steps. Firstly, the query phrases are turned into a list of all separate terms and the index for each term is looked up in the list orderly. Secondly, node lists for the corresponding terms are found according to the pointers. Thirdly, an intersection of the *NodeID* in the node lists of different terms is performed and then the offset of each

NodeID is verified to find the occurrences of the query phrases. The algorithm for the phrase search is as follows:

```
1.   Algorithm 1: Phrase Search Algorithm.
2.   Input: a query phrase
3.   Output: posting list of the query phrase found
4.   Separate the query phrase into a list of terms;
5.   Let L=list of terms;
6.   Find the first term in the L;
7.   Find the node list (NL) the pointer of the term is pointed to;
8.   While not EOF(NL) do
9.     Find the posting list (PL) corresponds to the NodeID in the NL;
10.    While not EOF (PL) do
11.      Record the entry of the posting list in PL₀[NodeID][Offset];
12.    End While
13.   End While
14.   While not EOF(L) do
15.    Find terms (t) in the term hash table in the inverted index;
16.    Find the node list (NL) the pointer (p) is pointed to;
17.    While not EOF(NL) do
18.      Find the posting list (PL) corresponds to the NodeID in the NL;
19.      While not EOF(PL) do
20.        Record the entry of the posting list in PL[NodeID][Offset];
21.        Check the matching of PL[NodeID][Offset] with PL₀[NodeID][Offset];
22.        If there is no matching then
23.          Break;
24.        Else
25.          Record the matching of PL[NodeID][Offset] in the PL₀[NodeID][Offset];
26.        End If
27.      End While
28.    End While
29.   Output the posting list of the query phrase which has been found;
```

3.3 Retrieval Stage

After the introduction of the preprocessing stage in the above section, we now describe our whole retrieval process. Before we go through the detail of the whole retrieval process, we first introduce some notions in scored pattern tree apart from scored pattern tree's definition in [22].

Firstly, the notions of *IR-node* are introduced. The nodes, for which a scoring function is defined and where an *IR-style* predicate is applied to, are called *direct IR-nodes* here. In addition, the inclusion of *direct IR-nodes* in the sub tree of a *non IR-node* will automatically converts the *non IR-node* into *IR-node*. This kind of *IR-node* is called *indirect IR-node*. Secondly, we introduce the notions of *pattern IR-node* and *data IR-node*. The *IR-node* in the scored pattern tree is called *pattern IR-node* while the corresponding *IR-node* in the result data tree matching the nodes in the scored pattern tree is called *data IR-node*. Thirdly, after *IR-nodes* in the scored pattern tree get their relevance scores, the scored pattern tree are turned into *initial scored data tree (ISDT)*. And after removing the irrelevant data *IR-nodes* in the initial scored data trees, the filtered data trees are produced. The root nodes of the filtered data trees will be scored using our relevance score method to finally create the *result data trees*.

The whole retrieval process is introduced through the following subsections.

3.3.1 Parsing of EXIR and Query Processing

In this section, the construction process of the *initial scored data trees* is analyzed, which are used for further filtering in later sections.

First of all, the *EXIR* expression is parsed to get a conjunction of predicates, which is consisted of tag constraints (*TCs*), value-based constraints (*VBs*) and node constraints (*NCs*). Secondly, these predicates should be applied to the corresponding nodes in the scored pattern tree according to EXIR. But not all the predicates can be explicitly assigned to nodes in the scored pattern tree. The traditional scored pattern tree can not handle the predicates which are not explicitly assigned to any node. Unlike the traditional scored pattern tree in previous works, we assign the predicates, which are not explicitly assigned to any node as expressed in the EXIR, to "*" node in the scored pattern tree.

Tree pattern query algorithms use an XML document and a scored pattern tree as input, and then output a collection of initial data trees. Meanwhile, the corresponding *TFs* and *IDFs* and *DPNums* of the nodes will be got.

The structure of each initial data tree in the output collection of result trees could be different, because in our scored pattern tree there could be the existence of "*" node, which gives to a good variety of the structures of the initial data tree in the output collection. "*" node can match different label path in the XML document without explicitly saying where the constraints should be applied. The following is the definition for initial data tree.

Definition 1 (Initial Data Tree): An initial data tree is a node labeled acyclic rooted tree, which satisfies the following four properties:

1. The root of the initial data tree is the same as that of the corresponding scored pattern tree;
2. The *data IR-nodes* in the initial data tree satisfy all the node constraints applied to the corresponding *pattern IR-nodes* in the scored pattern tree;
3. In the case of non -"*" node in the scored pattern tree, the relationship (parent child relationship or ancestor descendant relationship) of two nodes in the initial data tree is the same as that in the scored pattern tree;
4. In case of "*" node in the scored pattern tree, the relationship of two nodes in the initial data tree is not in accordance to that in the scored pattern tree.

3.3.2 Construction of Initial Scored Data Trees

With the *DPNums* and *TFs* and *IDFs* of the nodes, the relevance scores of nodes in the initial data trees can be computed by the scoring methods introduced in this subsection.

TF and *IDF* of each node in the initial data trees will traditionally be computed and then be used to compare with the threshold value to evaluate whether the node is relevant or irrelevant. The non-"*" node in the scored pattern tree, to which predicates are explicitly assigned to, is called *initially assigned node*. The node in the initial data trees, which includes the predicates initially assigned to the "*" node in the scored pattern tree and takes place of the "*" node, are called *lately assigned nodes*. The *lately assigned nodes,* which take place of the same "*" node, are called *brother lately assigned nodes* here. For the *initially assigned node* the *TF* and *IDF* methods are sufficient. But when it comes to the *lately assigned nodes*, the *TF* and *IDF* methods are not suitable enough. The relationship between different predicates, which are initially assigned to "*" in the scored pattern tree and lately assigned to the *lately assigned nodes*, are not taken care of by the *TF* and *IDF* methods, which isolate the *brother lately assigned nodes* from each other without taking into consideration of the structural relevance between the terms occur in them. The *TF* and *IDF* of the term in the node itself too low to be considered irrelevant, but structural relevance of the term in this node with those in other nodes may be very high. So only making use of *TF* and *IDF*, the structural linkage among the nodes will be broken and the relevance score will be inaccurately computed. The score of Article1\Section\Title node in Fig.3 (a) may be considered irrelevant, but it may have strong structural relevance with the Article1\Section\Text node, because they have the same parent node Article1\Section. It is possible Article1\Section\Title node is discarded, since it is irrelevant according to the *TF* and *IDF*. So for more accurate computation of the relevance score for *lately assigned nodes*, we import the concept of structural relevance.

$$RS_{te}^{t} = \frac{1}{1 + TD(te, te_i)} * TF(pa(te_i), te_i) \quad i \in (1, n) \tag{1}$$

The structural relevance scores for two terms *te* and *te_i* in the *brother lately assigned nodes* can be computed by this formula. Function *pa(te_i)* returns corresponding parent element of the term. *TF(te,n)* is the term frequency of the term *te* in node n, while *TD(te,te_i)* denotes term distance between *te* and *te_i*. Variable *i* falls into the range of *(1,n)*, *n* is the number *brother lately assigned nodes*, which originate from one "*" node.

There maybe many different terms occur in one XML node. Based on this formula, the average structural relevance score of the terms in a node can be computed by the following formula.

$$\overline{RS^{t}} = \frac{1}{m} * \sum_{i=1}^{m} RS_{te}^{t} * TF(te, n) \tag{2}$$

where m denotes the number of distinct terms occur in node *n*. Consequently, the *TF* and *IDF* scoring methods and the above structural relevance scoring method can be integrated to obtain the initial filter score for the "*" node by the following formula. The simple method of integrating these two scoring methods is weight sum.

$$IFS(n) = \alpha * TF(te, n) + (1 - \alpha) * \overline{RS^{t}} \tag{3}$$

where *IFS(n)* means the relevance score of node *n* in the initial data tree. *TF(te,n)* and $\overline{RS^{t}}$ denote the *TF* and *IDF* and structural relevance score respectively. α is a parameter, which can be selected by users or experts.

Given a *indirect IR-node v*, suppose there are m different child node of *v*, say v_1', v_2', ..., v_m'. The filter score of *v* can be obtained by the following formula.

$$IFS(n') = \sum_{i=1}^{m} IFS(n) * WS(v_i') \tag{4}$$

where *IFS(n')* denotes the relevance score of *indirect IR-node* and *WS(v_i')* is the semantic weight of each node v_i'. In Fig.3 are three results obtained by applying Query 1 to example XML document. The scores of the *IR-nodes* are indicated in the square brackets.

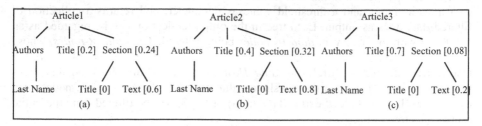

Fig. 3. Three initial scored data tree from Query 1

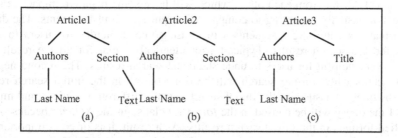

Fig. 4. Three filtered data trees from the initial scored data trees in Fig. 3, after filtering the irrelevant *IR-nodes*, which have the relevance score of less than the given threshold value

3.3.3 Result Filtering

When the initial scored data trees are constructed, the irrelevant nodes of them will be firstly filtered and then the filtered fragments will be formatted.

In order to get the irrelevant nodes filtered, a threshold value is usually chosen to evaluate whether a node is relevant or not. And making a decision on the threshold value poses much of a challenge. Typically, there are two kinds of method as to how the value of the threshold should be, namely, optimistic method and pessimistic method. In the optimistic method a predefined threshold value will be given, no matter how many nodes are relevant or irrelevant. May be in the extreme case, according to the given threshold value all the nodes are relevant or irrelevant. In the pessimistic method on the contrary, the concern is that whether there will be nodes returned or not. If the threshold is too high, then all the nodes will be irrelevant. As a result, relatively not too high threshold should used in the initial filtering process so as to prevent the extreme situation from happening.

The following formula is used to dynamically compute the threshold value after the filter scores have been obtained.

$$Threshold \quad value \ = \frac{1}{2} * (MIN \ (IFS \ ) + MAX \ (IFS \ )) \tag{5}$$

where *MIN(IFS)* and *MAX(IFS)* are the minimum value of the relevance score and maximum value of the filter score respectively, which are computed during the process of the construction of initial scored data trees in Section 4.3.2.

With the above threshold, the irrelevant nodes can be filtered by using filter function, which operates on initial scored data trees. It takes a collection of initial scored data trees as input, and a threshold value as parameter, and returns a collection of filtered data trees as output. Each tree in the output collection can be regarded as an input tree with the irrelevant *data IR-nodes* being filtered. The *data IR-nodes*, whose score are less than the threshold value, are considered irrelevant. Fig. 4 shows the result after filtering the irrelevant *data IR-nodes* in three initial scored data trees (nodes with score less than the threshold value are filtered). As a result, nodes such as, Article1/Title, Article1/Section/Title in (a) of Fig.3, will be filtered from the initial scored data tree of Fig.3(a).

Meanwhile, controlling the formats of the returning XML fragments is also very challenging. In the present XML IR systems, the formats of the returning XML fragments usually differ from each other, which will have some negative impacts. First of all, users are usually expecting to compare different returning fragments. The disunified formats of returning fragments will cause some difficulty for users to better understand the search results. Typically, an identical format for all the result fragments is very helpful for users to understand the search results. The second negative impact occurs when further search will be done based on the initial search results. When the query is issued on the initial search results, different meanings and mishandling of the query will be caused in the *for* or *let* clause in the *XQuery* because of the disunified formats in the initial search results. As a result, it is of great importance to propose a mechanism to create a unified format of the returning XML fragments.

There are two kinds of result fragment to be supported in our system, namely, disunified format and unified format. Disunified format here means that the structure of each returning XML fragment keeps heterogonous to each other, while unified format means each returning XML fragment keeps isomorphic to each other. In other words, the returning XML fragments in the disunified format approach will probably not have the same structure, while unified format approach will keep the returning XML fragment with the same structure.

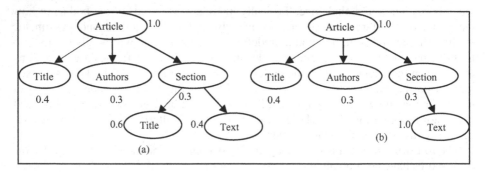

Fig. 5. The summary tree with weights is shown in (a), while in (b) is the reconstructed summary tree with renewable semantic weights

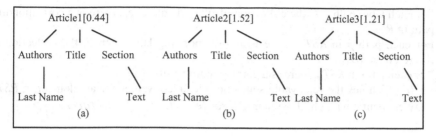

Fig. 6. Reconstruction of the filtered data trees in Fig.4 in accordance to the reconstructed summary tree with renewable semantic weights in Fig.5 (b). The score of the trees are showed in the brackets.

3.3.4 Unified Format Approach

As in HiD [21] index, a node n in an XML document can be uniquely represented by *(ln, dn)*, where *ln* and *dn* are the label path number and data path number of a node *n* respectively. Both the structure information and data instance information of node are needed to identify a node. The label path number is used to represent structure information, while the data path number is used to get the data node position.

In order to provide user a unified format approach, the structural information of the filtered data trees as the output of the filter function should be dynamically obtained. Here the summary tree of the XML document can no longer be used to get the structural information because the filtered data trees are no longer in accordance to the original structure of the XML document, which is summoned up by the summary tree of the document. The structural information of the filtered data trees must be resummoned up. With this intention, we propose a mechanism to reconstruct a new summary tree, which in the absence of XML DTD or schema for the XML document can provide sufficient information for the filtered data trees.

Firstly, the label paths of all the nodes in the filtered data trees will be recorded down. And in the summary tree every node also has its label path number. Secondly, the recorded label paths of the nodes in the filtered data trees will be used to check with that of the summary tree. Thirdly, based on the checking result, the summary tree can be reconstructed by removing the nodes in the summary tree, which do not occur in the filtered data trees. With this reconstructed summary tree, the structure information of the filtered data trees can be dynamically tracked. Due to the filtering of the irrelevant nodes in the initial scored data trees, the reconstructed summary tree will probably not be the same as the summary tree before reconstruction. Consequently, after recreating the summary tree, the semantic weights of all tags should also be renewed in order to follow the rule of semantic weights in [22] and better reflect the actual weight distribution requirements. We now define this kind of reconstructed summary tree as follows:

Definition 2(Reconstructed Summary Tree with Renewable Weights (RST_w))

A reconstructed summary tree with renewable weights RST_w for a collection of filtered data trees of an XML document D is a tree *(v, root, children)* originated from summary tree *(ST)*, where v is a finite set of element coming from the label set of D; root belong to v is the root of D; *children* is a mapping from elements to a sequence of child elements. This tree satisfies the following three properties:

1. For each label path p in the collection of filtered data trees, there is a corresponding path in RST_w;
2. For each path p in RST_w, there is a corresponding label path in the collection of filtered data trees;
3. For each path in RST_w, there is a corresponding path in ST;

$sew(e_i)'$ denotes the renewable semantic relevance weight for an element in RST_w. For any element e with child elements, $\sum sew(e_i)'=1$, $\forall e_i \in children(e)$

$$sew(e_i)' = sew(e_i) + \frac{sew(e_i)}{1 - \sum sew_d(e_i)} \times \sum sew_d(e_i) \qquad (6)$$

where $sew(e_i)$ denotes the original semantic weights of the summary tree, while $\sum sew_d(e_i)$ denotes the sum of the weights of the nodes being removed from the summary tree before reconstruction in the same level.

Now with RST_w, the filtered data trees can be reconstructed. The unified filtering function operates on the filtered data trees. It takes a collection of filtered data trees as input, and the RST_w as parameters, and returns a collection of unified data trees. Each data tree in the output collection, which is called unified data tree, will match the RST_w. We define unified data tree as:

Definition 3 (Unified Data Tree): A unified data tree, which is generated from a collection of filtered data trees and a reconstructed summary tree with renewable weights, includes the following properties:

1. For each label path p in the unified data tree, there is a corresponding path in the RST_w;
2. For each path p in the RST_w, there is a corresponding label path in the unified data tree;
3. For each path p in the unified data tree, there is a corresponding path in some of the filtered data trees;

Take a collection of filtered data tree and a reconstructed summary tree with updated weights as input and output a collection of unified data tree.

We take an example to explain the process of unifying the filtered data trees. According to Fig.4, Article1/Section/Text of (a), Article2/Section/Text (b), Article3/Title of (c) will be recorded down. The reconstructed summary tree with renewable semantic weights for the Fig.4 is shown in Fig.5 (b), while Fig.5 (a) is the summary tree before reconstruction. In Fig.6, the filtered data trees in Fig.4 are reconstructed according to the reconstructed summary tree in Fig.5 (b). Finally, score of the root node of each tree in Fig.6 will be computed using the relevance scoring methods we describe in [22]. Then the scores can be used to rank the unified data trees. Here the ranking is Article2, Article 3 and Article 1.

3.3.5 Disunified Format Approach:

After removing the irrelevant *data IR-nodes* in the initial scored data trees, which are showed in Fig.4 and the score of root node of each tree will be directly computed using the relevance scoring methods we describe in [22].

4 Evaluation of the System

In order to evaluate the effectiveness of the system, we choose SIGMOD Record in XML and sample data collection for XQuery 1.0 and XPath 2.0 Full-Text Use Cases. Experiments were run on a P4 1.8 GHz PC with 256 MB of RAM running Windows 2000 Server.

The data collection for XQuery 1.0 and XPath 2.0 Full-Text Use Cases consists of a collection of three books. All contain metadata and full text. For more competitive purposes, we extended this data collection to 100 books; the extended Data collection has more than 10000 elements with total size of 40MB. We selected 5 topics from XQuery 1.0 and XPath 2.0 full-text use cases and run 3 queries over Sigmod Record in XML. Table 2 lists these queries.

We use 'ftcontains' phrase to represent the full text retrieval, and use '&' and '|' to respectively represent the 'and' and 'or' relationship between two terms.

Query 1 to query 5 ran over the data collection. Query 6 to query 8 ran over Sigmod Record in XML. There is a common feature for these queries, i.e. these queries include full text constraints. For objectively evaluating the effectiveness of our algorithm, we investigated the Precision and Recall measure, which are often used to evaluate different algorithms of IR system. Precision represents the fraction of the retrieved documents which is relevant and recall means the fraction of the relevant documents which has been retrieved.

Figure 7 (a) and 6 (b) display the Precision and the Recall measurement of the query results respectively.

Now we simply analyze the query results. Overall, the precision and recall results are above 0.7, which means that SSRS system could effectively run XMLIR style query in average. Among these queries, the query 3, 5, 6 and 8 are more complex than others, but the precision and recall listed in Figure 7 (a) and 6(b) show that the corresponding precision and recall for these queries also are better, so SSRS system can effectively handle complex queries with respect to query structure and full text conditions.

Table 1. Queries for Experiment

Query	Content
Q1	/books/book//subject[.ftcontains "usability"]
Q2	/books/*[.ftcontains "usability testing"]
Q3	/books/book[./* [.ftcontains "goal" & "obstacles" \& "task"]]/title ftcontains "Software"
Q4	/books/book[./*[.ftcontains "usability testing"]]/Authors/*[.ftcontains "Software"]
Q5	/books/book[/metadata/*[."usability testing"]]//content/*[.ftcontains "Software"]
Q6	/IndexTermsPage//Author[.ftcontains "Han"]
Q7	/IndexTermsPage/*[.ftcontains "XML" $|$ "Information Retrieval"]
Q8	/IndexTermsPage[/Author/*[.ftcontains "Wang"]]//abstract ftcontains "XML"

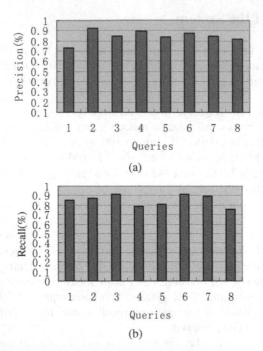

(a)

(b)

Fig. 7. Query running SSRS performance, (a) shows precision for different queries and (b) is recall for different queries

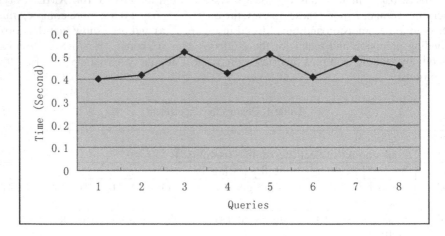

Fig. 8. Elapsed time for different queries

Fig. 8 shows the query processing performance for 8 queries running under SSRS system. It is obvious that SSRS system performs much better for all queries. For complex queries, query 3, 5 and 8, the elapsed time do not increased dynamically, from this point, we can know SSRS system can handle all kinds of queries efficiently. Elapsed time for query 7 is more than used by query 8, because the query result for query 7 include much more nodes than query 8.

5 Related Works

XML information retrieval is an active area of research. Now, the focus of research include, extending XML query languages with full text search, relevance scoring, ranking and index structure for information retrieval, etc.

In order to address the problem of XML IR, a lot of IR systems based on XML have been designed [1, 3, 4, 5, 6, 8, and 9]. However, none of the existing proposals accounts for structural and semantic relevance while computing the relevance score. Timber system [1] aims to integrate information retrieval and database queries by using XML as a means of representing structured text. In [6] an algebra, which is called TIX and based on the notion of a scored tree, is proposed. TIX can be used as a basis for integrating information retrieval techniques into a standard pipelined database query evaluation engine. And TIX also provides a way of specifying how to select appropriate elements among all the relevant ones in an XML document. Another system called XRANK [8] System is designed for ranking keyword search over XML documents. One of the design goals of XRANK is to naturally generalize a hyperlink based on HTML search engine such as Google.

At the same time, the so called full-text search with structured querying has also received a lot of attentions. A lot of research work, such as [2, 7, 11, 16] focus on XML query languages with full text search. TeXQuery [7] is powerful full-text search extension to XQuery and provides a rich set of fully composable full-text search primitives, such as Boolean connectives, phrase matching, proximity distance, stemming and thesauri. And TeXQuery satisfies the FTTF requirements specified in [17]. In [2], a language XIRQL is presented, which integrates IR related features, such as weighting and ranking, relevance oriented search, data types with vague predicates, and semantic relativism. However the language presented in this paper is based on XQL but not on XQuery [18] and XPath [19]. Other language, such as [16], is also not based on XQuery.

Relevance score is the core of XML IR system. A lot of attention has been put into this research field. Some researches [3, 10] still use the traditional vector space model to evaluate relevance score. Ranking approach [3] is assigning weights to individual contexts by extended vector space model. Main contribution of [10] is to dynamically derive the vector space that is appropriate for the scope of the query from underlying basic vector spaces. Researches [12–15] focus on how to query full text on XML documents. Most of these researches show that combining structure index and inverted list can have better performance for querying tasks.

6 Conclusion and Future Work

In this paper we have presented a novel system for dealing with information retrieval on XML documents. We propose the EXIR, which is a full text search extension to

XQuery. Based on the improved inverted index structure, the processing of EXIR can be efficiently supported. In order to better handle the occurrence of "*" in the scored pattern tree, we propose the structural relevance instead of using the traditional methods to compute the relevance score. Beside that, we propose novel methods to support two kinds of formats of the returning fragments.

Ongoing research on Top-K mechanism has attracted a lot of attentions. Realizing Top-k can largely improve the performance of the XML IR system. We also plan to develop our Top-k algorithm in our system in the future.

References

1. Yu, C., Qi, H., Jagadish, H.V.: Integration of IR into an XML Database. In: INEX (2002)
2. Fuhr, N., Grobjohann, K.: XIRQL: A Query Language for Information Retrieval in XML Documents. In: Proceedings of the 21th Annual ACM SIGIR Conference on Research and Development in Information Retrieval (2001)
3. Mass, Y., Mandelbrod, M., Amitay, E., Carmel, D., Maarek, Y., Soffer, A.: JuruXML-an XML retrieval system at INEX 2002 (2002)
4. Grabs, E., Schek, H.-J.: Flexible Information Retrieval from XML with PowerDBXML. In: INEX (2002)
5. Schlieder, T., Meuss, H.: Result Ranking for Structured Queries against XML Documents. In: DELOSWorkshop on Information Seeking, Searching and Querying in Digital Libraries (2000)
6. Al-Khalifa, S., Yu, C., Jagadish, H.V.: Querying Structured Text in an XML Database. In: Sigmod (2003)
7. AmerYahia, S., Chavdar Botev, J.: TeXQuery: A FullText Search Extension to XQuery. In: WWW (2004)
8. Guo, L.L., Shao, F., Botev, C., Shanmugasundaram, J.: XRANK: Ranked Keyword Search over XML Documents. In: Sigmod (2003)
9. AmerYahia, S., Lakshmanan, L.V.S., Pandit, S.: FleXPath: Flexible Structure and Full-Text Querying for XML. In: Sigmod (2004)
10. Grabs, T., Schek, H.-J.: Generating Vector Spaces On-the-fly for Flexible XML Retrieval. In: Proceedings of the ACM SIGIR Workshop on XML and Information Retrieval, ACM Press, New York (2002)
11. Florescu, D., Kossmann, D., Manolescu, I.: Integrating Keyword Search into XML Query Processing. In: Proc. of the Intern. WWW Conference, Amsterdam (2000)
12. Sacks-Davis, R., Dao, T., Thom, J.A., Zobel, J.: Indexing Documents for Queries on Structure, Content and Attributes. In: DMIB 1997 (1997)
13. Kamps, J., de Rijke, M., Sigurbjornsson, B.: Length. Normalization in XML Retrieval. In: ACM SIGIR (2004)
14. Williams, H.E., Zobel, J., Bahle, D.: Fast Phrase Querying with Multiple Indexes. ACM Transactions on Information Systems 22(4), 573–594 (2004)
15. Kaushik, R., Krishnamurthy, R., Naughton, J.F., Ramakrishnan, R.: On the Integration of Structure Indexes and Inverted Lists. In: Sigmod (2004)
16. Chinenyanga, T.T., Kushmerick, N.: An expressive and efficient language for XML information retrieval. J. American Society for Information Science & Technology 53(6), 438–453 (2002)
17. The World Wide Web Consortium. XQuery and XPath Full-Text Requirements http://www.w3.org/ TR/xmlquery-full-text-requirements/

18. The World Wide Web Consortium. XQuery 1.0: An XML Query Language.
    http://www.w3.org/TR/xquery/
19. The World Wide Web Consortium. XML Path Language (XPath) 2.0
    http://www.w3.org/TR/xpath20/
20. Initiative for the Evaluation of XML Retrieval (INEX) http://www.is.informatik.uni-
    duisburg.de/projects/inex03/
21. Han, Z., Xi, C., Le, J.: Efficiently Coding and Indexing XML Document. In: Zhou, L.-z.,
    Ooi, B.-C., Meng, X. (eds.) DASFAA 2005. LNCS, vol. 3453, pp. 138–150. Springer,
    Heidelberg (2005)
22. Han, Z., Shen, B.: A New Effective Relevance Scoring Algorithm for XML IR. In: Bres-
    san, S., Küng, J., Wagner, R. (eds.) DEXA 2006. LNCS, vol. 4080, pp. 12–21. Springer,
    Heidelberg (2006)

Readability Factors
of Japanese Text Classification

Lukáš Pichl and Joe Narita

International Christian University
Osawa 3-10-2, Mitaka, Tokyo, 181-8585, Japan
lukas@icu.ac.jp
http://cpu.icu.ac.jp/~lukas/

Abstract. Languages with comprehensive alphabets in written form, such as the ideographic system of Chinese adopted to Japanese, have specific combinatorial potential for text summarization and categorization. Modern Japanese text is composed of strings over the Roman alphabet, components of two phonetic systems, Japanese syllabaries hiragana and katakana, and Chinese characters. This richness of information expression facilitates, unlike from most other languages, creation of synonyms and paraphrases, which may but do not need to be context-wise substantiable, depending not only on circumstance but also on the user of the text. Therefore readability of Japanese text is largely individual; it depends on education and incorporates life-long experience. This work presents a quantitative study into common readability factors of Japanese text, for which thirteen text markers were developed. Our statistical analysis expressed as a numerical readability index is accompanied by categorization of text contents, which is visualized as a specific location on self-organizing map over a reference text corpus.

1 Introduction

Data mining, or Knowledge-Discovery in Databases is a well-established field of computational data analysis [1], built on information retrieval tools such as statistics, machine learning, pattern recognition by associative neural networks, and other algorithms. Projected onto the textual data, data mining comprises of text categorization, summarization, and clustering, concept extraction, attitude analysis, and semantic data modeling [2]. Text mining of ideographic Asian languages has been, however, left aside the mainstream research, in part because of the language barrier in communicating research results internationally, and in part because of associated economic implications. Japanese shopping search and price comparison engines, for instance, master text mining for product comparison purpose, which results in increased rates of business competition.

Computational processing of Japanese text including readability categorization is of relevance not only because of data mining in Japan, but also for its relation to computational processing of Chinese language, which represents the fastest growing online population. Chinese characters, or "kanji" were incorporated to Japanese as early as the 5th century, but were subjected to different

S. Bhalla (Ed.): DNIS 2007, LNCS 4777, pp. 132–138, 2007.

Table 1. Thirteen readability markers for five groups of Japanese text

abbr.	readability factors	I	II	III	IV	V
cv	coefficient of variation of sentence length	0.72	0.66	0.64	0.55	0.58
sk	special kanji over 15 strokes	0.89	0.88	0.88	0.88	0.92
fk	frequently used kanji (ratio of)	0.72	0.63	0.69	0.78	0.76
pa	percentage of roman letters in the text	0.09	0.02	0.01	0.02	0.14
pk	percentage of katakana letters in the text	0.11	0.03	0.04	0.08	0.08
pc	percentage of kanji letters in the text	0.26	0.33	0.24	0.44	0.39
ph	percentage of hiragana letters in the text	0.55	0.62	0.71	0.47	0.39
cp	tooten to kuten ratio (commas to dots)	1.49	1.74	1.78	1.31	2.35
ra	relative frequency of runs for alphabet	0.59	0.90	0.87	0.82	0.45
rh	relative frequency of runs for hiragana	0.35	0.37	0.30	0.45	0.47
rc	relative frequency of runs for kanji	0.59	0.62	0.69	0.48	0.45
rk	relative frequency of runs for katakana	0.29	0.33	0.25	0.31	0.29
ls	average number of letters per sequence	60.85	52.99	45.47	45.84	70.13

policy on character simplification in the second half of 20th century. Modern Japanese text typically consists of kanji, two (phonetic) syllabaries (hiragana and katakana), and the Latin alphabet. Hiragana and katakana are phonetic systems, consisting of 46 basic symbols (characters of "gojuon"[1]) corresponding to particular syllables each, which evolved from simplified characters, but include extra concepts for syllable mixing (addition of "ya", "yu", "yo" to syllables ending with "i") and voicing ("ha", "pa", "ba"); the syllabary system thus consists of data, some of which can be subjected to voicing commands (dakuten and handakuten); interestingly, the number of characters included in the syllabaries varies as 41, 46, 48 or 50, depending on the point of view, and definition of obsoleteness. This might appear as if the Japanese syllabary system was avoiding the principle of counting; this is also obvious in case of general characters, usable amount of which simply depends on individual education level. Let us note here that UNICODE tables of hiragana and katakana have 96 characters (full-width syllabary version including delimiters).

The short account above illustrates how understanding of Japanese writing system ties closely with Japanese history and education policy. Teaching methods of Japanese vary from those used in Japanese schools to those designed for foreign learners, with immediate implications on evaluation of Japanese text readability. For the sake of present analysis, "Japanese text" in what follows refers to a string of UNICODE characters in the region of the character encoding table assigned to all four writing subsystems, including delimiters.

2   Markers of Readability

By readability we mean the possibility to make sense of text contents [3,4,5]. In English, various readability indices have been studied, for instance Simple

---

[1] Literally meaning 50 sounds.

Measure Of Gobbledygook [6], Gunning fog index [7], Flesch-Kincaid Readability Test [8], or Coleman-Liau Index [9]. The ability of declamation can be one of the aspects of readability; however, ideographic Japanese text can be understood to considerable extent even without any knowledge on its pronunciation. Let us summarize the role of the four writing subsystems used in Japanese text. Roman letters usually describe concepts foreign to Japanese culture, which commonly accompanies text specialization.

The emphasizing role is also typical for the syllabary katakana, which sometimes fulfills the role similar to italics in western text, or is simply used for phonetic transcription of foreign proper names. Hiragana is common in grammar constructs, transcription of pronunciation (known as furigana) or as a replacement of kanji (if complicated or to indicate emotional load). While a modest use of hiragana simplifies text readability, if used excessively, there arise problems with disambiguation of homonyms, and also difficulties with increased length of text (lower information density accompanied by higher requirements on mental concentration). Finally, about as many as two thousand kanji characters are used in the text of newspapers on regular basis.

A typical noun consists of a pair of kanji; however, long sequences of characters without intermission by any other writing subsystem are unusual, and their meaning is difficult to discern. The aforementioned factors motivate our selection of readability markers as listed in Table 1. Included herewith are also typical values of these markers for categories of (i) blog, (ii) textbook, (iii) translation, (iv) newspaper, and (v) research article text, for which the cumulative variance of principal component data subspaces raises as 54.79%, 78.49%, 90.42% and 100.00%; the principal component loadings are shown in Table 2.

The numerical values of text markers listed in Table 1 for all five text categories illustrate not only typical features of each genre, but also some limitations of written text as a whole. For instance, sentence length varies most in case of blogs, relative frequency of kanji runs is the highest for translations, and special kanji characters are most often used in research papers. In blog text, however, text delimiters are quite often missing, which gives rise to phantom sentences after data preprocessing; moreover, some characters are used to express emotional state of the author (or meant as projection of emotional state onto the reader); in such evolving area of linguistics, definition of readability has not been available yet. Except for the last six entries in Table 1 [4], the readability markers were deployed for the sake of present analysis. In other words, the extension of readability markers has not been established in the literature.

3   Readability Index

In order to present the results of readability categorization, we introduce the following compact notation for functions defined on readability markers above,

Table 2. Correlation of principal components with readability markers

0.88	0.18	-0.93	0.84	0.76	0.86	-0.48	0.73	0.61	-0.97	0.47	0.65	-0.85
-0.45	-0.51	0.36	0.53	0.58	0.43	0.68	-0.28	0.17	0.20	0.51	-0.74	-0.51
0.14	0.23	0.03	-0.01	0.15	0.24	-0.25	-0.48	-0.76	0.06	0.68	0.16	0.15
0.07	0.81	-0.04	0.09	0.27	-0.11	0.49	-0.40	0.14	-0.10	-0.25	0.04	-0.01
0.89	0.40	-0.94	0.79	0.74	0.76	-0.39	0.66	0.65	-0.98	0.31	0.69	-0.79

$\alpha = (-0.12\text{ls} - 1.37\text{ra} + 7.4\text{rh} - 23.18\text{rc} - 5.4\text{rk} - 4.67\text{cp} + 115.79)/100$;

$\beta = 1$ for $\alpha \geq 1$ and $\beta = \alpha$ otherwise;

$\gamma = 0$ for ph$=0$, $\gamma = |\text{pc/ph} - 3/7| - (\text{pc/ph} - 3/7)^2$ for $1/4 \leq \text{pc/ph} \leq 2/3$, $\gamma = 7/10 - (3/40)(\text{pc/ph})$ for pc/ph$>2/3$, and $\gamma = 3\text{pc/ph}$ otherwise;

$\delta = 1$ for pk$>3/20$, and $\delta = (3/20)\text{pk}$ otherwise;

$\epsilon = 1$ for pa$>3/20$ and $\epsilon = (3/20)\text{pa}$ otherwise;

$\nu = -5/3(\text{fk} - 0.7) - 0.5-$ for fk>0.7, and $\nu = 1 - 5/7\text{fk}$ otherwise;

$\zeta = |(10/3)(\text{sk} - 17/20) - 0.5|$ for sk$>17/20$, $\zeta = 1 - (10/17)\text{sk}$ otherwise.

The readability index of Japanese text is then composed as follows,

$$\text{Readability index} = (\beta + \gamma + \delta + \epsilon + \nu + \zeta + \text{cv})/7, \tag{1}$$

which is normalized to the interval $< 0, 1 >$.

The above formula goes far beyond the readability index formulated by Tateishi et al. [4], which is based only on the counts of writing subsystem runs, punctuation, and sentence length. In particular, the number of strokes of kanji (approximately corresponding to school level of kanji), overall frequency for all writing subsystems, and coefficient of variation of sentence length have been added as variables. The functional form of readability index was adjusted for the present reference text corpus. It should be, however, considered a definition, in line with other readability indices. Numerical experiments in the next section demonstrate the plausability of the functional form.

The readability index in Eq. (1) was implemented online as a Java applet at http://inf.icu.ac.jp/jri/, default appearance of which is shown in Fig. 1.

By means of the online application, users can immediately obtain values of readability markers, stratify kanji used in the text to seven school grades (according to kanji education system), and find the location of the input text on a two-dimensional self-organizing map of reference corpus. The grey-scale coloring of the map can be used to visualize either the readability index itself or the readability markers, as shown in Fig. 2.

4   Self-Organizing Mapping

Although the readability index discussed in previous section is capable of ranking text from easily to hardly readable, the thirteen readability markers can be further exploited for finer categorization analysis. One obvious tool for that

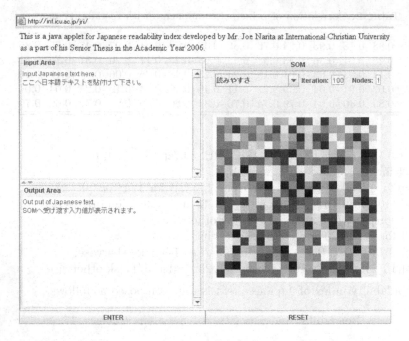

Fig. 1. Online java applet for readability index of Japanese text

purpose widely used in literature is the self-organizing map (SOM), developed for two-dimensional visualization of data clusters [10,11]. We adopt this tool here based on the corpus of reference text introduced above. The algorithm starts with nodes on a square grid, each of them (total count S) being associated with an N-dimensional reference vector m_i; the training set for SOM consists of T vectors (identical dimension N). The map organizes as follows: given an input vector $x(t)$, the nearest vector $m_c(t)$ is found by means of Euclidian distance (or in principle any other measure in the N-dimensional vector space could be used), $|m_c(t) - x(t)| = \min_{i=1...S} |x(t) - m_i(t)|$.

Next, the vectors m_i are updated according to simple learning rule $m_i(t) = m_i(t) + hc_i(t)(x(t) - m_i(t))$, where $hc_i(t)$ is non-zero only in a neighborhood of the node c with a radius given as $R(t) = R(0)(1 - t/T)$ (in the unit of grid mesh size), and $hc_i(t)$ for a node i inside the circle with center c and radius $R(t)$ has the value $h(t) = h(0)(1 - t/T)$. In other words, the learning neighborhood and speed of learning shrink as the set of training input vectors becomes exhausted. Starting with a random initial population of reference (feature) vectors $m_i(t)$, the 2D map organizes into clusters that correspond to various categories of the text.

If the SOM result is taken as a point of departure, one could in principle motivate readability index by assigning a scale to the resulting data clusters, and optimize its parameters with a restrain of some prescribed boundary conditions. In this way, the functional form of the readability index could be derived from data categorization (or clustering) method, although we did not pursue such approach here.

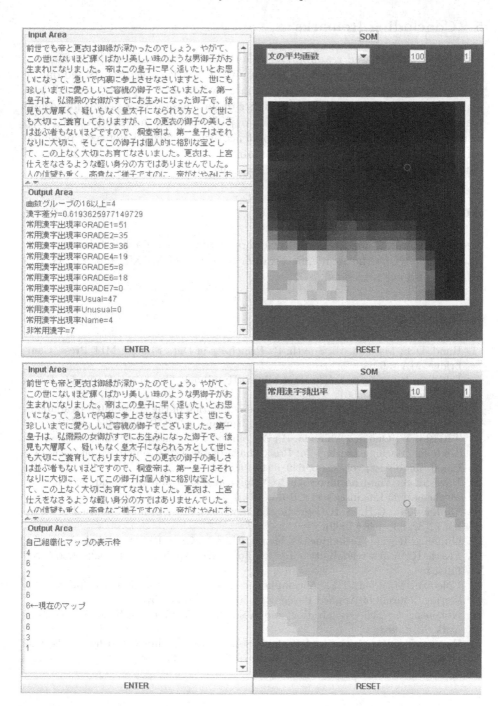

Fig. 2. Location of input text on self organizing map of reference corpus (color scale can be adjusted either for readability index or one of readability markers)

5   Concluding Remarks

Thirteen literally based readability markers were developed and expressed numerically for a corpus of reference text in five thematic categories. Based on the results, we proposed a definition of readability index of Japanese text. A java applet analyzing the literal characteristics of Japanese text was developed and made available online (http://inf.icu.ac.jp/jri/) including a self-organizing map visualization module. Direct correspondence was confirmed between principal components of reference corpus in readability marker space and boundaries of data categories in the two-dimensional visualization space.

Acknowledgements

The authors would like to thank reviewers of the article for helpful comments. Partial support by the Academic Frontier Program of MEXT and ICU Grant-in-Aid is also acknowledged.

References

1. Hand, D.J., Mannila, H., Smyth, P.: Principles of Data Mining. MIT Press, Cambridge, MA (2001)
2. Feldman, R., Sanger, J.: The Text Mining Handbook. Cambridge University Press, Cambridge, UK (2006)
3. Smith, E.A., Kinkaid, P.: Derivation and Validation of the Automated Readability Index for Use with Technical Materials. Human Factors 12, 457–464 (1970)
4. Tateishi, Y., Ono, Y., Yamada, H.: A Computer Readability Formula of Japanese Texts for Machine Scoring. In: Proceedings of the 12th Conference on Computational Linguistics, vol. 2, pp. 649–654 (1988)
5. Hayashi, S.: Yomi no nouryoku to Yomiyasusa no youin to Yomareta kekka to. Mathematical Linguistics 11, 20–33 (1959)
6. Mc Laughlin, G.H.: SMOG grading: A new readability formula. Journal of Reading 12(8), 639–646 (1969)
7. Gunning, R.: The Technique of Clear Writing. McGraw-Hill, New York, NJ (1952)
8. Flesch, R.: A new readability yardstick. Journal of Applied Psychology 32, 221–233 (1948)
9. Coleman, M., Liau, T.L.: A computer readability formula designed for machine scoring. Journal of Applied Psychology 60, 283–284 (1975)
10. Kohonen, T.: Self-Organization and Associative Memory. Springer-Verlag, New York, NJ (1988)
11. Roussinov, D.G., Chen, H.: A Scalable Self-organizing Map Algorithm for Textual Classification. Artificial Intelligence Journal 15, 81–111 (1998)

Differences and Identities in Document Retrieval in an Annotation Environment

Paolo Bottoni, Michele Cuomo, Stefano Levialdi,
Emanuele Panizzi, Marco Passavanti, and Rossella Trinchese

Department of Computer Science
University of Rome "La Sapienza"
Via Salaria 113, 00198, Rome, Italy
{bottoni,levialdi,panizzi}@di.uniroma1.it
{cuomom,marcopassavanti,trinchese}@gmail.com

Abstract. Digital annotation of web pages presents two types of problems which are unknown to traditional annotation and which are connected to the dynamicity and the openness of the Web. The first problem is related to the possibility of replicating a document over multiple sites, so that it can be retrieved over the Web at different URLs or with different queries. This poses the need to associate to a web page all the annotations pertaining to its content, even if they were created while accessing the same content under a different URL. The second problem is related to the dynamics of individual HTML pages that often consist of insertions, deletions or movement of page segments. Annotations related to portions of the page that have moved within the page itself should be retrieved and shown to the user. To reduce the impact of these phenomena on the usefulness of the annotation process, our annotation system MADCOW incorporates two algorithms which assess the identity of two pages under two different URLs, and the differences between two versions of a page under the same URL, taking the proper actions in order to retrieve all the pertaining annotations.

1 Introduction

Systems for digital annotation are inspired by the annotation activities that a person usually performs when reading or perusing a paper document or text of any sort, in professional, leisure, or private contexts.

The annotation process is typically composed of two aspects: 1) the identification of the portion of document to which reference is made, and 2) the production of the content to be associated with that portion.

While paper annotation modifies the physical document on which the process is performed, digital annotations can be separately stored and linked to the document, without modifying its digital content. This requires the use of specific software to be added to or integrated with the software which is used to explore the original content.

In a series of previous papers [3, 1, 4, 2], we have presented the development of such a software system, named MADCOW (for Multimedia Annotation of Digital

S. Bhalla (Ed.): DNIS 2007, LNCS 4777, pp. 139–153, 2007.

Content over the Web), based on a client-server architecture in which the client is a plugin which gets integrated in some existing Web browser, and the server side maintains both the actual content of the annotation, a *webnote* in MADCOW parlance, and an index relating the URLs of the annotated documents and those of the associated webnotes.

MADCOW supports the possibility of producing webnotes which are not only composed of texts, but possibly of images, videos, audio files and in general multimedia content. Analogously, the content to be annotated can be any type of multimedia object embedded in a HTML page. For some type of formats, it is also possible to refer to portions of this embedded materials. For example, besides annotating an image in its entirety, one can produce annotations for structures or groups of structures which are considered relevant in it, while for continuous media such as video or audio, annotations can be associated with intervals.

With MADCOW a user can both produce annotations and retrieve annotations on an already explored page, whether produced by him/herself, or by other users who have chosen to make their annotations public. The presence of a webnote referring to the currently loaded page on the MADCOW server is signalled by small icons, called *placeholders*, in correspondence of the position of the annotated content.

While adding the advantage of preserving the physical document when adding an annotation to it, digital annotation presents two types of problems which are unknown to traditional annotation and which are connected to the dynamicity and the openness of the Web.

The first problem is related to the possibility of replicating a document over multiple sites, so that the same document can be retrieved over the Web at different URLs. Moreover, the same physical document can be retrieved in response to different queries. If the document is not associated with a unique *Digital Object Identifier* (DOI), different annotations could be associated with different URLs, which embody parameters of the original query. Hence, a user interested in the annotations on a specific object would retrieve different webnotes for each download of the same page.

The second problem is related to the dynamics of individual HTML pages. Even if the pages are not created on the fly by some content management system, a same URL can refer over time to different versions of a HTML source, as produced by its owner. These modifications can produce a complete rewriting of the page, but more often consist of insertions, deletions or movement of page segments. Even if restricted, these modifications can impact the accuracy with which a placeholder marks the location of an annotated segment. Moreover, to avoid the phenomenon of *orphan annotations*, if the expected content does not match with the current content for the stored location in the page, the placeholder cannot be shown.

To reduce the impact of these phenomena on the usefulness of the annotation process, MADCOW incorporates two algorithms which assess the identity of two pages under two different URLs, and the differences between two versions of a page under the same URL. In the first case, the webnotes produced on the same document, under whichever URL, are retrieved and their placeholders presented to the user. In the second case, MADCOW is able to move a placeholder from its original location to the new location of the annotated fragment, if it exists in the same page.

The rest of the paper proceeds as follows. In Section 2, we revise existing approaches in literature to the management of these two problems. Section 3 gives an overview of MADCOW with particular reference to the organisation of its server side. Sections 4 and 5 illustrate the solutions implemented in MADCOW to manage identity and variation among web pages. Finally Section 6 draws conclusions and points to ongoing developments of MADCOW.

2   Related Work

Many studies have been performed on the problems of duplicate content identification by the use of signatures and similarity functions.

During this work, we referred to both methods that evaluate similarity of entire documents by evaluating the frequency of terms contained in them [11], and methods that consider selected portions of documents, code them into a number using a coding function, and rely on the equality of corresponding couples of coded values to assess the similarity between two documents [8, 13, 12, 7, 6, 9].

3   Digital Annotations and the MADCOW System

MADCOW exploits the possibility of directly accessing the DOM (Document Object Model) tree managed by the browser, both to locate the annotated content and define suitable rendering effects on it – e.g. highlighting of marked portions or positioning of place-holders – and to match the current content at a given node with the original content for that node.

MADCOW capability of annotating static pages has been expanded to deal with dynamic pages composed of different frames, recursively scanning a page content looking for frames, up to the annotation of pages not containing frames.

In order to do so, the document to be managed is not statically obtained from `Explorer.Document`, but is passed as a parameter to the methods for annotation management.

All actions take place only after the page and all its parts have been completely loaded and the corresponding DOM tree created.

Once a selection is made in a page, and the button triggering the actual annotation process is clicked on, the document attribute `activeElement` is used in the method `elementoAttivo(IHTMLDocument2 p)` to follow the path in the tree which leads to the document containing the selected element. When this is within a frame, `activeElement` does not contain an object of type `HTMLBodyClass`, but a frame object. This becomes the value of `p` for a new invocation of `elementoAttivo()`. When a `HTMLBodyClass` object is reached, the process resumes its normal behaviour.

If the user first activates the annotation button and then selects a portion of the document, after the first click MADCOW notices that the current selection is empty and starts waiting for a selection event. Once this occurs, MADCOW runs the same procedure as before.

An analogous strategy is used to position placeholders for annotations done within frames. The method `downloadAll()`, after performing some validity checks, including those on the user's identity, calls `searchFrame(IHTMLDocument2 doc)`, which recursively looks for the frames contained in the document referred by `doc`. Whenever a document with a BODY, rather than FRAMESET, tag is reached, the method `downloadAll2(IHTMLDocument2 p)` is called, which performs the actual download of the placeholders, as for a normal HTML page.

The annotations are composed in a new lightweight dialog window such as that in Figure 1.

Fig. 1. The window to write the content of an annotation

The retrieved webnotes are presented as normal HTML pages, dynamically generated from the database content, and are associated with context metadata information on the page in which the original content was present, the author and the time of the annotation, as shown in Figure 2.

4 Duplicate Source Documents

Creation and publication of content is one of the relevant aspects of the Internet. While in the first years of the web content creation was left to companies, universities and public organizations, with the advent of blogs and other forms of facilitation of web page creation many web users publish new web pages at an impressive rate.

Content in such pages is often adapted or copied from other existing pages, so it is more and more common that the same content is available completely or partially under different URLs. If we add the effect of mirror sites that deliver

Fig. 2. A webnote presented as a HTML page

exactly the same content of their original counterparts, and the one of formatting (when the same content is published in different formats in order to be accessed from different devices, or with different connection bandwidth or by people with different abilities), we can explain why only 70% of web pages have a unique content [13]. More precisely, many reasons exist for duplication of web contents, including:

- Product descriptions in ecommerce sites. Many ecommerce sites sell the same products and describe them on their pages using the same text provided by the manufacturer.
- Printer friendly versions of web pages
- Identical pages reached by different URLs. Many websites admit information for the webserver in the URL, such as referral information, session information, etc. For example: many websites include session IDs in URLs in order to trace visitors; pages that are dynamically generated according to query-string (URL) parameters, are insensitive to parameter permutations, thus they have the same content under different URLs.
- template content: template based websites repeat the same content in header, footer and often right and left columns of all their pages;
- content syndication: many websites allow other websites to reproduce their articles for free;
- ecommerce affiliation: many smaller ecommerce websites are affiliated to larger ones; they obtain content from them and publish it under a different graphical appearance;

- mirrors and replication of popular documents: many websites use mirror technology to provide better performance; many popular documents, for example software manuals, are replicated on several (50-100) different sites;
- unauthorized copies. Many websites or blogs, in order to increase the number of users and the revenues from advertisement, include unauthorized copies of material from popular sites, so that it can drive users from search engines.

Identification of duplicate documents consists in comparing documents to evaluate their syntactical similarity. It is either possible to compare all the couples of documents in a collection in order to cluster them into subsets of similar documents (*n-to-n problem*) or to compare one document with all the documents in a collection to assess its similarity to one or more of them (*one-to-n problem*).

Applications of duplicate identification have different purposes that also imply slightly different definitions of what is a duplicate. Typical applications are in the fields of search engines, copyright violation, broken links management, anti-spam email filtering, search of related documents, etc.

4.1 Duplicate Identification Methods

The goal of most duplicate identification methods is that of finding pages that are similar to others. For this reason, an obvious method of coding web-pages with a single value and comparing such values is not sufficient, as it would recognize as duplicates only identical pages. Most typical approaches are *ranking* and *fingerprinting*. Both of them are applied after that a processing of the original document is performed in order to reduce occurrence of words that do not differentiate documents from each other (*stopping*), to allow comparison of words with the same prefix reducing the number of different words (*stemming*) and to better organize the data to improve algorithm efficiency (*term parsing*).

The ranking approach consists of computing a similarity measure between the *query document* and each document in the collection. Documents in the collection are then sorted using such values and those that rank higher are considered duplicates. The similarity functions are generally based on the occurrence of the same terms in the documents, often normalized by the length of the documents. Classical similarity functions are *inner product*, in which the ranking of a document in the collection is proportional to the frequency of a term in the document and in the whole collection, and *normalized inner product* and *cosine measure* which avoid the effect of assigning higher ranking to longer documents. An example of ranking can be found in [11].

In the fingerprinting approach, on the other hand, the similarity between two documents is obtained by applying a function to selected portions of each document thus obtaining a tuple of values called signature, and then comparing values in the two signatures, counting the number of identical values found. There are four relevant aspects in the fingerprinting implementation:

- the granularity, i.e. the size of each selected portion (string) of the document
- the resolution, i.e. the number of values in the signature

- the selection strategy used to chose strings within the document
- the function used to code each selected string

Several authors proposed algorithms and systems for fingerprinting. See [8, 13, 12, 7, 6, 9].

4.2   Duplicate Documents in MADCOW Digital annotations

When a document or a portion of it is replicated many times, it is possible that different users access the duplicates and annotate them. The bad effect derived from this is twofold: i) two users that annotate two different duplicates of the same page, are not aware of each other's annotation, while they could benefit from knowing it and they could share their thoughts; ii) when a user annotates a page whose URL may change at each visit, like in the session-ID case described above, he may not be able to retrieve his annotation anymore when accessing the page again.

For the above reasons, it seems important to associate an annotation to any page that is a duplicate of the one which was originally annotated and thus to provide the user with all the annotations related both to the current page and to its duplicates.

The association of a web-note to a URL, as it was in MADCOW before this work, is not sufficient to this purpose because it allows to retrieve only those annotations created accessing the page with that precise URL. On the other hand, it would be useful to relate, conceptually, the web-notes to the page and to visualize them any time the page content is accessed using any of its URLs.

A duplicate, for MADCOW, is defined by similar content and similar structure. In fact, the structure of the DOM is used to associate a web-note to a specific portion of the web page, and it is not possible to associate the web-note if the structures of the two pages do not match at all. Most duplicates in MADCOW are related to mirror, session IDs in URLs, and URLs related to parts of the same pages (named anchors).

We decided for a fingerprint approach to identify duplicates in MADCOW, in order to match, besides identical pages accessed with different URLs, also pages with some differences in their content, but with a sufficiently similar structure.

We tested two of the algorithms found in literature: the one proposed by Broder, [6] and the one proposed by Pugh and Henziger [9]. We studied the behavior of these algorithms with different values for their parameters as well as their applicability to MADCOW.

Broder. In our implementation we visit the DOM and extract the text contained in it. Then we identify *shingles*, i.e. sequences of g contiguous words (where g is the granularity parameter), and we code each shingle with Rabin's coding function [10]. We sort the values obtained, and choose the r larger ones as the signature of the page (where r is the resolution of the method). We first analyzed the effect of varying the granularity in the range 5..10 as proposed by Broder. It did not show any relevant difference, so we chose g equal to 8 for our subsequent study. Then we varied the resolution, and that implied a variation of complexity for the algorithm, but we found that with low values of r, we had too many false positives, while a value of 84 used by the author, would lead to an unsustainable complexity for MADCOW . Moreover, we found this method unsuitable for MADCOW as we could not cope with page structure.

Pugh. We coded each node in the DOM that was related to text (and not to formatting), i.e. <P>, <TD> etc. We create r lists of DOM nodes, by using the coding value *modr*. We then code the text of each list, obtaining r different values, our signature. Two pages are considered similar if at least one couple of values matches. We tested the implementation with different resolutions, i.e. different number of lists. The higher r, the higher the number of false positives.

We also tackled the case of pages that have large nodes (with more than 100 characters) and that differ just for one node. We applied the algorithm in [5] in order to better code the node in the signature.

On the other hand, small pages with very few content may be considered duplicates because single DOM nodes will be represented in each list, and thus a single node, containing a usual word or sentence, can affect duplicate identification. We addressed this problem by fusing together lists that contain very few text. This is similar to a variable resolution, and proved to work well in our study.

Finally, we tested the problem of pages in a web site that have a common structure and content due to the use of a *template* to generate them. We tested 120 pages from the site www.trovacinema.it and found that a resolution of 7 worked well in distinguishing different pages, although some false positives could not be avoided.

Our results suggested that this method is more suitable for MADCOW thus we chose it for our final implementation.

5   Modified Source Documents

When considering dynamicity of web pages, we can distinguish between modifications in the structure, identifiable in MADCOW as a modification of the set of paths of labels in the DOM tree, and content modifications.

An extensive study has been conducted to identify the major sources in variability of Web pages.

5.1   A Study of Web Page Modifications

About fifty *Welcome pages* from ten categories of websites (e-commerce, search engines. e-papers, art, television networks, provincial administration, university, Internet providers, sport and science) have been visualised and saved with a MSIE 6.0 Web broser each day, over a period of two weeks.

These pages have subsequently been studied exploiting UltraEdit32 Professional Text/HEX Editor Version 11.20a, which allows the comparison of two text files, highlighting their differences, and the DOM inspector incorporated in Mozilla Firefox Version 1.5.0.1, which allows the examination of the DOM tree loaded in the browser. This way, we could observe the evolution of both the structure and the content of the selected pages.

The analysis has shown the similarity in the structure of all pages, which is based either on tables (exploiting the <TABLE> tag) or containers (<DIV> tag, or combinations of the two, possibly with several nesting levels. A layout organization similar to that of Figure 3 has observed to be common.

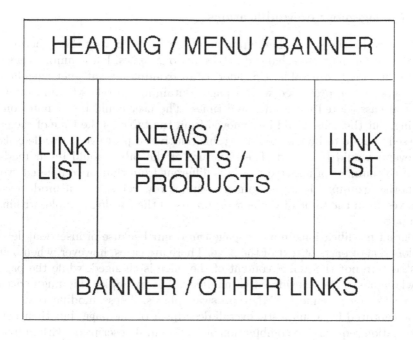

Fig. 3. The typical organisation of a portal Welcome page

This layout provides a sort of exoskeleton, in which the upper part of the page contains a heading, or a menu, or some banner, the left and right columns present links, typically organised into categories to access internal parts of the Website or external and related links, typically in the right column, while the central column hosts the most relevant contents. The bottom row may be present or not, to contain other banners or links.

The portion which is most subject to changes is therefore the central part. Its internal structure is based on table cells (<TD>) or <DIV> blocks, which change their position as the days pass, as new ones are inserted, until they are removed altogether. Moreover, some elements in this structure may vary their content in time, especially those which refer to external links (tags <A>) or images (tags <IMG>, or formatting tags, such as <STRONG>, <B>, <CENTER>, <FONT>, <SPAN>, etc.)

One can however observe that the most significant nodes in the hierarchical structures of the central part are <DIV>, <TABLE> (or <TD>) and <P> (paragraph). The latter is less frequent in these Welcome pages, and more largely used in pages deeper in the organisation of the Website, containing mostly text.

Some less relevant changes within the content of a page are quite frequent and used for special purposes. For example comment tags may vary if they are used to indicate the page creation date, or some attributes of a tag may change without a change in the actual content of the tag.

5.2   Management of Modifications

As observed by the study there are a number of reasons why the structure of a Web document may have changed between two accesses. For example, cells may have been added to a table. Consider as an example an intranet agenda such that users can be presented with a page containing a table where meetings or events are associated with dates and times. The user could insert notes on the meeting, but the cells would be removed from the table as the time of the event is passed, maybe to be inserted in a table relative to past events, while cells for new events would be added. The structure of the table would result modified only if the number of inserted events is different from that of completed events. As another example, in an e-commerce site, a row relative to a product could be moved from the table of the new proposals to the catalogue table, within the same page.

Content modifications in a web page can occur because of insertion, deletion or relocation of parts of text of the page. There are cases, however, when neither the structure nor the global content of the page is changed, while the page on the whole has to be considered as modified. For example, this might occur if two blocks of the same HTML type swap places. These modifications would not be captured by a simplistic overall descriptor of the page, but their correct identification requires the combination of a structural description with individual descriptions of the content of each node.

This notion of modification can be extended to any type of digital content, for example in situations where a page presents a "Picture of the day" feature, which would associate different images with the same URL at subsequent visits.

The features for DOM access and manipulation incorporated within MADCOW offer several possibilities for managing modifications both of structure and of content within a dynamic environment.

In particular, with respect to the management of modifications of the page content or structure, we obtain the following features.

- **Frame management.** The DOM of the current main page is traversed looking for nodes of type FRAME.
- **Dynamicity management.** Each DOM node is accessed and coded through a hash function, whose value is saved together with the annotation content. Once the page is reloaded, the hash codes of the nodes for which an annotation is found are matched with the codes for the loaded

page, so that an annotated node can be retrieved even if its position in the DOM has changed.

- **Placeholder placement.** The DOM of the loaded page is traversed according to the path stored at webnote creation time, in order to look for the position in which to insert the icon. The inserted icon is treated as an additional node.
- **Text highlighting effect.** A text node for which an annotation has been produced is analysed to look for the portion of text which has actually been selected. This part is highlighted through a new FONT node, created on the fly, set to a specific color.

We now describe the problems related to the location of content in a modified structure, before discussing the use of hash codes for identification of variations or restructuring of the content.

Exploiting these features, there are two fundamental operations to be performed:

1. To assess whether the position indicated by the webnote still contains that part of content which was the actual subject of the annotation activity.
2. If a modification of the original page is detected, to try and identify the current position of that content, in order to allow the correct placement of the placeholder or of the highlighting.

To this end, at loading time a hash code is computed for each node of the DOM tree. The value of the code depends on the node type, its content, attributes and values. The values of the nodes for the current page can thus be matched against the values associated with the webnotes retrieved for that same page.

To check whether the zone in which a retrieved webnote is in the original position, a path corresponding to the stored XPath is traversed in the tree for the current page. If the path cannot be followed to its final node, this means that page has incurred a structural change. However, even if a node of the requested type is found, we still have to check whether a content modification has occurred, by evaluating the hash code for the reached node, and matching it against the code associated with the original annotation. If the match succeeds, we have found the exact originally annotated content. Otherwise, we search the DOM for the correct position of the annotated content, if it has not been eliminated from the page altogether. To this end, two methods have been devised, one based on a normalised form of the DOM tree, and the other on a linearisation of the tree via an ArrayList. We are going to present these two methods in the following subsections.

5.3   Management of Trees

Two methods to reduce the computational cost of locating and retrieving the annotated part have been considered, one based on normalised trees and the other on linearised ones and a comparison has been performed.

Normalised trees. A normalised tree is one in which only significant nodes appear, i.e. those which refer to the structure of the page, such as DIV, TD, TABLE, or P, without considering tags pertaining to formatting information. This way, all content is lifted to its nearest enclosing significant tag.

Now, the quest for a corresponding node can be done more effectively in the normalised tree. However, if a match is found in the normalised tree, the exact position in the original DOM tree has to be retrieved. This second search can, however, be performed only in the subtree rooted in the last significant node found in the match for the normalised tree.

It is also possible that nodes with the same coding are found in different positions in the page, if some structure and content is replicated over the page. In this case heuristics have to be adopted, such as picking the node with the least distance from the original position.

Linearised trees. As for the case of the search on the normalised tree, the use of a linearised structure, in this case an ArrayList, allows a more effective retrieval.

The list is constructed by visiting the tree and inserting at each position a pair $(hash, XPath)$ identifying the associated code and path for each node. After completing the construction of the list, it is sorted according to the code values, so as to make subsequent searches faster.

This structure also simplifies the task of identifying the presence of repeated elements. Moreover, once an element is found, this directly gives us the path to the original node, without having to revisit the current DOM.

Discussion. The method based on normalised trees is more efficient in construction, $O(n)$ where n is the number of nodes, against $O(n^2)$ for the linearised version, due to the ordering process. On the other hand, the linearisation allows a faster retrieval, $O(log\ n)$ against $\Theta(m) + k * O(n)$, where k is the number of significant nodes in the linearised tree. Supposing a prevalence of retrieval over annotation processes (annotate once, retrieve many times), the method based on linearisation has been chosen.

5.4   Hashing

The hash code for a HTML document is calculated immediately after the page has been loaded and the DOM tree created, before any placeholder has been presented on the page, as this would alter the tree with the introduction of new HTML code.

An HashTableIndex is created, in which each element associated with the code is a structure Hash_DOM, containing the URL string for the page, and two pointers to ArrayList. The hashtableC list contains, for each node, its code and a pointer to the corresponding DOM node. The hashtableP list contains the code and the PATH to the node.

The resulting memorization structure is shown in Figure 4.

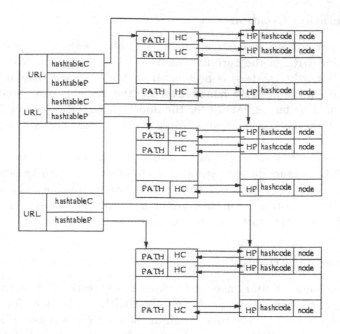

Fig. 4. The organization of the hashing structure in the MADCOW client

The two `ArrayList` referred to by `hashtableC` and `hashtableP` maintain references to one another. In the rest of the paper, we assume that the `HashTableIndex`, `hashtableC` and `hashtableP` have been sorted according to URL, hashcode and PATH, respectively.

The hashing algorithm is based on a function of type $f(s) = \sum_{i=0}^{n-1} B_{n-i-1} s_i$, with $B = 2^k$ for some k. As the range of $f(s)$ is thus unlimited and we need to constrain the results of the function within a bounded interval, we adopt the adapted function $f(s) = (\sum_{i=0}^{n-1} B_{n-i-1} s_i) mod\ W$, for $W = 2^w$, w the size of a memory word. This choice has the problem that, since both W and B are powers of 2, the value of the hash function depends only on the last $log(W)/log(B)$ characters of the string.

Now, if we consider the 7-bit ASCII set, only 97 of the characters in it are printable, while the remaining ones are control characters, which rarely occur in a string. Of these 97, the commonest are alphanumeric characters, which are coded by 62 ASCII codes. Moreover, as characters are contiguous in the set, and similarly are digits, the last six and four bits are sufficient to identify alphabetic characters and digits, respectively. Hence a choice of $k = 6$ should be adequate. This hashing scheme works if strings mostly differ in their final letters. However, it may happen that strings mostly differ in their first characters, as is often the case for domain names. To compensate for this, we consider the first 6 bits, which would be lost when the variable *result* is shifted to the left, and XOR them with the shifted result variable.

5.5   Annotation Creation

When the user triggers the activation process, after selecting a portion of the document, the URL of the current page is used to look in HashTableIndex if an element with that address is present. If such an element is found, let it be *HD*, one looks in the table HD.hashtableP for the element corresponding to the PATH returned by build_Path() for the node closing the selection. It is worth noticing that the path is built, discarding possible tags with id="madcow", as these are created by previous annotation interactions.

Let *HP* be the element found in hashtableP. The hashcode for this node is found at HP.HC.hashcode, exploiting the redundancy between hashtableC and hashtableP. This information is then maintained throughout the interaction to produce the annotation for the current selection, and stored on the server together with the other metadata for the webnote.

5.6   Webnote Retrieval

To retrieve an annotation, we assume to have already retrieved from the database the metadata relative to an annotation. The URL of the page for which the notes must be shown is a key to a search in the HashTableIndex. If an element corresponding to the URL is found, let it be *HD*, the hash code for the found annotation is looked for in HD.hashtableC and, if found, stored in a variable *HC*. The adjacent elements are checked to see if they have the same hash code. If more occurrences of elements with the same code are found, one of them is selected as the "true" content for which the note was produced. Namely, the closestNode(ArrayList HCODE, string PATH, int index) method chooses the node whose position is the closest one, counting the DOM nodes to reach the node indicated in the path returned by the server. This method works on hashtableC, so that each time a node with the searched code is found, the value of the attribute HC.HP.PATH is used to found the node which is closer to the annotation source.

Once an element of hashtableC is found which satisfies the different criteria for identification, it contains a direct reference to the DOM node in which the placeholders has to be inserted.

6   Conclusions

We have studied two problems related to annotating digital documents over the web: the need to retrieve annotations related to portions of the page that have moved within the page itself, and the need to associate to a web page all the annotations pertaining to its content, even if they were created while accessing the same content under a different URL. We found solutions to both these problems and implemented them in our system, MADCOW.

References

1. Bottoni, P., Civica, R., Levialdi, S., Orso, L., Panizzi, E., Trinchese, R.: MADCOW: a Multimedia Digital Annotation System. In: AVI 2004, pp. 55–62. ACM Press, New York (2004)
2. Bottoni, P., Levialdi, S., Panizzi, E., Pambuffetti, N., Trinchese, R.: Storing and retrieving multimedia web notes. IJCSE (to appear)
3. Bottoni, P., Levialdi, S., Rizzo, P.: An analysis and case study of digital annotation. In: Bianchi-Berthouze, N. (ed.) DNIS 2003. LNCS, vol. 2822, pp. 216–230. Springer, Heidelberg (2003)
4. Bottoni, P., Civica, R., Levialdi, S., Orso, L., Panizzi, E., Trinchese, R.: Storing and retrieving multimedia web notes. In: Bhalla, S. (ed.) DNIS 2005. LNCS, vol. 3433, pp. 119–137. Springer, Heidelberg (2005)
5. Brin, S., Davis, J., García-Molina, H.: Copy detection mechanisms for digital documents. In: SIGMOD 1995, pp. 398–409. ACM Press, New York (1995)
6. Broder, A.: On the resemblance and containment of documents. In: SEQUENCES 1997, vol. 00, p. 21. IEEE Computer Society Press, Los Alamitos, CA, USA (1997)
7. Chowdhury, A., Frieder, O., Grossman, D., McCabe, M.C.: Collection statistics for fast duplicate document detection. ACM Trans. Inf. Syst. 20(2), 171–191 (2002)
8. Manber, U.: Finding similar files in a large filesystem. In: 1994 Winter USENIX Technical Conference, pp. 1–10 (1994)
9. Pugh, W., Henzinger, M.H.: Detecting duplicate and near-duplicate files. US Patent 6658423 (December 2003)
10. Rabin, M.O.: Fingerprinting by random polynomials. Report TR-15-81, Center for research in computing technology, Harvard University (1981)
11. Sanderson, M.: Duplicate detection in the Reuters collection. Technical Report TR-1997-5, Department of Computer Science, University of Glasgow (1997)
12. Shivakumar, N., Garcia-Molina, H.: Scam: a copy detection mechanism for digital documents. In: Proc. International Conference on Theory and Practice of Digital Libraries (1995)
13. Shivakumar, N., Garcia-Molina, H.: Building a scalable and accurate copy detection mechanism. In: DL 1996, pp. 160–168. ACM Press, New York (1996)

Towards Scalable Architectures for Clickstream Data Warehousing

Peter Alvaro, Dmitriy V. Ryaboy, and Divyakant Agrawal

ASK.com
555 12th Street, Suite 500
Oakland, CA 94607, USA
{palvaro,dmitriy.ryaboy,divy.agrawal}@ask.com

Abstract. Click-stream data warehousing has emerged as a monumental information management and processing challenge for commercial enterprises. Traditional solutions based on commercial DBMS technology often suffer from poor scalability and large processing latencies. One of the main problems is that click-stream data is inherently collected in a distributed manner, but in general these distributed click-stream logs are collated and pushed upstream in a centralized database storage repository, creating storage bottlenecks. In this paper, we propose a design of an ad-hoc retrieval system suitable for click-stream data warehouses, in which the data remains distributed and database queries are rewritten to be executed against the distributed data. The query rewrite does not involve any centralized control and is therefore highly scalable. The elimination of centralized control is achieved by supporting a restricted subset of SQL, which is sufficient for most click-stream data analysis. Evaluations conducted using both synthetic and real data establish the viability of this approach.

1 Introduction

Click-stream data warehousing has emerged as a monumental information management and processing challenge for Internet scale enterprises such as Web search companies, E-commerce enterprises, and other Web-based companies. In the old brick-and-mortar days data warehousing typically involved tracking and analyzing sales transaction data against other business processes in an enterprise. Even though the scale of the problem expanded and became more complex with the rapid expansion of retail conglomerates such as Wal-Mart and others, the data at the granular level still remained manageable. The scale of data is constrained since it corresponds to a line item in every sales transaction which is physically bounded. In such environments, customer interaction with an enterprise is tracked at the point of actual sales. The main challenge in traditional data warehousing is integration of data perhaps from hundreds of store-fronts, and in general the solution suites are often based on commercial RDBMS technologies [2,1,3].

The nature of interaction between a customer (or user) and an enterprise has undergone a dramatic change in the context of Web-based enterprises. First, the

S. Bhalla (Ed.): DNIS 2007, LNCS 4777, pp. 154–177, 2007.

physical constraint on the number of geographical locations and the capacity of a store is no longer present. The consequence of this is that the number of customers who can interact with an enterprise is virtually unlimited. A second and perhaps more important change is that Web-based businesses have moved away from sales-based tracking of their clients to click-based tracking, which is akin to retailers such as Wal-Mart tracking their customers as they walk down the shopping isles and eye-ball different products in the shelves. The main reasons for these paradigm shifts are several. First is the enabling nature of the current information technology that allows us to track user interactions at every mouse-click and keyboard input. Second, the potential usage and benefit of this information is significant. For example, a Web-based retailer can exploit this information to personalize its *virtual store* to suit the preferences of every one of its clients (or each client category). Imagine, walking into a grocery store which reorganizes the products based on ones preferences and individual tastes.

Although the above paradigm is well-intentioned from a business perspective, it poses significant information management and information processing challenges. First and foremost among them is that there is a huge explosion of data. A part of this data explosion arises due to the sheer scale of the Internet which expands the potential customer-base enormously for any Web-based enterprise. Much worse, however, is the data explosion due to the change in data granularity from sale-based user transactions to click-based user interactions. Due to the fine granularity of this data, it needs to be aggregated and summarized in meaningful ways before it can be put to use. The current focus of most clickstream data warehousing systems is to aggregate and summarize highly granular data to detect meaningful patterns. Typically, such analysis is done in terms of fairly standardized attributes such as *time, geography, business entity*, and *product/service*. Most of these attributes are hierarchical in nature and some of them have multiple hierarchies associated with them (e.g., time and geography). The performance of the enterprise is tracked in terms of different types of metrics or measures such as: user counts, user queries, user visits/sessions, and user selection/clicks, etc.

The click-stream information is generated as log files at the enterprise's Web site and are delivered to the IT infrastructure responsible to maintain the click-stream data warehouse[1]. In general, the log information is cleaned, parsed, and organized so that it can be structured and stored using commercial RDBMS systems. However, considering the data explosion as well as the 24×7 nature of Web-based enterprises it is becoming increasingly evident that the current RDBMS technology is stretched to its limits both in terms of storage as well as in terms of processing. Furthermore, commercial RDBMS evolved primarily to handle workloads for OLTP systems and therefore are heavily bloated in terms of technology components such as ACID transaction support and write-ahead logging to deal with failures and restarts. Click-stream data warehousing systems, on the other hand, warrant a RISC analog of DBMS technology that

---

[1] Note that the log data is generated as a continuous stream. However, for administrative purposes it is stored in terms of finite-sized log files.

supports consistent appends instead of atomic writes, redundant storage instead of write-ahead logging, fast (and perhaps highly parallelized) aggregation and summarization instead of ACID transactions. This realization is shared by the broader research and development community. Several efforts are underway to develop research and production systems to enable ad-hoc analysis of massive data-sets. Notable among these efforts are MapReduce [6], SAWZALL [12], and related projects [7,5] from Google, the Pig Project from Yahoo [11], and the Dryad project from Microsoft [9].

A common theme among all these efforts is to exploit data-parallelism to enable ad-hoc analysis of massive data-sets over a large cluster of thousands of computers. In each of these proposals, the programmers need to specify and structure their computation in terms of a proprietary data-parallel programming language. In return, the run-time system relieves the programmer from the mundane and complex details related to concurrency, synchronization, and failures. Although these proposals make a powerful and compelling case, their availability and applicability is limited due to their proprietary nature. In that, the underlying software is not available that can be used to build a click-stream data warehousing substrate on top of these systems. Furthermore, these proposals require that business analysts and application programmers learn a new language other than SQL which is not completely declarative.

In this paper, we take a different approach. We start by making an argument that the full power of relational SQL is not necessary for supporting most of the routine reporting and analysis that are needed to support click-stream data warehousing. This is especially the case when the data warehouse schema adheres to the dimensional structure and is based on the star schema. We then identify a restricted version of SQL that can support click-stream data warehousing. Next, we argue that given that most Web-based enterprises collect their click-stream logs on a large cluster of computers, aggregation and summarization of this data should exploit the inherent parallel processing capacity available in the system. This recommendation is based on our observation and experience that collating the log data from multiple machines and pushing it upstream to be stored in an RDBMS storage creates severe storage bottlenecks and processing latencies. In particular, due to the streaming nature of logged data even minor failures result in a domino effect in terms of system downtime. In effect, we propose to create a middleware storage engine that allows Business Intelligence Application Programmers to specify their queries in terms of restricted SQL syntax which in turn are executed directly on the distributed log data stored on a large cluster of computers. The middleware converts the SQL specification into data-parallel execution tree by using the data distribution information about the underlying data according to the algorithm presented below, runs the job on the entire cluster, and returns the results of the computation.

We conclude by presenting some real-world examples and noting that while the system is designed for aggregating and reporting on click-stream data, it can work well for other applications, and has the added benefit of being easy to mount on top of pre-existing distributed data, in essence turning any well-defined dataset distributed over a large number of machines into an ad-hoc database.

The rest of the paper is organized as follows. In Section 2, we identify the necessary subset of SQL that must be preserved to enable click-stream analysis as well as is amenable to distributed and parallel processing. In Section 3, we present the details of the different components of our runtime environment and the design of query rewriting and data redistribution for executing user queries. An evaluation of overall performance of our system using real data and actual implementations, as well as comparison to technologies proposed by [12] and [11] is presented in Section 4. Section 5 concludes the paper.

2   Restricted SQL and Click-Stream Analysis

2.1   Typical Data Warehouse Queries

In a data warehouse environment, the longest running queries usually involve large sequential scans of raw data and deal with filtering and aggregation. In this phase, the data is projected into a smaller dimensional space, and aggregate functions are applied to the projected tuples. If any joins are to be performed, they are typically executed as a later step when the number of keys has been significantly reduced.

Figure 1 describes a very common mapping workflow. The objective is to aggregate a very large set of records down to a meaningful digest of the relevant

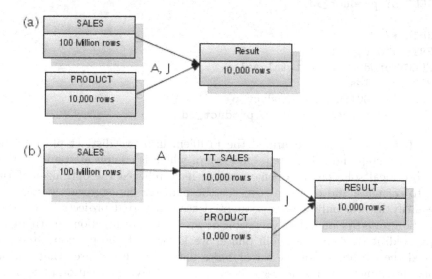

Fig. 1. Mapping a large dataset to a significantly smaller set of relevant dimensions can be done by performing the Aggregation (A) and Join (J) at the same time, or in two phases. In practice, database programmers find that explicitly specifying the second approach yields significantly better performance.

dimensions, in this case a 'product' dimension. To make the digest meaningful to a user, it is often necessary to dereference certain fields to meaningful names, as illustrated here in the mapping of a numeric product_id to a product name. A simple query in ANSI standard SQL syntax achieves this goal in a single step:

```
INSERT INTO report
  SELECT product_name,sum(total_sales)
  FROM sales s
    INNER JOIN  product p
      ON s.product_id = p.product_id
  GROUP BY product_id
```

Tranforming this statement into an efficient parallel execution plan is another matter altogether. There are literally dozens of different methods available to the optimizer for performing the join, aggregation, and projection steps of the query. Instead of relying on the optimizer to make the best decision, many data warehouse programmers will choose to apply their own domain knowlege of the relative size of the *Product* entity to *Sales*, and deconstruct the problem in this way:

```
/* Aggregation */
INSERT INTO tt_sales
  SELECT product_id,sum(total_sales)
  FROM sales
  GROUP BY product_id

/* Join */
INSERT INTO report
  SELECT product_name,total_sales
  FROM tt_sales s
        LEFT OUTER JOIN product_p
          ON s.product_id = p.product_id
```

By doing so, we have separated the problem into two distinct phases – an aggregation step and a join step – neither of which is complicated and neither of whose ideal solutions is ambiguous. We have relieved the optimizer of the need to calculate and compare costs, as the plan for the aggregation step will necessarily involve only a large sequential scan, a sorted projection (to evaluate the group-by) and the evaluation of an aggregate function (in this case simple addition). As for the join phase, we know that the previous step has reduced the cardinality of the set of *Sales* data to such a degree that performance should not be an issue, regardless of the underlying execution plan of the join.

Figure 2 describes a problem more specific to clickstream data warehousing. The objective is to report on the use of a button on a web page. An unqualified

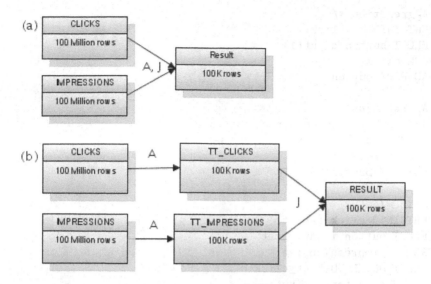

Fig. 2. Computing the intersection of several large entities is a common task in data warehouses, and one that has even more significant repercussions for using a one-phase approach due to the size of the datasources. By separating the filtering and aggregation step from the join, database engineers apply domain knowledge to achieve better performance.

count of the number of times the button was pressed is not particularly useful in itself, as this count will tend to trend along with traffic. Instead, we want to track the ratio between the number of times the button was pressed and the number of times that the button was presented to a user. This 'click to impression ratio' can then be used to gauge whether the button is being presented in an optimal fashion. A naive approach joins the raw entities on the button id, and carries out both aggregations with one statement:

```
INSERT INTO click_report
  SELECT button,count(distinct i.user_id),count(1)
  FROM impression
    LEFT OUTER JOIN click c
      ON c.button = i.button
  GROUP BY button
```

Here, the optimizer is presented with a problem similar to the previous example, but with much more severe consequences. If it chooses a poor plan for the join (executing nested loops for each comparison of buttons from both entities, for example), the aggregation step might never complete. Even with current statistics and an ideal optimizer, the join is unlikely to perform well, so this problem is nearly always approached in two phases:

```
/* Aggregation */
INSERT INTO tt_click
  SELECT button,count(1)
  FROM click
  GROUP BY button

/* Aggregation */
INSERT INTO tt_impression
  SELECT button,count(distinct user_id)
  FROM impression
  GROUP BY button

/* Join */
INSERT INTO click_report
  SELECT button,i.cnt,c.cnt
  FROM tt_impression i
    LEFT OUTER JOIN tt_click c
      ON c.button = i.button
```

As in the first example, the aggregation phases are now clean and unambiguous, relieving the optimizer of burden of cost evaluation and plan building. Our domain knowledge of the deliberate reduction of click-level data (hundreds of millions of rows) to an aggregate of buttons (probably well under one million) shows us that the join phase on the resulting entities, while still an ambiguous problem, will have a low cost even in the worst case, and may be performed with a fraction of the resources needed to perform the aggregation steps.

2.2   Restricted SQL

A declarative language differs from an imperative language in that a programmer must specify the outcome that is desired, rather than the steps to take towards that outcome. Such a framework inherently provides a level of abstraction between the language and the execution plan or algorithm, allowing the latter to change frequently, even at runtime, while the former remains a consistent standard. In RDBMS, for example, a (mostly) standard SQL syntax generates vastly different query plans not only from database to database (Oracle, MySQL, SyBase, etc.) and version to version, but from query instance to query instance as data statistics change over time.

   In these traditional RDBMS, the translation layer between the declarative query language and the underlying filtering, projection and aggregation plans is the optimizer, which evaluates a number of different execution paths based on estimated cost. A parallel-aware relational optimizer presents very complicated problems that are outside the scope of this discussion. A significant body of work exists in the area of distributed and parallel query optimization [15,8,10]. The work reported here is also closely related to the issue of scalable access to data in peer-to-peer systems [4] and grid environments [14].

The high cost of joins on large entities, poor optimizer choices and the un-wieldiness of indexes undermine the implicitly permissive nature of SQL, which allows large statements even if they are unlikely, if not incapable, of performing as well as an equivalent batch of statements separating the aggregation and relational phases. Though the temptation exists to write a single relational algebra statement that encapsulates the filtering, grouping, intersection and ordering phases of an analysis workflow in one statement, most code tends toward separating and simplifying the aggregation steps for practical reasons.

SQL is an attractive query language because its ability to express a majority of data warehousing and data mining problems is well established, and because it is widely adopted. If we extend the data warehousing best practice described above into a rule – that is, define a limited SQL syntax which excludes joins – we are left with a familiar language that still adequately satisfies the majority of data warehouse aggregation query requirements. This limited syntax also provides a model whose execution plans are easy to parallelize and optimize.

In our restricted model, the filtering phase is encapsulated by the familiar *Where* clause. It can be of an arbitrary complexity, but because joins are out of scope, the where clause may only refer to columns in the given row or the results of functions evaluated on them. Just as Google's Sawzall ensures a highly granular filter phase by enforcing a single row of state, this restriction guarantees that the filtering phase of queries can be carried out in a highly parallel fashion, local to the storage of the raw data.

The *Group By* clause handles the aggregation phase. Unlike the severely limited *Where* clause, this is a workalike to the SQL equivalent. A combination of column references and the results of functions project tuples to which aggregate functions may be applied. A limited number of aggregate functions are provided: specifically, we select those that are the most valuable and are easily parallelizable. This short list includes *Count(),Sum(), Min(), Max()*, and *Count Distinct()*. The efficiency of a parallel strategy for evaluating the *Group By* and aggregate functions will depend on the the underlying table. Although the list of aggregate functions will be small and relatively static (like Sawzall's Aggregators [12], or SQL's aggregate functions, for that matter), single-row functions can be freely created and added, allowing data transformation capabilities identical to that of standard SQL.

The familiar *Having* and *Order By* clauses are also provided. These query attributes are by their nature difficult to parallelize, but are indispensable to analysis of this kind.

The resulting thin language can interpret the aggregation phase SQL into highly efficient parallel execution plans. The semantic structure of the restricted SQL (where clause, group by and aggregate function) maps cleanly to the phases of parallel data retrieval (filtering, projection and aggregation), allowing a very simple set of rules to drive the optimization. When used in conjunction with a local RDBMS to handle the join phase, we may implement all of the workflows described above with a massively parallel aggregation step.

3   The Run-Time Environment

In this section, we start by describing the static components of the system that are needed for data storage and data description. We then provided a complete overview of how queries are executed on a cluster of servers. Finally, we describe the underlying rules that are employed for taking the source SQL query and rewriting it to execute in a distributed storage infrastructure. We also describe how the data is redistributed in the case of mis-alignment of the query with respect to the partitioning scheme in order to facilitate the query execution.

3.1   Storage Engines and Data Definition Language

In order to hide the details of I/O, field extraction and column ordering, we place two levels of abstraction between the files on disk and the the parallel query system. In addition to modularizing those aspects of code and configuration that are likely to change frequently, this separation vastly simplifies our design. With the extraction and labeling issues external to it, the main algorithm may treat all input as a series of key-values pairs with composite values, whose field values may be extracted by name.

Storage Engine. We first define a storage engine API as providing at minimum three functions, open(), next() and close(). This model is inspired by the MySQL storage engine abstraction, which likewise provides a simple but quite complete API for row-level data access. Assuming indexed data, MySQL also provides a seek() function, but we are only concerned at this time with sequential data access. With this in place, the query engine may assume that while there is data left to fetch, next() will return a row each time it is called.

A querying system with an SQL front end can only deal with 2-dimensional data sets, as all the functionality of the language assumes them. Often, we find a tabular data structure on the raw data files already (Comma-Separated Value text files, for example), so the mapping is one-to-one and the storage implementation is quite simple. We can, however, pass any arbitrary data through the parallelized aggregation system if we can first express the raw data as 2-dimensional records. A storage engine provides this layer of abstraction.

Storage engine implementations will differ based on existing file types. Typically, the bulk of their code will deal with details of directory scanning, file I/O, and extraction (most often string tokenization). Because our file model might allow for multiple calls to next() per file read (as in the case of nested data, transformed into 2 dimensions by a storage engine), some state may be used to hold data buffers. The basic storage engines include a CSV engine and a parallel storage engine, used internally for communication among branch nodes. More inventive abstractions could include a web storage engine that implements a pagegrabber and an HTML parser, supporting something like "select * from ask.com".

DDL – Data Definition Language. While a storage engine defines how to fetch rows from a data source, it does not tell us what is in those rows or how to interpret and manipulate the record columns. This configuration belongs at a different level, since a given storage engine could have an infinite number of record types stored within it. We want to avoid writing code when new record types are implemented, so we choose an already familiar language (DDL SQL) to define our objects.

Our Data Definition Language takes form of an extended SQL *Create Table* statement. Unlike the RDBMS equivalent, it does not actually create any objects or storage, but merely defines an existing structure, mapping column names to ordinals. It is here that the user may associate a particular storage engine library with a table, possibly passing parameters to that engine.

```
CREATE TABLE queries (
dt date PHYSICAL PARTITION,
site varchar(128) DISTRIBUTION PARTITION,
query_txt varchar(1000),
cnt int)
STORAGE CSV BUCKETS (10);
```

The DDL always indicates which column or columns comprise the *Distribution Partition*, meaning they are aligned with the first level of host-wise data partitioning. In the example above, we know that every data value in the *site* column will hash to a value of 0-9, and that any given value of *site* will occur in one and only one partition. It is precisely this knowledge that will allow the system to optimize its query plan to maximize parallelism and minimize data movement.

The *Physical Partition* flag is an optional keyword indicating that the given column is the sub-(or physical) partition of the table. Subpartitions are usually implemented at the storage tier as separate files within a directory. Unlike the primary partition, which is the result of uniform hashing on a particular column value or values, the value of a subpartition column will map to one of these files, representing a point along an axis (i.e. date) which is sequential and for which range queries are likely. Indicating in the DDL that a table is physically partitioned by the values of a particular column will allow a subpartition-aware storage engine to open and scan only relevant files.

In order to support retrieval of nested datatypes without needing to implement new storage engines, the STORAGE clause of a DDL statement may reference an expanded version of another defined table. These references can be nested so as to represent the chain of transformations to a 2-dimensional structure necessary to query the data in an SQL framework.

```
CREATE TABLE words (
word varchar(128) FLATTEN(0))
STORAGE EXPAND(queries,query_txt,split(/\s/))
```

This DDL statement defines a table inheriting the (dt,site,cnt) elements from the queries table defined above. In place of the query_txt column, which has been expanded by calling the split() function, we have the word column, which contains a value from one iteration of the split() function. *Flatten()* takes a subscript, as the expansion function will often produce multiple values from which to choose. Some sample source data, the rows returned when querying the queries table, and the rows returned when querying the words table are shown in Tables 1-3.

Table 1. Data stored on disk

"'03/03/2007'","'ask.com'","'25th president'","'23'"
"'03/04/2007'","'ask.com'","'Best pizza in Boston'","'12'"
"'03/05/2007'","'ask.co.uk'","'John Taylor'","'10'"

Table 2. same data as interpreted by the CSV Storage Engine with the Queries DDL

DT	SITE	QUERY_TXT	CNT
03/03/2007	ask.com	25th president	23
03/04/2007	ask.com	Best pizza in Boston	12
03/05/2007	ask.co.uk	John Taylor	10

Table 3. same data as further interpreted by the *Flatten()* Storage Engine with the Words DDL(c)

DT	SITE	WORD	CNT
03/03/2007	ask.com	25th	23
03/03/2007	ask.com	president	23
03/04/2007	ask.com	Best	12
03/04/2007	ask.com	pizza	12
03/04/2007	ask.com	in	12
03/04/2007	ask.com	Boston	12
03/05/2007	ask.co.uk	John	10
03/05/2007	ask.co.uk	Taylor	10

3.2   Parallelization Algorithm

Having described the data access mechanism, we now present the method used to evaluate queries on this data. In developing this algorithm, we focus on maintaining maximum scalability and resource availability. Cross-communication and message-passing must be minimized, reducing interdependencies of sub-processes on different nodes. This is achieved by ensuring that any transformations to the queries, temporary data structures, and execution plans are deterministic – given a query and a DDL, any node can infer the complete plan without consulting any other node.

A client can connect to any node on the cluster to initiate the computation for a given click-stream analysis query. This node begins a recursive process that builds an execution tree with itself as the root and several child nodes elected based on resource availability[2] as intermediate nodes. The nodes recruit their own child nodes if necessary, and the process repeats until a leaf node is created for every physical data partition[3]. The leaf nodes apply the query to their slices of data, and report on results of computation to their parent nodes. Pseudo-code for the algorithm is presented in Figure 3 and described in detail below.

At the beginning of the query execution process, the root node examines the query and the underlying DDL to determine if temporary storage will be needed; if so, it prepares storage space on the cluster. At non-leaf nodes the query received from the parent is passed down to the child, and the local query and ddl are rewritten in accordance with the query rewrite algorithm presented in the next section. The leaf nodes examine the raw data and apply the filters found in the *Where* clause and perform any necessary aggregations.

If redistribution is not required, the leaf nodes pass the resulting data up to their parents, which apply their rewritten queries and pass the data further up, until the results arrive at the root node and can be sent back to the querying entity.

If redistribution is required, the leaf nodes redistribute the filtered tuples across the cluster in accordance with the rewritten DDL. Since rewrite is deterministic, leaves can send the data out to the storage cluster autonomously. When all of its children return successfully, the root node then starts the process again, applying its rewritten query and the new DDL to the temporary table. This repeats until the requested projection is aligned with the table distribution key and the recursive execution process is complete.

The actual algorithm we use in production is slightly more complicated, as batches of data can pre-aggregated to reduce network traffic.

A sample execution tree is illustrated in Figure 4. In the diagram, edges are labeled with the SQL and DDL pairs that are passed from a parent node to a child node. The boxes are labeled with the sql and ddl applied to the data that is either read from disk, or returned by their children. The nodes are represented as separate boxes, but in reality they are not necessarily different machines, just different instances of the program. Naturally, the nodes that read data from disk can only be on the machines that have access to the data.

In the illustrated case, the original (sql, ddl) pair, hereafter referred to as the *query package* is sent to a node, which then becomes the head node for this query. It parses and converts the query package according to the rules found in Section 3.3, and initiates the parallelization algorithm. Since the head node can see that temporary storage will be required for some repartitioned data, it prepares the storage for it on the cluster, elects some nodes to function as its children, and sends them the query package. As they each have more than 1 partition to work

---

[2] Note that we assume that resource management and fault-tolerance are managed via a cluster management layer such as the system described in [13].

[3] A single machine can function as multiple nodes in the tree, as necessary – in fact, the whole process could run on a single machine, although of course one would lose all the benefits of parallelization in this case.

INIT_QUERY($sql, ddl, depth$)

```
 1    (new_sql, child_sql, new_ddl) ← QUERY_REWRITE(sql, ddl)
 2    repartition ← SHOULD_REPARTITION(sql, ddl)
 3    if repartition
 4        then create temp table on cluster using new_ddl
 5    result ← EXEC_QUERY(sql, ddl, depth, ddl.all_partitions)
 6    if repartition
 7        then
 8                ▷ reissue the query on the temporary table
 9                return INIT_QUERY(new_sql, new_ddl, depth)
10        else return result
```

EXEC_QUERY($sql, ddl, depth, partitions$)

```
 1    (new_sql, child_sql, new_ddl) ← QUERY_REWRITE(sql, ddl)
 2    data ← NEW
 3    if depth = 1 or length(partitions) = 1
 4        then
 5                ▷ Leaf Node
 6                data ← APPLY(sql, ddl, partitions)
 7                if SHOULD_REPARTITION(sql, ddl)
 8                    then
 9                            REPARTITON(data, new_ddl)
10                            return Done
11                    else
12                            return data
13        else
                 ▷ number of children depends on total partitions and desired depth
14            for n ← 1 to f(partitions, depth)
15                do in parallel
16                    child ← SPAWN
17                    data + = child .EXEC_QUERY(child_sql, ddl, depth − 1, SLICE(partitions))
18            wait until children are done
19            if SHOULD_REPARTITION(sql, ddl)
20                then return Done
21            return APPLY(new_sql, new_ddl, data)
```

Fig. 3. Parallelization algorithm

on, and are not yet at the maximum allowed depth, they branch out to other nodes. These nodes receive the query package, apply the original SQL to raw data, and cast their slices of resulting data out to the network in accordance with the rewritten DDL. The second layer of nodes waits for the third to finish, and passes the *Done* status to its parent. The master node checks that all the children completed successfully, and, since it knows the query required repartitioning, starts the process over, this time with the rewritten query package. When the leaf nodes read the data, they apply the rewritten query to the physical (filtered and redistributed) data, and pass the results to their parent nodes, which then aggregate the results using a rewrite of the rewritten query package – a second and, in this case, final rewrite. They then pass the results to the head node, which applies the third query package to these intermediate results, and returns the final result to the user.

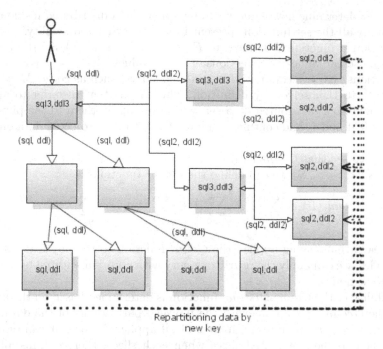

Fig. 4. Sequence of query package rewrites and data flow through the distributed system

A careful reader will notice that the two intermediate nodes created on the first pass (prior to repartitioning) do almost no work, and seem unnecessary. There are several reasons for their creation. The number of children a single node can have is often limited by external constraints, such as a maximum number of sockets that can be open on a single machine. By controlling the desired depth, the system is able to achieve two things – stay within the external constraints, and distribute the task of aggregating the final results, reducing load on the head node.

3.3   Query Rewriting and Data Redistribution

Because we do not support the concept of joins on distributed data, the requirements for an optimizer are drastically simplified. The *Where* clause is a static list of operations that evaluates to a truth value for each raw record without reference to other entities. Therefore, at each step of the parallel execution tree, we need only to project the requested tuples (*Group By*) and evaluate the aggregate functions. The main challenge lies in making sure the data is distributed in such a way that projections from different physical partitions can be combined without loss of data, and that set functions may be evaluated on a maximum number of nodes, taking full advantage of available parallelism.

In order to determine how a query can be applied in a distributed fashion, we must examine all the set functions present in the query for reducibility.

We say that a function F reduces to F' if $F(a, b) = F'(F(a), F(b))$ where a and b can be scalar values or collections of scalar values.

All functions that are commutative and associative, such as Sum, Min, and Max, reduce to themselves, by definition. This means that in order to apply these functions to a distributed data set, we simply apply them to the physical data on the leaf nodes, and continue applying them to the projected data on the parent nodes.

It is easy to see that $Count$ reduces to Sum, since given a set P partitioned into subsets $S_1, S_2, ..., S_n$ we can define the $Count$ functions as

$$Count(P) = \sum_{\forall p \in P} 1 = \sum_{\forall s \in S_1} 1 + \sum_{\forall s \in S_2} 1 + ... + \sum_{\forall s \in S_n} 1$$

This means that having applied the $Count$ function to the physical data on the leaf nodes, we can apply the rewritten query with Sum substituted for $Count$ on the parent nodes.

Reducibility of the $Count$ $Distinct$ function is determined based on the alignment of the $Group$ By columns with the physical partitioning of the data. Because $Count$ $Distinct$ requires awareness of all applicable rows, it can only be safely applied to the distributed slices when each slice either contains all the applicable data, or when the applicable data is guaranteed not to overlap on the distinct column across slices. In other words, in order to successfully apply the $Count$ $Distinct$ function the data must distributed on some subset of the $Group$ By columns, on the combination of the $Group$ By columns and the argument to the $Count$ $Distinct$ function, or only on the argument to the $Count$ $Distinct$ function. Any other distribution can lead to double-counting of rows when the $Distinct$ operation is performed on leaf nodes.

Given these facts, we can present the basic pseudo-code for the REDUCE function.

REDUCE(f, sql, ddl)

```
1    if IS_ASSOCIATIVE(f) and IS_COMMUTATIVE(f)
2        then return f
3    if f = count
4        then return sum
5    if f = count_distinct
6        then
7                if f.argument = ddl.distribution_key or
8                ddl.distribution_key ∈ sql.grouping_columns or
9                f.argument + sql.grouping_columns = ddl.distribution_key
10                   then return sum
11   return null
```

Although a limited number of core aggregators are provided, a number of other functions, such as *Average* and *Stddev*, can be derived. For example, a query involving *Average* can treated as a query for $Count(x)$ and $Sum(x)$, and a query involving *Stddev* can be considered to ask for $Count(x)$, $Sum(x)$, and $Sum(x^2)$, with the final computation being performed by the root node.

The following examples illustrate the results of the query rewrite applied to the same query given different data distribution scenarios. In all cases, the goal is to count the total number of visits and the number of **unique** users for every country, region, city combination. These examples exhaustively cover all possible distribution scenarios.

```
SELECT   country, region, city,
count(visit_id), count(DISTINCT user_id)
FROM site_visits
WHERE eyes = 'blue' and hands = 'large'
GROUP BY country_id, region_id, city_id
```

Case 1: Distribution Key is part of the Group-By tuple. The data is distributed on the *Group By* tuple (country_id, region_id, city_id) or any non-empty subset of the tuple such as (country_id, region_id). It is easy to see that the results of this query run on the whole dataset are equivalent to applying the filters to individual partitions, performing the aggregations locally, and concatenating the results. Distribution by the tuple (country_id, region_id) guarantees that all rows where this tuple is the same – including all the applicable values of city_id – go into a single bucket , so we can safely compute the unique counts. Rows returned by leaf nodes to intermediate nodes require no further aggregation, since the slice data contains all relevant records.

Case 2: Distribution Key is the Group-By tuple + the unique aggregator parameter. Let us consider a slightly more difficult case, where the distribution is on partition key (country, region, city, user_id). The filtering and $Count()$ aggregation can be performed on the local partitions. The distinct count applied on the local partitions would not be accurate, since records with the same country, region, city combination and different user ids are guaranteed to be distributed across multiple buckets. By the same token, it is clear that no unique combination of (country_id, region_id, city_id, user_id) can occur in more than one partition. This means that we can apply the original query to the slices, and have the intermediate nodes perform a small bit of aggregation using the following rewritten query:

```
SELECT   country, region, city,
sum(count_visit_id), sum(distinct_count_user_id)
FROM data_from_children
GROUP BY country_id, region_id, city_id
```

Case 3: Distribution Key is the unique aggregator parameter. This case is equivalent to case 2 – since no records containing the same user_id can exist in multiple buckets, it is safe to apply the original query to the data slices, and aggregate at intermediate nodes.

Case 4: Distribution Key is not aligned with the query. Consider the case of the table being distributed on some other partition key. We can execute the filtering phase on the local partitions, reducing the cardinality of the data set, and the reducible aggregation operations, but we cannot perform the unique aggregation, since the same tuple can occur on multiple partitions. Instead, we project the tuples (*country_id*, *region_id*, *city_id*, *user_id*) into a temporary distributed table that is partitioned by (*country_id*, *region_id*, *city_id*, *user_id*). We can now apply the rewritten query below to the temporary table and proceed as in case 2.

```
SELECT country, region, city, sum(visit_counts), count(distinct user_id)
FROM tt_site_visits
GROUP BY country, region, city
```

The generalized query package rewrite algorithm is as follows. First, add the grouping columns from the original query as columns to the new DDL and SQL. Then for each measure, add its reduced function to the SQL measure list if it is reducible; otherwise add the measure's arguments to the grouping column list of the SQL, and the columns in SQL and DDL. The pseudocode is below.

QUERY_REWRITE(*sql*, *ddl*)
1   *new_ddl* ← NEW*DDL*(*columns* : *sql.grouping_columns*)
2   *new_sql* ← NEW*SQL*(*columns* : *sql.grouping_columns*)
3   **for** *measure* ∈ (*sql*.*measures*)
4   **do**
5   *reduced_function* ← REDUCE(*measure*, *sql*, *ddl*)
6   **if** DEFINED(*reduced_function*)
7   **then** *new_sql*.*measures*+ = *reduced_function*
8   **else**
9   *new_sql*.*grouping_columns*+ = *measure.arguments*
10   *new_sql*.*columns*+ = *measure.arguments*
11   *new_ddl*.*columns*+ = *measure.arguments*
12   **return** (*new_sql*, *sql*, *new_ddl*)

Given these facts, it is easy to determine if repartitioning will be required based on the query package alone. Repartitioning is needed when the query contains irreducible measures, and the distribution key does not match one of the cases 1, 2, and 3 above. Repartitioning is never needed when the query does not contain irreducible measures.

SHOULD_REPARTITION(sql, ddl)

```
1  for measure ∈ sql.measures
2      do
3          if
4              not DEFINED(REDUCE(measure)) and
5              not sql.grouping_columns ⊃ ddl.distribution_key and
6              ddl.distribution_key ≠ (measure.arguments ∩ sql.grouping_columns) and
7              sql.grouping_columns ≠ measure.arguments
8                  then return TRUE
9  return FALSE
```

With this final procedure, we have presented a complete description of the parallelization engine required for the reduced instruction set SQL we propose.

4 Evaluation

4.1 Performance Evaluation: Synthetic Data

To evaluate the performance of the described system, we demonstrate several benchmark tests using a table that contains approximately 3 million URLs and some associated data. In all cases, the numbers shown are average run times of several executions of the query.

First, we show that the system scales well by distributing a sample dataset on 2, 4, 6, and then 10 nodes and performing the same query involving a *count distinct* operation on the URL, which is the distribution key. As shown in Figure 5, the time required to perform the query is inversely proportional to the number of servers in the cluster, as expected. There are theoretical complications that arise when the number of nodes becomes extremely large; we expect them to be handled by the cluster job management layer [13,7,6], and they are outside the scope of this paper.

In Figure 6 we compare the speed of the system to an enterprise class, parallel commercial RDBMS with a comparable number of CPUs. The performance on two typical queries is shown – a count of distinct URLs, and a count of distinct top-level domains derived from the URLs using the *substring* and *position* functions. The parallel query system described in this paper vastly outperforms the RDBMS, even when the nodes are under heavy load.

4.2 Performance Evaluation: ASK.com datasets

Figure 7 demonstrates the results of applying the proposed technique to a real-life dataset at Ask.com. A basic metric used by business analytics is the number of unique sessions and unique users different properties receive in any given time period. The same query was applied to the enterprise-level, parallel, commercial, 64-bit RDBMS and to a cluster of 16 32-bit Linux computers running the Knuckles software. The data was organized on the cluster in 256 evenly

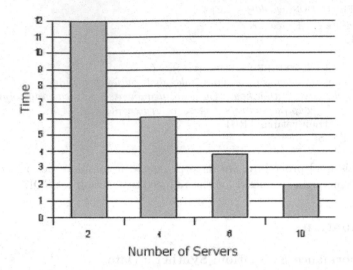

Fig. 5. Execution time is inversely proportional to the number of nodes on the cluster

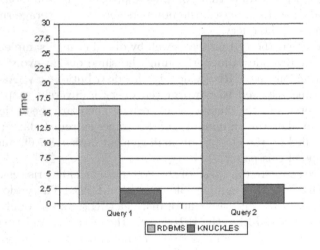

Fig. 6. The Knuckles system vastly outperforms an enterprise-level commercial RDBMS installation

distributed partitions, which were sub-partitioned (using the *Physical Partition* attribute) by day. BerkeleyDB was used for storage, with the record fields packed into a complex, delimited value of the BDB key-value pairs. It is worth noting that this distribution wasn't chosen specifically for the task of demonstrating Knuckles performance – the partitioned dataset already existed on the cluster for other purposes. We simply wrote a Storage Engine

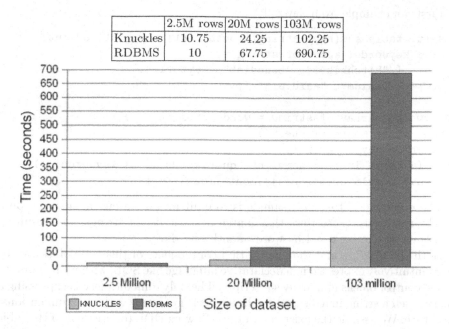

	2.5M rows	20M rows	103M rows
Knuckles	10.75	24.25	102.25
RDBMS	10	67.75	690.75

Fig. 7. Knuckles performs significantly better than a commercial RDBMS as the size of the dataset increases

that knew how to access the data. This demonstrates the additional benefit of Knuckles – that of being able to bring the database directly to the data, which is often already distributed in the enterprise due to unrelated storage or processing considerations.

While performance is comparable for *small* datasets – in fact, the RDBMS is slightly faster on average – Figure 7 shows that the benefits of explicit parallelism and lack of overhead of the Knuckles system quickly demonstrate themselves as the size of the datasets increases.

4.3 Comparison to Other Efforts

Sawzall [12] and PigLatin [11] authors describe their approaches to attacking the same basic problem of simplifying parallel access to large data stores. Unlike the approach presented here, they have found that new languages are needed to express the aggregation tasks at hand, arguing that their solutions lead to programs that are easier to read, write, and express. Here we present examples from both papers alongside our restricted SQL solutions.

The examples address the issue of gathering statistics on term frequency within user queries – a common enough problem in the search analytics space that both Google and Yahoo use word count tasks as canonical examples of parallelizable data retrieval scenarios.

First, an example from Sawzall:

```
result: table sum[key: string][month: int][day: int] of int;
static keywords: array of string =
        { "hitchhiker", "benedict", "vytorin",
        "itanium", "aardvark" };

querywords: array of string = words_from_query();
month: int = month_of_query();
day: int = day_of_query();
when (i: each int; j: some int; querywords[i] == keywords[j])
    emit result[keywords[j]][month][day] <- 1;
```

The objective of this code sample is to emit into a stateful aggregator, on a row-by-row basis, those rows that match a static list of strings. The resulting aggregate should be keyed by word, month and day.

Both the filtering and the aggregation components of this code block can be more intuitively expressed in a declarative language like SQL, without recourse to loops, control blocks and array subscripts. The only difficult piece is expressing a query, itself a string in a discrete field, as a vector of terms within a 2-dimensional structure. We assume the existence of the following DDL (or create it, if the table has not been defined, and we are dealing with pre-existing distributed raw files):

```
/* define the underlying data file structure */
CREATE TABLE queries (
        dt date,
        pn number DISTRIBUTION PARTITION,
        query varchar(1000))
 STORAGE CSV BUCKETS(10)
```

For the actual solution, we employ the built-in *Flatten* storage engine, which treats the *query* field of *Queries* as a nested record and instantiates a pseudo-table to express this abstraction.

```
CREATE TABLE words (
        word varchar(100) FLATTEN(0))
          STORAGE EXPAND(queries,query,split(/\s/));

SELECT word,trunc(dt,mm),trunc(dd,date),count(word)
FROM words
  WHERE word IN ('hitchhiker', 'benedict', 'vytorin',
                  'itanium', 'aardvark')
GROUP BY word,trunc(mm,date),trunc(dd,date)
```

The PigLatin paper presents a wordcount example that differs from Sawzall's in that no filter is applied, and the words are counted in a raw fashion, with reference to date:

```
input = LOAD 'documents' USING StorageText();
words = FOREACH input GENERATE FLATTEN(Tokenize(*));
grouped = GROUP words BY $0;
counts = FOREACH grouped GENERATE group, COUNT(words);
```

Although it may appear obscure at a glance, this approach is more like ours, in that it takes a declarative approach to the retrieval language. Built-in transformation functions allow the programmer to change the dimensional space at query time rather than as a part of the data definition, though this makes the structure and transformations somewhat difficult to visualize. Our restricted SQL equivalent would be the same as above, without the *Where* clause and the date functions in the *Group By*:

```
CREATE TABLE words (
        word varchar(100) FLATTEN(0))
STORAGE EXPAND(queries,query,split(/\s/));

SELECT word,count(word)
FROM words
GROUP BY word
```

PigLatin authors state that "programmers don't 'like' SQL." The language's longevity and widespread adoption belie this statement. In our experience, the semantics of a declarative language are not a stumbling block for most programmers; rather, it is the unpredictability of a black-box process that translates the declarative statement into a procedural execution plan that confounds them. By reducing the available instruction set we are able to retain the intuitive and familiar framework while eliminating the need for a complex optimizer.

5   Conclusion

In order to support the requirements of their logging infrastructures, web enterprises that perform clickstream data warehousing and analysis already implicitly support a high level of data parallelism. Largely as a result of the ubiquity of relational databases as datastores, such enterprises rely heavily on SQL to express analysis problems, though experience has shown that a restricted style of SQL is best suited to data warehouse queries. Existing database systems – even MPP systems – provide an SQL environment over clickstream data at a high cost, as loading into databases creates a bottleneck that defeats the implicit parallelism of the enterprise's computing resources. Cluster-aware database alternatives focus on taking better advantage of these resources, but require users to learn new procedural data query languages. We consider the wide adoption of SQL among programmers and analysts to be a resource in itself: we seek a system which squanders neither the ubiquity of SQL as a data retrieval language nor the data-parallel infrastructure already in place in so many enterprises.

To better support indefinite scaling, our system is based on deterministic rather than data-relative plans, and on local rather than global process control. The RISC analog of our restricted SQL engine participating on each node in the system carries out its operations independently, while the collective set of these decisions in each query instance replaces the conventional optimizer decision. It is a shared-nothing system in the most extreme sense, in that even statistics and data dictionaries are abandoned. A modular storage engine framework and Data Definition Language allow tabular access to arbitrary data formats on disk. All optimizations follow a simple rule: try to push as much of the computation as possible as far down the execution tree as possible. We are left with a framework ideally suited for filtering and aggregating distributed data by exploiting the inherent parallel potential in its distribution.

When the data is not partitioned in such a way as to maximize the resources participating in the computation, we redistribute a projection of the requested data in order to fulfill this property – and in order to fulfill the property of determinism, we always perform this redistribution under these circumstances. The cost of loading the network in order to better distribute the data load across CPUs is justified by the observation that network speed is continually increasing at a lowering cost, while enhancements to processors in the coming years are predicted to be in the form of increased numbers of cores rather than faster chip speeds. In a tree-based execution model, we can better afford to send data across the network than we can afford to push computations up the tree, where parallel resources are more scarce.

We conclude by observing that our proposed design is in essence a P2P analog of click-stream data warehousing. Most earlier efforts on distributed or parallel database systems made *query optimization* decisions by relying on some form of centralized control. By relying on deterministic rewrite of queries with completely localized control, we have developed a system that can be easily scaled to hundreds or thousands of processors.

Acknowledgments. We would like to thank the following people at Ask.com for supporting this work: Nick McCann, Manager, Data Engineering Development, Erik Collier, Director of Product Management, Shane McGilloway, VP of Research & Analytics, and Bob Zweig, VP of Information Technology.

References

1. Ibm db2 product family,
   http://www-306.ibm.com/software/sw-bycategory/subcategory/SWB30.html
2. Oracle products & services, http://www.oracle.com/products/index.html
3. Teradata: A division of ncr, http://www.teradata.com/t/
4. Androutsellis-Theotokis, S., Spinellis, D.: A survey of peer-to-peer content distribution technologies. ACM Comput. Surv. 36(4), 335–371 (2004)

5. Chang, F., Dean, J., Ghemawat, S., Hsieh, W.C., Wallach, D.A., Burrows, M., Chandra, T., Fikes, A., Gruber, R.: Bigtable: A distributed storage system for structured data (awarded best paper!). In: OSDI, pp. 205–218. USENIX Association (2006)

6. Dean, J., Ghemawat, S.: Mapreduce: Simplified data processing on large clusters. In: OSDI, pp. 137–150 (2004)

7. Ghemawat, S., Gobioff, H., Leung, S.-T.: The google file system. In: SOSP, pp. 29–43 (2003)

8. Graefe, G.: Volcano - an extensible and parallel query evaluation system. IEEE Trans. Knowl. Data Eng. 6(1), 120–135 (1994)

9. Isard, M., Budiu, M., Birrell, A., Fetterly, D.: Dryad: Distributed Data-Parallel Programs from Sequential Building Blocks. In: EuroSys 2007, Lisbon, Portugal (March 2007)

10. Kossmann, D.: The state of the art in distributed query processing. ACM Comput. Surv. 32(4), 422–469 (2000)

11. Kumar, R., Olston, C., Reed, B., Srivastava, U., Tomkins, A.: Research project: Pig, http://research.yahoo.com/project/pig

12. Pike, R., Dorward, S., Griesemer, R., Quinlan, S.: Interpreting the data: Parallel analysis with Sawzall. Scientific Programming Journal 13(4), 277–298 (2005)

13. Shen, K., Yang, T., Chu, L., Holliday, J., Kuschner, D.A., Zhu, H.: Neptune: Scalable replication management and programming support for cluster-based network services. In: USITS, pp. 197–208 (2001)

14. Venugopal, S., Buyya, R., Ramamohanarao, K.: A taxonomy of data grids for distributed data sharing, management, and processing. ACM Comput. Surv. 38(1), 3 (2006)

15. Yu, C.T., Chang, C.C.: Distributed query processing. ACM Comput. Surv. 16(4), 399–433 (1984)

Localized Mobility-Aware Geometric Graphs for Topology Control in Heterogeneous Mobile Ad Hoc Networks

Khoriba Ghada[1], Jie Li[1], and Yusheng Ji[2]

[1] Graduate School of Systems and Information Engineering University of Tsukuba, Japan
[2] Information Systems Architecture Science Research Division, National Institute of Informatics, Japan
Khoriba@osdp.cs.tsukuba.ac.jp, lijie@cs.tsukuba.ac.jp, kei@nii.ac.jp

Abstract. This paper studies the topology control in heterogeneous mobile ad hoc networks (MANETs), where mobile nodes have different transmission ranges, and moving speeds. We consider symmetric graph where two nodes can communicate directly if and only if they are within the transmission range of each other. We propose mobility-aware geometric graphs in which the mobility parameter has been used instead of distance parameter to minimize nodes transmission power. The proposed graphs are localized in which they depend on collecting information from the one-hop neighbors only. In the proposed graphs, each node transmission range is tuned according to its neighbors moving speeds. Nodes then update their neighbors list according to their new transmission range. We introduce locale algorithms for updating nodes transmission range and the mobile network graphs according to symmetric and cone-based graphs. To study the performance of the proposed graphs, a mobile ad hoc network environment has been simulated. The simulation results show that the proposed graphs reduce the average transmission power by more than forty to fifty percent and reduce the average nodes degree while preserving network connectivity property.

Keywords: Geometric graphs, Graph connectivity, Heterogeneous Transmission range, Mobile Ad Hoc Networks, Topology control.

1 Introduction

In wireless ad hoc networks, nodes are allowed to communicate directly to each other using wireless transceivers without the need for a fixed infrastructure. The features of a typical ad hoc network are as follow [1]:

- Heterogeneity: A typical ad hoc network is composed of heterogeneous nodes which may have different maximum transmission ranges.
- Mobility: In a typical ad hoc networks most of the nodes are mobile.
- Relatively dispersed network: The adoption of the ad hoc networking paradigm is justified when the nodes composing the network are geographically dispersed.

S. Bhalla (Ed.): DNIS 2007, LNCS 4777, pp. 178–188, 2007.

A mobile ad hoc network (MANET) is a kind of heterogeneous wireless ad hoc network formed by a set of mobile nodes in a self-organizing way without the need for a fixed infrastructure. In MANETs all nodes cooperate to achieve certain goal tasks, such as fast traffic information delivery on highways and urban areas, ubiquitous Internet access, and delivery of location-aware information. MANETs raised many challenging research problems as it has some unavoidable limitations as memory, energy, and processing elements. One of challenges for MANETs is to minimize the energy consumption. Topology control protocols concern with nodes' transmission range coordination to reduce energy consumption and interference while preserving connectivity and/or increasing network capacity. MANETs usually suffer from the topology connectivity problem. Connectivity problem appears when any message sent at one time slot between two connected nodes and one of the nodes moved before the topology update, so the topology will be out of date, and may be disconnected. The localized solution has the advantage that nodes depend on collecting information from the one-hop neighbors only. In MANETs, topology control protocols can be seen as a layer between the routing layer and medium access control (MAC) layer [1]. The topology control protocol sets the new transmission range to the MAC layer, and updates the neighbor-list used for the routing layer.

Most prior research on network topology control assumed that wireless ad hoc networks are modeled by unit disk graphs (UDG) [2][3], where nodes are static (i.e. not moving). In UDG, nodes can communicate with each other as long as their Euclidean distances are no more than a constant threshold. UDG cannot be perfectly used with heterogeneous wireless ad hoc networks, in which the maximum transmission range of wireless devices may vary and each node may have its own transmission range. To solve the problem with UDG, the mutual inclusion (MI) graph is proposed in which, two nodes can communicate directly if and only if they are within transmission range of each other [4]. Also Yao graph [5] is proposed for topology control in heterogeneous wireless ad hoc network. In Yao graph, each node divides its transmission range to equally size sectors then communicate with the nearest neighbor in each sector. Yao graph and its extensions [2] [3] [4] [6] are proved having good characteristics in the topology control in respect of sparseness, bounded node degree, and can be constructed locally in an efficient way. However, nodes in heterogeneous wireless ad hoc networks modeled by MI graph, Yao graph and Yao graph extensions are assumed being static in the previous work. Another interesting related work [7] has studied the topology control for MANETs where nodes may move but have the same maximum transmission ranges.

In this paper, we study the topology control problem for heterogeneous MANETs where nodes may move and may have different maximum transmission ranges. The main contribution of this paper is to propose novel mobility-aware extensions to MI graph [4], and Yao graph [5], respectively for MANETs. The proposed graphs mobile mutual inclusion (MMI)graph and mobile Yao ($MYao$) graph are mobility-aware extensions to mutual inclusion (MI) graph, and Yao graph, respectively. In the two proposed graphs, MMI graph and $MYao$ graph starting with MI graph as the initial graph, the nodes choose their neighbors depending on mobility parameter instead of distance parameter. MMI graph and $MYao$ graph is constructed locally as each node depends only collecting information from the one-hop neighbors only. We conduct comprehensive computer simulation to study the performance of the proposed MMI and $MYao$ graph. The simulation results show that the MMI proposed graph can reduce the physical node degree by at least four degree *(in case of 100 nodes)* regardless of nodes' speed. Also MMI graph and $MYao$ graph reduces the average transmission power by more than fifty percent and forty percent, respectively while preserving network connectivity property.

The remaining of the paper is organized as follows. Section 2 introduces some basic definitions and related work. Section 3 explains the proposed graph in details. Simulation results and performance metrics are presented in section 4. Conclusions and possible future research work are introduced in section 5.

2   Preliminaries

2.1   Geometric Graphs

In the theory of geometric random graphs, a set of points are distributed according to some probability distribution in a certain region. Points are then connected according to some rule. In this paper, MANET can be defined as a geometric graph, where nodes are connected according to MI graph or Yao graph, respectively as in Fig.1.

A heterogeneous mobile ad hoc network consists of a set N of n mobile nodes distributed in 2-dimension plan. Each node u_i has its own transmission range r_i and its own moving speed ν_i. Any signal sent by node u_i can be received by any node u_j if $d_{i,j} \leq min(r_i, r_j)$, where $d_{i,j}$ is the Euclidean distance between u_i, u_j.

MI graph should be used to model heterogeneous wireless networks, as there is no assumption for equally transmission ranges. MI graph is defined as follow [4].

Definition 1. Mutual inclusion (MI) graph. In MI graph two nodes can communicate directly if and only if they are within transmission range of each other. Formally, $MI = (N, E)$ defined as follows. There is an edge $(u_i, u_j) \in E$ if and only if $d_{i,j} \leq min(r_i, r_j)$, where $d_{i,j}$ is the Euclidean distance between nodes u_i and u_j.

Another basic geometric graphs used in topology control problem research work is Yao graph. The basic Yao graph is defined as follow [1].

Definition 2. Yao Graph (YG). Given a set N of points in the plane, and an integer $k \geq 6$, the Yao Graph of parameter k is the directed graph $YG_k = (N, E_k)$ defined as follows.

At each node $u_i \in N$, divide the plane with radius r_i into k equally sized sectors originating at u_i. Denoting by $S_i^1, ..., S_i^k$, the sectors for node u_i, the edge $(u_i, u_j) \in E_k$ if and only if u_j is the closest neighbor of u_i in S_i^l.

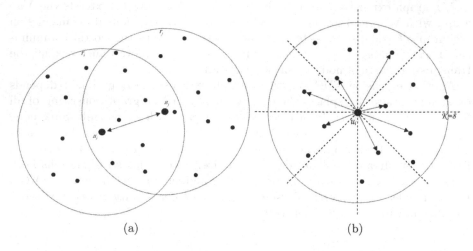

Fig. 1. (a)In MI graph, node u_i and node u_j can communicate directly if and only if there are within the transmission range of each other. (b)Yao graph, node u_i, divide the plane with radius r_i into 8 equally sized sectors, then choose the nearest node in each sector.

2.2 Related Work

In this section, we will refer to some topology control research work which depend on geometric graphs. Many graphs extensions have been proposed for topology control problem as [4] [6] [7] [8]. The basic used graphs are Relative Neighborhood Graph (RNG), Gabriel Graph (GG), and Yao graph (YG). Generally for any set of points N, RNG is a subset of or equal to the GG graph and Yao graph, i.e, ($RNG(N) \subseteq GG(N)$), ($RNG(N) \subseteq YG_k(N), k \geq 6$). These graphs are sparse, bounded average node degree, and can be constructed locally in an efficient way. The power stretch factor of Gabriel Graph is always one, Yao graph is bounded by a constant, and Relative Neighborhood Graph could be as large as the network size minus one [9] [10].

Mobility sensitive approaches are proposed in [7]. The research depends on solving the problem of inconsistency that emerged from node mobility. This research extends the relative neighborhood graph (RNG), minimum spanning tree (MST), and shortest path tree (SPT) based protocols by two different mechanisms. The first was consistent local views that avoid inconsistent information and delay. The second was the mobility management that tolerates outdated information.

The Directed Relative Neighborhood Graph ($DRNG$) [8] is a localized topology control algorithm for heterogeneous networks depending on RNG. Ordered Yao graph $OrdYaoGG$ topology control structure applies the ordered Yao Structure on Gabriel graph structure [10]. In $OrdYaoGG$, all nodes have same maximum transmission ranges and uses only 1-hop information. The final topology is planar graph, whose node degree is bounded.

MI graph extensions are introduced in [4]. EYG_k graph, extends the Yao graph with MI initial graph, in which each node u_i partitions its transmission region into k equal-sized sectors. In each sector, u_i keeps the closest communication neighbor u_j, if the transmission range of u_j is more than or equal the transmission range of node u_i, preserving bidirectional links.

Another extension is EYG_k^* graph which extends EYG_k graph. It depends on sink structure. It replaces the directed star in a Yao graph consisting of all links toward a node u_j by a directed tree $T(u_j)$ with u_j as sink. Sink node u_j constructs a Tree $T(u_i)$ rooted at u_i, where u_i is some in-neighbor of u_j in EYG_k graph. Then it informs the end nodes of the selected links to keep such links if they already exist or add it otherwise. EYG_k and EYG_k^* graphs both are sparse and have constant bounded power and length stretch factors. EYG_k^*, has a bounded node degree. These extensions assume a static network without mobility in wireless ad hoc networks.

3   Proposed Graphs

To deal with the node mobility for MANETs, we here introduce two extensions to MI graph and Yao graph, respectively.

3.1   Proposed Mobile Mutual Inclusion Graph

MI graph should be used to model heterogeneous wireless networks, as there is no assumption for equally transmission ranges. The Mobile MI graph MMI proposed graph is mobility-aware extension to MI graph to deal with motion parameter. MMI graph is defined as follow.

Definition 3. MMI graph. Let $MI_T(N,E)$ be the maximum power MI graph at time T, where T is updated when the node changes its position, and $MI_T'(N,E')$ be the sub-graph generated by MMI graph as follow:

Each node $u_i \in N$, with transmission range r_i and speed ν_i chooses the slowest neighbor u_j that satisfies $r_j \geq r_i$, and $\nu_j \leq \nu_i$.

Algorithm 1 shows how to construct the mobile mutual inclusion graph. From MMI) construction steps, we can see that MMI is a localized protocol as it depends only on speed information from only first-hop neighbors.

3.2   Proposed Mobile Yao Graph

The proposed extension for Yao Graph, Mobile Yao graph $MYao_k$, will be constructed using MI graph as initial graph, considering mobility parameter. Yao graph based on MI graph is connected if MI is connected. Yao graph extensions in wireless ad hoc networks [4], have bounded node degree and energy efficient. Here we propose a mobility-aware extension for Yao graph to deal with mobility parameters instead of distance parameters. The proposed $MYao_k$ can be defined as follow.

Algorithm 1. *Constructing MMI*

1: Each node u_i, with transmission range r_i and speed ν_i locally broadcasts a message with ID_i and it's speed ν_i to all first-hop neighbors $Neigh(u_i)$ in its transmission range.

2: When node u_i updates its position: u_i checks for all neighbors, and chooses u_j as new neighbor if:
   - u_j has the minimum speed ν_j among all neighbors of u_i.
   - The transmission range $r_j \geq r_i$ which ensures bi-directional links, and
   - The speed $\nu_j \leq \nu_i$ which ensures the connectivity.

3: Each node u_i updates its neighbor list, and updates the transmission range according to the new neighbors.

4: MMI graph, will be the union of all chosen links.

Definition 4. $MYao_k$ graph. Let $MI_T(N, E)$ be the maximum power MI graph at time T, where T is updated when the node changes its position, and $MI'_T(N, E')$ be the sub-graph generated by $MYao_k$ as follow:

Given an integer $k \geq 7$, each node $u_i \in N$, with transmission range r_i and speed ν_i, divides its plane into k equally sized sectors $S_i^1, ..., S_i^k$ originating at node u_i. In each sector node u_i chooses the slowest neighbor u_j that satisfies $r_j \geq r_i$, and $\nu_j \leq \nu_i$. Assuming the network consists of N nodes and initial topology is MI graph.

Algorithm 2 shows how to construct $MYao$ graph. From $MYao_k$ construction steps, we can see that $MYao_k$ is a localized protocol as it depends only on direction and speed information from only first-hop neighbors.

Algorithm 2. *Constructing MYao$_k$*

1: Each node u_i, with transmission range r_i divides the disk $d(u_i, r_i)$ of π angle around the node, into k equal sized sectors each with angle $2\pi/k$. These sectors can be represented with its number and center node as $S_i^1, ..., S_i^k$.

2: Each node u_i locally broadcasts a message with ID_i and it's speed ν_i to all first-hop neighbors $Neigh(u_i)$ in its transmission range.

3: When node u_i updates its position: u_i checks for all neighbors, and add u_j to S_i^l, if:
   - u_j is inside the lth sector,
   - u_j has the minimum speed ν_j among all neighbors of u_i in this sector,
   - The transmission range $r_j \geq r_i$ which ensures bi-directional links, and
   - The speed $\nu_j \leq \nu_i$ which ensures the connectivity.

4: Each node u_i updates its neighbor list, and updates the transmission range according to the new neighbors.

5: $MYao_k$ graph, will be the union of all chosen links.

3.3 Proposed Graph Properties

To show the benefits of the two proposed graphs for the topology control in MANETs. We study the properties of the proposed graphs in this section.

For N nodes, Let $MI_T(N, E)$ be the maximum power MI graph at time T, and $MI_T'(N, E')$ be the sub-graph generated by $MYao_k$ or MMI graphs.

Properties 1: $MI_T'(N, E')$ results from $MYao_k$ has linear number of links at most kn links, where k is a constant represents the number of sectors and n is the number of nodes.

For example, when dividing the range around node u_i into k sectors, it chooses only one node in each sector. So node u_i will get at most k neighbors. For n nodes network, links will be at most kn.

In the case of MMI, the number of links for each node can reach $(n-1)^2$ in the worst case, but in average each node choose only one logically neighbor node so the number of links can be only n.

Properties 2: $MI_T'(N, E')$ results from $MYao_k$ or MMI preserves connectivity, if the initial MI graph is connected.

If the nodes are static, the condition $r_j \geq r_i$ that ensures the bidirectional links can preserve the connectivity condition, as proven in [4].

If the nodes are mobile, the condition $\nu_j \leq \nu_i$ can preserve the connectivity. As node u_j will not change its position before node u_i. Each time node u_i changes its position it will recall the graph to be update.

4   Performance and Simulation

4.1   Performance Metrics

Topology control problem aims to reduce the energy consumption and interference while preserving the connectivity. The major performance metrics, for topology control, are node degree, connectivity, transmission power and throughput. Connectivity and throughput parameters measures how the proposed graphs preserves network properties. The graphs were evaluated in terms of the following metrics.

- *Physical Node degree:* Let $G_P = (N, E_P)$, the communication graph generated by a certain topology control protocol P. For a given $u \in N$, the physical node degree of u in G_P is the number of nodes within $u's$ transmission range when it transmits at the broadcast power [1]. To measure the effect of the structures on the interference, average physical nodes degrees were computed.
- *Connectivity:* A Graph $G = (N, E)$ is connected. If for any two nodes $u_i, u_j \in E$, there exists a path from u to v in G [1]. The connectivity ratio computed as the ratio of all connected pairs to all possible pairs in N.
- *Transmission power:* average power used for every packet transmission is computed.

- *Throughput:* network throughput computed as total number of received bits per second.

4.2   Simulation Parameters

The simulation for the proposed graphs has been implemented in C++ using ns2.30 simulator [11]. We implement EYG_k graph, which uses the distance parameter [4], and the proposed $MYao_k$ graph both with $k = 8$, and the proposed MMI graph. MI graph, the maximum power graph, was tested to indicate the effect of different topology control on the network properties.

The initial position for the nodes and its transmission range are randomly generated. Each graph is updated when node updates its position. The domain, in which 100 mobile nodes are distributed, is a square area $1000 \times 1000m^2$. Each node has random initial transmission range uniformly distributed between $200m \sim 250m$. In our simulation, we assume that the minimum transmission range for nodes after graph update is 180m. Nodes move with random speed between 10 to 50m/sec in 500sec. The traffic between nodes is generated as constant bit rate (CBR) with packet size 512Byte and 60 connections. Each result is repeated 10 times and associated with a 95 percent confidence interval.

4.3   Results

Node degree performance. To measure the effect of the structures on the interference, average physical nodes degrees were computed. Fig.2 shows that the proposed MMI graph has the lowest node degree among the others. Although $MYao$ graph is the nearest to the initial graph. From the throughput results, the lower node degree affects the network throughput, so the proposed $MYao$ graph can preserve better throughput ratio than MMI graph.

Connectivity performance. The connectivity ratio computed as the ratio of all connected pairs to all possible pairs in N. We compute the average ratio of all connected pairs to all possible pairs in the network. The simulation results show that the proposed graphs preserve connectivity, if the maximum power MI graph is connected. The connectivity ratio for these graphs may differ than the MI basic graph by 0.01 to 1 percent.

Transmission power performance. Average power used for every packet transmission is computed. As shown in Fig.3, The lower value is for the proposed MMI graph. It has the smallest average transmission range. MMI graph reduce the transmission power with more than fifty percent than MI. $MYao_k$ graph, and EYG_k graph, for $k = 8$, reduce the transmission power by at least forty percent than MI graph. The initial graph MI has a constant transmission power as initially defined for the node (the maximum one), as it has no criteria to adjust the transmission power according to nodes distance or speed parameters.

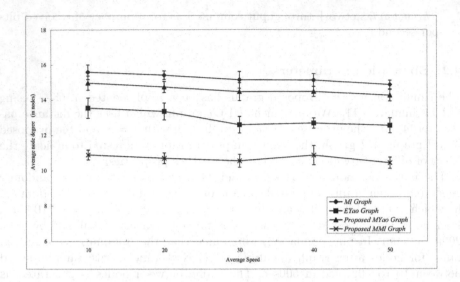

Fig. 2. Average node degree for 100 nodes

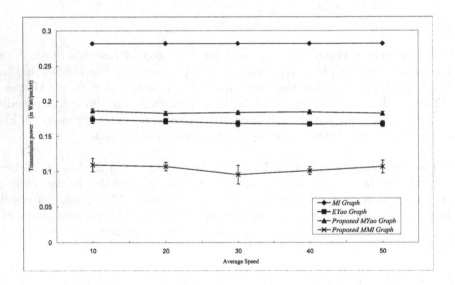

Fig. 3. Average transmission power for 100 nodes

Throughput performance. Network throughput computed as total number of received bits per second. Fig.4 shows that the throughput (bit/sec) for the maximum power MI graph is maximum. The proposed MMI has the lowest throughput value while $MYao$ has higher throughput. It is clear that the throughput level is related to transmission power and node degree parameters.

As when transmission power is minimized to lower level, the physical node degree will be also minimized while the probability of loosing packets will be higher.

The network throughput is affected by other protocols as the routing protocols and MAC Layer protocols. From this point, we can conclude two things. First, topology control problem can be studied as large scale protocol changing in routing and MAC protocols. Second, it is required to study topology control as optimization problem to minimize the energy consumption while maximizing the network throughput.

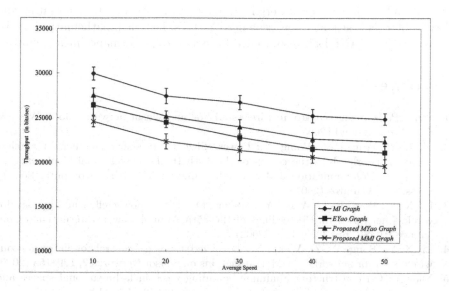

Fig. 4. Throughput for 100 nodes

5   Conclusion

Topology control problem is one of most important issues for MANETs. The main target of topology control is to decrease power consumption and interference while preserving network desired properties as connectivity. Using geometric graphs for this problem proves efficiency along many researches. Most of researches deal with wireless network in static state in the case of sensor network studying. In this kind of networks, the main target is to minimize the energy consumed. While in MANETs, the problem of preserving the connectivity with various node speed is added. Various transmission range and speed should be considered.

In this paper, a mobility-aware solution is proposed. It depends on using the speed parameter instead of only distance parameter. Using of speed parameter add a new dimension in dealing with the mobile networks. It improves the preservation of the connectivity parameter along with highly dynamic network.

$MYao$ and MMI are proposed as mobility-aware extensions to Yao graph and symmetric mutual inclusion graph.

We can see that the proposed graphs are constructed locally using only first-hop information. The simulation results show that the MMI proposed graph can reduce network interference as it reduce the physical node degree by at least four degree (in case of 100 nodes) regardless of nodes' speed. The proposed MMI and $MYao$ graph reduces the average transmission power by more than fifty percent and forty percent respectively while preserving network connectivity property.

The study of topology control problem using more parameters as motion direction can be a future study. Merging topology control solution with routing and MAC protocols is still an open issue in MANETs. Detailed mathematical model framework for MANETs is also needed for accurate performance measurement.

References

1. Santi, P.: Topology control in wireless ad hoc and sensor networks. John Wiley & Sons Ltd. England (2005)
2. Li, X.-Y., Wan, P.-J., Wang, Y.: Power efficient and sparse spanner for wireless ad hoc networks. In: Proceedings of the 10th IEEE International Conference on Computer Communication and Networks, pp. 564–567. IEEE Computer Society Press, Los Alamitos (2001)
3. Li, X.-Y., Wan, P.-J., Wang, Y., Frieder, O.: Sparse power efficient topology for wireless networks. In: Proceedings of the 35th Annual Hawaii International Conference HICSS, pp. 3839–3848 (2002)
4. Li, X.-Y., Song, W.-Z., Wang, Y.: Localized topology control for heterogeneous wireless sensor networks. ACM Transactions on Sensor Networks 2, 129–153 (2006)
5. Yao, A.: On constructing minimum spanning trees in k-dimensional spaces and related problems. SIAM Journal on Computing 11, 721–736 (1982)
6. Alzoubi, K., Li, X.-Y., Wang, Y., Wan, P.-J., Frieder, O.: Geometric spanners for wireless ad hoc networks. IEEE Transactions on Parallel and Distributed Systems 14, 408–421 (2003)
7. Wu, J., Dai, F.: Mobility-sensitive topology control in mobile ad hoc networks. IEEE Transactions on Parallel and Distributed Systems 17, 522–535 (2006)
8. Li, N., Hou, J.C.: Localized topology control algorithms for heterogeneous wireless networks. IEEE/ACM Transactions on Networking (TON) 13, 1313–1324 (2005)
9. Santi, P.: Topology control in wireless ad hoc and sensor networks. ACM Computing Surveys 37, 164–194 (2005)
10. Song, W.-Z., Wang, Y., Li, X.-Y., Frieder, O.: Localized algorithms for energy efficient topology in wireless ad hoc networks. In: Proceeding of MobiHoc 2004, Roppongi, Japan, pp. 98–108 (2004)
11. Fall, K., Varadhan, K.: The ns Manual. The VINT Project, UC Berkeley, LBL, USC/ISI, and Xerox PARC January (2007), http://www.isi.edu/nsnam/ns/

Using SCTP to Implement Multihomed Web Servers

Md. Nurul Islam and A. Kara

Department of Computer Science and Engineering,
The University of Aizu, Aizu-Wakamatsu, Fukushima-ken, Japan
{d8072201,kara}@u-aizu.ac.jp

Abstract. Stream Control Transmission Protocol (SCTP) is an emerging transport protocol for sending data from one point to another over the Internet. SCTP provides innovative features beyond TCP, such as multihoming and multistreaming in a single SCTP association. With the explosive growth of Internet users, it remains an unsecured environment; hence the data protection from the malicious interception becomes a crucial issue. In this paper, we present how the multihoming feature of SCTP can be exploited to implement a multihomed web client and server for providing resilience to network failure and a certain level of data protection against the malicious interception as well as achieving high throughput by utilizing the multiple paths that exist between the multihomed web client and server. We discuss the potential threats during data transmission and the way of defense using multihoming in this aspect.

Keywords: Multihoming, Stream Control Transmission Protocol, Transport Protocol.

1 Introduction

At present we use various applications such as Internet telephony, IP TV, pay per view, video conferencing, e-banking and some other transactions over the Internet. With the explosive growth of Internet, it remains an unsecured environment as the attackers can extract information from the wire and intercept the communications by traffic analysis; hence, the data reliability and protection from the malicious interception becomes a crucial issue. Most of the case, end to end encryption is widely used for this concern; however there are still some problems in this encryption infrastructure, as for example, it increases the network overheads and degrades the application throughput. In addition, end to end encryption requires a proper key management mechanism [1], [2]. Moreover, most of the current endpoints do not implement all required encryption algorithms due to expensive and prohibitive reason. For example, a client host may implement them, but the host that the client host wants to communicate with may not.

Another encryption approach is a technique of encrypting some parts of a compressed data file whereas leaving rests unencrypted that has been proposed for several multimedia applications. For example, Spanos et al. [3] proposed a scheme to reduce the computationally intensive processes of encryption and decryption of video as well as to improve the enciphering/deciphering throughput. Droogenbroeck et al.

S. Bhalla (Ed.): DNIS 2007, LNCS 4777, pp. 189–202, 2007.

[4] claimed that it is not compulsory to encrypt all data in the image encryption and described several partial encryption techniques to provide security mechanisms as well as an overall visual check that might be desirable in some applications like searching from a shared image database. Reference [5] also presents the suitability of the partial encryption of images instead of the use of full encryption. However, this type of partial encryption scheme does not strive for maximum data protection but decrease the computational complexity [6]. In this paper, we present how the multihoming feature of Stream Control Transmission Protocol (SCTP) can be exploited to implement a multihomed web client and server for providing resilience to network failure and a certain level of data protection against the malicious interception as well as achieving high throughput by utilizing the multiple paths that exist between the multihomed web client and server.

The performance of SCTP has been investigated in some previous papers. The suitability of SCTP in a wide area network, especially when competing with TCP has been studied by Jungmaier et al. [7]. The authors conclude that SCTP can share the link equally in terms of link layer load with other SCTP connections and friendly with the TCP protocol. Ravier et al. [8] have investigated the multihoming feature of SCTP through experimental studies. The authors have shown that SCTP is a viable solution of multihoming for link optimization and redundancy. Recently, Islam et al. [9] have shown the impact of data chunk transmissions with different sizes of payload on the throughput of a multihomed SCTP association in both network stable state and packet loss environment. Fu et al. [10] have proposed an analytical model to evaluate the performance of SCTP multihoming, and showed that it provides the steady state throughput of both primary and alternative path of a multihomed SCTP association. Reference [11] shows that a link failure causes negligible loss in the throughput of a SCTP association because in this case the SCTP sender quickly reroutes the packets through an alternate path. Kashihara et al. [12] have proposed a multipath transmission algorithm using multihoming to improve the goodput during handover across heterogeneous wireless access networks by sending the same packets along multiple paths. However, to the best of our knowledge, there is no existing work that explains a multihomed web client and server running on SCTP. Mutihoming feature of SCTP can provide web server a certain level of network redundancy. In this fault tolerant approach, the web server can be reachable even when one of its path fails, another path can be used to communicate without interruption. In this paper, we explain how routing paths can be separated between a multihomed web client and server running on SCTP.

The rest of the paper is organized as follows. The basics and features of SCTP are explained in section 2. The security threats, network monitoring attacks are explained in section 3. Separation of routing paths between a multihomed web client and server is presented in section 4. We show our experimental setup and results in section 5. Comparison with related work is shown in section 6. Finally, the conclusion is provided in section 7.

2   Overview of SCTP

IETF Signaling Transport (SIGTRAN) Working Group primarily developed SCTP to transport signaling messages over IP. However, recent studies demonstrate that SCTP is a robust general-purpose, message-oriented, reliable transport protocol that can be useful for many applications at the same time [13]. Transmission Control Protocol (TCP) [14] is most commonly known transport layer protocol on the Internet but limited to the changing needs of signaling transport and data transfer intensive applications. On the other hand, SCTP satisfies these needs by using features like multi-streaming and multihoming along with a message- oriented and a partial ordering service [7], [15]. TCP provides both reliable data transfer and strict order-of-transmission delivery of data but not all applications need both. Head-of-line blocking by TCP causes unnecessary delay. Moreover, TCP is relatively vulnerable to denial of service (DoS) attacks. The most well known attack is called SYN flooding attack [16]. In response to these limitations, IETF SIGTRAN Working Group designed the new protocol SCTP to provide additional performance and reliability for the signaling transport over IP networks [17].

2.1 SCTP Multihoming

An SCTP association supports multihomed hosts. In fault tolerant approach, when one path fails, another interface can be used for data delivery without interruption. Multihoming allows two endpoints to setup an association with multiple addresses for each endpoint. During association initialization, each endpoint lists its IP addresses as well as its port number. Hence, the SCTP sender or receiver has a list of transport addresses that share the same port number. The SCTP sender selects a primary destination address and transmits all data chunks through this address and rest addresses are considered as alternate destination addresses. This built-in support for multihomed endpoints allows high availability applications to perform switchover to an alternate destination address without interrupting the data transfer during link failure situation.

The Fig. 1 shows two multihomed hosts X and Y. X1 and X2 represent two IP addresses for host X. Y1 and Y2 represent two IP addresses for host Y. The host X and Y can make an association using these four IP addresses. Then SCTP association is ({X1, X2}, {Y1, Y2}). However, according to current SCTP implementation such

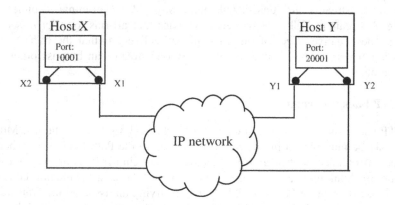

Fig. 1. Example of multihomed hosts

as [18], only one pair of IP addresses can be chosen for normal communication, called primary addresses. Here, primary destination address for host X and Y is Y1 and X1 respectively. And alternate destination address for host X and Y is Y2 and X2 respectively. The alternate destination addresses are used when primary fails. The primary destination address may become inaccessible, possibly due to an interface failure, server congestion, or due to slow route convergence (e.g. of BGP) around path outages. In this example, if the primary destination address Y1 becomes unreachable, multihoming keeps the SCTP association active by allowing host X to send data to host Y using alternate destination address Y2. The receiver endpoint sends the acknowledgements of data and control chunks back to the originating peer destination addresses; as a result, the sender endpoint can determine the reachability of the peer destination addresses.

When its peer is multihomed, an SCTP sender sets an error counter for each of the destination addresses of the peer endpoint. If, a retransmission timer expires on an outstanding DATA chunk sent to the primary destination address, then an error counter for this destination is incremented. When the error counter of primary destination address reaches an upper bound, Path.Max.Retrans [17], then it is declared unreachable. The endpoint monitors the reachability of the idle destination addresses of the peer by sending periodic HEARTBEAT chunks to all the idle destination addresses. During the setting up of an association, the periodic interval at which HEARTBEAT chunks are sent to a destination is provided. HEARTBEAT chunk has to be acknowledged by a HEARTBEAT-ACK. If the HEARTBEAT chunk is not acknowledged with a HEARTBEAT-ACK within the retransmission timeout (RTO), then an error counter for that destination is incremented. When the error counter of a destination address reaches an upper bound, Path.Max.Retrans, then that destination address is declared unreachable. However, when the HEARTBEAT-ACK is received from the destination, then the error counter is cleared and that destination address can be made active again.

2.2 SCTP Multistreaming

SCTP supports multiple streams in a single association. If message loss occurs in one stream, other streams are unaffected, that way, SCTP eliminates unnecessary blocking. A stream in TCP is a sequence of bytes that affirms to strict in-sequence delivery. The negative effect of this in-sequence delivery is that the bypass among streams is not permissible. However, in SCTP, one stream can by-pass others upon prioritization.

2.3 SCTP Packet Format

An SCTP packet is composed of a common header and a series of chunks. Multiple chunks may be multiplexed in one SCTP packet up to the Path MTU size. There are two types of chunks: control and data chunks. Data chunks contain user messages while control chunks contain control information. SCTP uses the same port concept as TCP and User Datagram Protocol (UDP) in identifying an association. Transmission error is detected through the use of 32-bit Adler checksum, which is more robust than the 16-bit checksum used by TCP and UDP. SCTP packets with an invalid checksum are silently discarded by the receiver. SCTP common header also contains a 32-bit verification tag. This verification tag is association specific and exchanged between endpoints at association startup. The verification tag provides a key that allows a receiver to verify that the SCTP packet belongs to the current association and is not an old or stale packet from a previous association. Each chunk begins with a type field, which is used to distinguish data chunks and different types of control chunks, followed by chunk specific flags and a chunk length field needed because chunks have a variable length. In addition, for the reliability and congestion control, each data chunk is assigned a unique Transmission Sequence Number (TSN), while Streams identifier and Stream sequence number are used for multistreaming support.

3 Network Monitoring Attacks

This section depicts a vision on possible structure of the potential threats during data transmission.

Now-a-days, hackers or intruders can extract information from the wire and intercept the communications by traffic analysis without either user being aware intrusion has occurred, called network monitoring attacks [19]. Thus, at present, these attacks are a serious problem in IP networks due to their stealthy nature. As for example, in VoIP, it gives an attacker the ability to listen or record private phone conversations. For e-banking systems, these attacks are the obvious threats to the customer's personal computers as they compromise the integrity of the communications between the e-banking system and the customer. Hence, communicating data may be vulnerable to such passive attacks during data transmission over the IP networks. In the network monitoring attacks, the intruders configure the respective network interface in promiscuous mode. The intruders then run a network monitoring tool that captures any packets sent on the network through that interface and stores or sends it for later usage by the attacker, thus, the attacker discovers valuable information in this way. There are various applications, listening devices or network analyzers including tcpdump [20], ethereal [21], Snort [22] and Snoop (bundled with the Solaris operating system) that can be used for capturing and reassembling data being transmitted over the network.

4 Separation of Routing Paths

This section explains how routing paths can be separated between a multihomed web client and server using multihoming feature of SCTP.

TCP involves one source and one destination IP address during the connection. It means that even if the TCP sender or receiver contains more than one physical address with multiple IP addresses, only one of these IP addresses per end will be utilized. Consequently, TCP connection is vulnerable with the network monitoring attacks occurred on the path followed by the connection, therefore it exposed to a potentially insecure environment. On the other hand, an SCTP association supports multihomed hosts; therefore, multiple distinct paths exist between two SCTP peer endpoints.

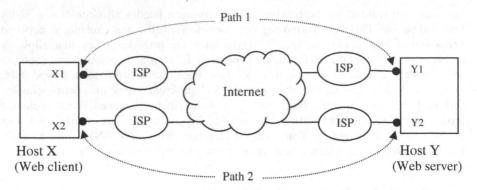

Fig. 2. An SCTP association with multiple paths

According to [17], definition of path is- "The route taken by the SCTP packets sent by one SCTP endpoint to a specific destination transport address of its peer SCTP endpoint". In the general IP network terminology, a host is called multihomed if it can be addressed by multiple IP addresses [23]. SCTP supports multihomed hosts; as a result, it may allow a sender to transmit data to a multihomed receiver through different destination addresses. The Fig. 2 shows an example of SCTP association between two multihomed hosts X (web client) and Y (web server). X1 and X2 represent two IP addresses for the client. Y1 and Y2 represent two IP addresses for the server. During association initialization, each endpoint lists its IP addresses as well as its port number. Hence, the SCTP sender or receiver has a list of transport addresses that share the same port number. According to standard SCTP [17], the SCTP sender selects a primary path and transmits all data chunks through this path and alternate paths are used for retransmission purpose to improve the probability of reaching the remote endpoint. In Fig. 2, we assume Y1 is the primary destination for the client, if Y1 becomes unreachable, multihoming keeps the SCTP association active by allowing the client to communicate the server using alternate destination address Y2. However, if we consider a TCP connection in Fig. 2 where the web client connects to the web server using (X1, Y1) and if Y1 fails, the web server will become

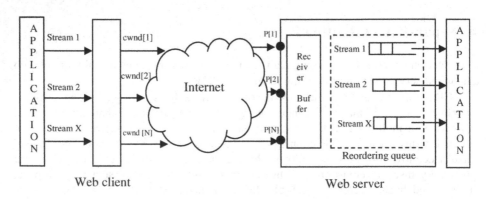

Fig. 3. Architecture of proposed load sharing method

unreachable to the client and the connection will be abort. Hence, a multihomed web client and server running on SCTP can provide resilience to such network failures and alternate path can be used to communicate without interruption.

We propose a load sharing method which enables an SCTP endpoint to split the source data to transmit over all available paths that can enhance the fail-over performance as well as increase the throughput for reliable data transmission. The Fig. 3 shows the architecture of proposed load sharing method.

Data_assignment ()
1. input number of paths, N;
2. input rwnd;
3. set sum_cwnd= 0, cwnd[N];
4. for i=1 to N
5.   input cwnd[i];
6. end for
7. for i=1 to N
8.   sum_cwnd=sum_cwnd+cwnd[i];
9. endfor
10.   for i =1 to N   // Here $\lfloor \rfloor$ means floor value.

11.   send data through path[i] equal to $\lfloor(cwnd[i]/sum\_cwnd* rwnd)\rfloor$;
12.   endfor
13. end

Fig. 4. Algorithm for data assignment across multiple paths

The Fig 4. states the data assignment algorithm. As we consider multihomed data receiver, the data sender needs to maintain a separate congestion window (cwnd) for each transport addresses of data receiver due to different network paths and their congestion status may not related. We assume that N number of paths are available between a multihomed web client and server running on SCTP; it means that the sender can reach to the receiver using N number of destination addresses and the sender maintains a separate cwnd for each path. The variable cwnd [i] represents the current congestion window of path [i], where i =1, 2, 3, ... , N. As for example, cwnd [1], cwnd [2], ..., cwnd [N] represents the congestion window (cwnd) maintained at the sender for path[1], path [2], ..., path [N] respectively.

The variable sum_cwnd represents the aggregate congestion window, which specifies the largest amount of data the sender can inject into the paths between a multihomed web client and server before causing path congestion. As we mentioned that SCTP sender maintains a separate congestion window (cwnd) for each destination addresses, if SCTP sender maintains cwnd [1], cwnd [2],...,cwnd [N] for path [1], path [2] ...., path [N] respectively, then the aggregate congestion window of these paths, sum_cwnd is,

cwnd [1] + cwnd [2] + ...+ cwnd [N] = $\sum$cwnd [i] ; where i =1, 2, ... , N

In this load sharing method, the sender uses sum_cwnd variable to control the flow of data transfer. The sender does not transmit new data to the receiver if receiver window (rwnd) indicates that there is no receiver buffer space left and SCTP sender's sending rate is min (rwnd, sum_cwnd). According to the data assignment algorithm states in Fig. 4, the sender assigns the data over all available paths proportional to each path's congestion window. Congestion window changes and updates dynamically by many causes like packet loss, congestion and other network conditions, so data assignment based on congestion window can be an intelligent decision even in fast variation network paths.

We argue that the risk of network monitoring attacks can be reduced by transmitting source data over all available paths because in this case an intruder needs to know the identity of multiple paths and then access, capture, decode and diagnose the network traffic to retrieve the data from multiple network interfaces. Hence this method can be used to minimize or even disable potential data interceptions during data transmissions because the attacker can not collect all data at a single point. Therefore, a multihomed web server and client can achieve high performance and reliable data transmission by using SCTP. In a load sharing approach, SCTP sender transfers data simultaneously over all available end-to-end paths to the receiver, so there is no primary path concept in this approach; as a result the congestion control information for each path fairly updated that can ensure better error detection and control as well as provide data confidentiality during data transmission between a multihomed web client and server.

Moreover, according to the SCTP specification in [17], the SCTP sender selects a primary path and transmits all data chunks through this path. The SCTP sender changes the primary path when it notices that the primary path becomes unreachable, it is called fail-over which allows a SCTP sender to send traffic to an alternate path [17], [24]. The primary path is switched to an alternate path when the error count on the data delivering path becomes larger than a certain threshold. In addition, an SCTP sender provides for application-initiated changeovers, so that the sending application can move the outgoing traffic to another path by changing the sender's primary destination address [17]. According to the current changeover mechanism, one path is active and all others are in idle and this changeover occurs from a single active path to a single idle path. However, an extension of SCTP can be developed by changing the primary path pretty often to route data traffic on different paths that exist between the multihomed web client and server which can also reduce network monitoring attacks. One simple method would be a round-robin fashion which means each time data transmission would be sent on a different path than was served last time. For instance, if the peer has three destination addresses Y1, Y2, Y3, then first transmission would be sent to Y1, the second transmission would be sent to Y2 and the third transmission would be sent to Y3. If a fourth transmission were required, the sender would rotate back to Y1. Hence, this advanced changeover mechanism could be use for data transmission between a multihomed web client and server running on SCTP to reduce network monitoring attacks by changing primary destination pretty often, thus sending data traffic to a potentially different path.

Many cryptography approaches are proposed to enhance data transmission reliability; however, it requires an additional exchange or transmission of information over some kind of secure channel which add some overheads in the application that decreases the throughput. We propose to utilize the multiple paths during data transmission between the web server and client running on SCTP to enhance the performance and reliability of data transmission against network monitoring attacks.

We claim that an intruder can do attacks easily if the path chosen by the web client and server upon initialization of the connection is used for data transmissions throughout the lifetime of the connection. Hence, we propose that multihoming feature of SCTP can be used to take the advantage of data protection against network monitoring attacks by splitting source data to transmit over multiple paths or changing the primary path pretty often to route data traffic on different paths; as a result it will be difficult for the intruders to carry out all tasks for the successful network monitoring attacks. Even if the attacker captures some parts of transmitted data, the chance of extracting the information is very little because it needs all or at least most of the data to decode and retrieve the original information. Thus, data transmissions over multiple paths prevent the intruder from intercepting information. Moreover, cryptography security approaches add overhead in the applications but our proposed method does not add any overhead, consequently, the applications running on multihomed web client and server can achieve high throughput for web transfers.

In addition, in a web server running on TCP, after receiving a SYN request, the server allocates resources assuming that the connection will be established soon. If there is no response to the server's SYN/ACK by a client's ACK in a certain time, the connection information expires. However, this causes unnecessary allocation of resources of the server. An attacker can send many TCP connection requests with spoofed source addresses to a web server, as a result, the target web server's resources can be exhausted and, no more incoming TCP connections can be established. Thus, the web servers running on TCP are vulnerable related to SYN attacks [16]. On the other hand, SCTP protects such SYN attacks using its four-way handshake with a cookie mechanism during association establishment. The cookie mechanism in SCTP keeps the web server stateless without allocating any resources until the web client that initiates a connection maintained state, thus an attacker's ability to perform a SYN attack is reduced.

Routing path separation near the multihomed web client and server can be accomplished by selecting different ISPs and sending data traffic over several network interfaces. However, to make better separation of routing paths over the core network, Multiprotocol Label Switching (MPLS) [25] could be used. When the packets go through a MPLS network or Internet, Label Switched Routers (LSRs) perform routing based on Label Switching, hence, this routing approach is simpler than a conventional IP router. The label is carried in a packet header, and which represents the packet's FEC (Forwarding Equivalence Class), a particular path through the network of LSRs is specified. That is, each FEC represents a group of packets that share the same requirements for their transport. LSRs only look up the labels and just forward each packet based on its label, they do not inspect the IP header. MPLS also provides the suitability to divert and route traffic during link failures or congestion.

5   Experimental Setup and Results

Our experiments are meant to show the separation of routing paths near the endpoints using multihoming feature of SCTP. The setup is constructed of two Linux computers where each of them has two network interfaces as shown in Fig. 5. The SCTP implementation, SCTP library (sctplib) [18] was installed in both hosts. To extract

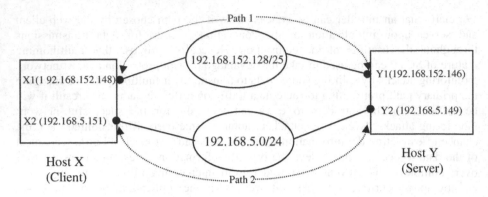

Fig. 5. Experimental setup

information while running the experiments, we used ethereal network protocol analyzer [21] and it simply captured all the packets that a certain network interface saw.

The technique we use in our experiments is to emulate a multihoming scenario by selecting two ISPs, one IP segment 192.168.5.0/24 which is connected to AINS (The University of AIZU Information Network System) and another segment 192.168.152.128/25 which is connected to Plala (Plala networks Inc.), one of Japan's

No.	Time	Source	Destination	Protocol	Info
1	0.000000	192.168.152.148	192.168.152.146	SCTP	INIT
2	0.000588	192.168.152.146	192.168.152.148	SCTP	INIT_ACK
3	0.001105	192.168.152.148	192.168.152.146	SCTP	COOKIE_ECHO
4	0.002058	192.168.152.146	192.168.152.148	SCTP	COOKIE_ACK
5	0.002563	192.168.5.151	192.168.5.149	SCTP	HEARTBEAT
6	0.002789	192.168.5.149	192.168.5.151	SCTP	HEARTBEAT_ACK
7	2.546126	192.168.152.148	192.168.152.146	SCTP	DATA
8	2.546762	192.168.152.146	192.168.152.148	SCTP	SACK DATA
9	2.564517	192.168.152.148	192.168.152.146	SCTP	SACK
10	3.002548	192.168.5.149	192.168.5.151	SCTP	HEARTBEAT
11	3.002936	192.168.5.151	192.168.5.149	SCTP	HEARTBEAT_ACK
12	4.557070	192.168.152.148	192.168.152.146	SCTP	DATA
13	4.557613	192.168.152.146	192.168.152.148	SCTP	SACK DATA
14	4.757909	192.168.152.148	192.168.152.146	SCTP	SACK

```
⊞ Frame 1 (136 bytes on wire, 136 bytes captured)
⊞ Linux cooked capture
⊞ Internet Protocol, Src: 192.168.152.148 (192.168.152.148), Dst: 192.168.152.146 (192.168.152.146)
⊟ Stream Control Transmission Protocol, Src Port: 59752 (59752), Dst Port: 7 (7)
     Source port: 59752
     Destination port: 7
     Verification tag: 0x00000000
     Checksum: 0x341d5af3 [correct CRC32C]
  ⊟ INIT chunk (Outbound streams: 10, inbound streams: 10)
    ⊞ Chunk type: INIT (1)
       Chunk flags: 0x00
       Chunk length: 88
       Initiate tag: 0x14501206
       Advertised receiver window credit (a_rwnd): 110592
       Number of outbound streams: 10
       Number of inbound streams: 10
       Initial TSN: 450171367
```

```
0000  00 00 00 01 00 06 00 e0  18 99 22 c2 04 e4 08 00   ........ .."....
0010  45 10 00 78 00 40 00 40  40 84 87 7a c0 a8 98 94   E..x.@. @..z....
0020  c0 a8 98 92 e9 68 00 07  00 00 00 00 34 1d 5a f3   .....h.. ....4.Z.
0030  01 00 00 58 14 50 12 06  00 01 b0 00 00 0a 00 0a   ...X.P.. ........
0040  1a d5 11 e7 c0 00 00 04  00 0c 00 08 00 05 00 06   ........ ........
0050  00 06 00 14 fe 80 00 00  00 00 00 00 02 02 b3 ff   ........ ........
0060  fe 03 ac 38 00 06 00 14  fe 80 00 00 00 00 00 00   ...8.... ........
0070  02 e0 18 ff fe 99 22 c2  00 05 00 08 c0 a8 98 94   ......". ........
0080  00 05 00 08 c0 a8 05 97                            ........
```

Fig. 6. Ethereal trace of path 1

leading Internet Service Provider. The Fig. 5 shows our network topology. The host X (client) communicates with the host Y (server). X1 and X2 represent two IP addresses for the client. Y1 and Y2 represent two IP addresses for the server. The client and server makes an association, ({X1, X2}, {Y1, Y2}) using these four IP addresses. X1of the client and Y1 of the server form path 1. Similarly, X2 of the client and Y2 of the server form path2. The network capacity is 100 Mb/s for each network interface. After initiating the association, a single stream of data chunks were sent between the client and server. The Ethereal trace in Fig. 6 shows the trace for the path 1 (192.168.152.148 → 192.168.152.146). Note that the receiver sends the Selective Acknowledgement (SACK) to the address which the corresponding SCTP packet originated.

No. ▾	Time	Source	Destination	Protocol	Info
1	0.000000	192.168.152.148	192.168.152.146	SCTP	INIT
2	0.000588	192.168.152.146	192.168.152.148	SCTP	INIT_ACK
3	0.001105	192.168.152.148	192.168.152.146	SCTP	COOKIE_ECHO
4	0.002058	192.168.152.146	192.168.152.148	SCTP	COOKIE_ACK
5	0.002563	192.168.5.151	192.168.5.149	SCTP	HEARTBEAT
6	0.002789	192.168.5.149	192.168.5.151	SCTP	HEARTBEAT_ACK
7	2.546126	192.168.152.148	192.168.152.146	SCTP	DATA
8	2.546762	192.168.152.146	192.168.152.148	SCTP	SACK DATA
9	2.564517	192.168.152.148	192.168.152.146	SCTP	SACK
10	3.002548	192.168.5.149	192.168.5.151	SCTP	HEARTBEAT
11	3.002936	192.168.5.151	192.168.5.149	SCTP	HEARTBEAT_ACK
12	4.557070	192.168.152.148	192.168.152.146	SCTP	DATA
13	4.557613	192.168.152.146	192.168.152.148	SCTP	SACK DATA
14	4.757909	192.168.152.148	192.168.152.146	SCTP	SACK

```
⊞ Frame 5 (80 bytes on wire, 80 bytes captured)
⊞ Linux cooked capture
⊞ Internet Protocol, Src: 192.168.5.151 (192.168.5.151), Dst: 192.168.5.149 (192.168.5.149)
⊟ Stream Control Transmission Protocol, Src Port: 59752 (59752), Dst Port: 7 (7)
    Source port: 59752
    Destination port: 7
    Verification tag: 0x6ebd5d04
    Checksum: 0x2d5df31e [correct CRC32C]
  ⊟ HEARTBEAT chunk (Information: 28 bytes)
    ⊞ Chunk type: HEARTBEAT (4)
      Chunk flags: 0x00
      Chunk length: 32
    ⊞ Heartbeat info parameter (Information: 24 bytes)
```

```
0000  00 00 00 01 00 06 00 02  b3 03 ac 38 04 e4 08 00   ........ ...8....
0010  45 10 00 40 00 02 40 00  40 84 ad ab c0 a8 05 97   E..@..@. @.......
0020  c0 a8 05 95 e9 68 00 07  6e bd 5d 04 2d 5d f3 1e   .....h.. n.].-]..
0030  04 00 00 20 00 01 00 1c  4b a2 48 e1 00 00 00 00   ... .... K.H.....
0040  83 13 c5 90 ef 17 b7 17  43 2a 62 b3 a1 d2 ca 78   ........ C*b....x
```

Fig. 7. Ethereal trace of path 2

The next Ethereal trace in Fig. 7 shows the trace for the path 2 (192.168.5.151 → 192.168.5.149). We can see from the trace that HEARTBEAT chunks are sent to the path 2 to monitor the reachability of the path and to update that path's RTT (Round Trip Time) where the HEARTBEAT information size is 28 bytes. It is worthwhile to note that by default, the SCTP sender sends periodic HEARTBEAT chunks to all idle destination addresses which are not currently using for data transmission. During the setting up of an association, the periodic interval at which HEARTBEAT chunks are sent to a destination is provided. Each HEARTBEAT chunks has to be acknowledged by a HEARTBEAT-ACK message. If the HEARTBEAT has not acknowledged with a HEARTBEAT-ACK within the retransmission timeout (RTO), then an error counter for that destination is incremented. When the error counter of a destination address reaches an upper bound, Path.Max.Retrans [17], then that destination address is declared unreachable. But when the HEARTBEAT-ACK is received from the destination, then the error counter is cleared and that destination address can be made active again.

6 Comparison with Related Work

References [26], [27], and [28] describe Onion Routing, a technique for reducing the network's vulnerability to traffic analysis. In this circuit-based routing approach, messages are transmitted from source to destination through a set of onion routers (proxies). The objective of onion routing is to make traffic analysis harder because in this case the intruder needs to have the ability to monitor every onion router in a network to discover the path of a message. However, this routing architecture does not provide any perfect privacy; it only provides a continuum in which the degree of privacy is a function of the number of participating routers vs. the number of compromised or malicious routers [29]. Furthermore, in Onion Routing approach, if any onion router fails or leaves the network then the corresponding path cannot continue functioning with the remaining routers or it cannot include the routers those are newly joined to the network. In our proposed multipath data transmission approach for a multihomed web client and server running on SCTP, if any communicating path fails, it may effect slightly on the data transfer, however it does not terminate the communication because the other available paths can be used for data transmission.

In an ad hoc network, data transmission is vulnerable to malicious attacks due to the absent of any central security gateway which can provide data protection service for mobile nodes. Moreover, IPsec protocol is also infeasible to such an ad hoc wireless network scenario [30]. Several security issues related to ad hoc networks have been discussed in the literature such as [31], [32], [33], and [34]. The aim of these works is to take advantage of multi-route between the communicating routes to achieve high performance and enhance data confidentiality. Recently, Li et al. [30] proposes a multipath routing algorithm called Multipath TCP Security (MTS) to reduce the passive attacks in ad hoc wireless networks. The author suggest some ideas such as on receiving the route check packets, the source node adaptively changes the current route to be the fastest one, which means that the routing path can be changed dynamically to improve the performance and data transmission reliability.

We believe that our proposed method yields better protection against network monitoring attacks than the existing approaches because the multihomed web client and server have enough information and control over the multihomed environment, as a result, built-in support for multihoming makes SCTP data sender feasible to accomplish the tasks such as monitoring available paths, failure detection and fail-over that allows an intelligent distribution of data traffic among multiple paths for reducing the network monitoring attacks. In addition, since the network monitoring attacks usually occur close to the endpoint by configuring the respective network interface in promiscuous mode, consequently, we propose that the sender distributes data traffic over multiple network interfaces near the endpoint, so that the attacker can not collect all data at a single point.

7 Conclusion

In this paper, we have presented that the multihoming feature of SCTP can be exploited to implement a multihomed web client and server for providing resilience to network failure and a certain level of data protection against the malicious interception as well as achieving high throughput by utilizing the multiple paths that exist between the multihomed web client and server. We have discussed the potential threats during data transmission and the way of defense using multihoming in this aspect. We argue that the benefits of SCTP multihoming can be exploited by the applications running on multihomed web client and server for reducing network monitoring attacks as well as achieving high throughput.

References

1. Lu, W.P., et al.: Secure Communication in Internet Environments: A Hierarchical Key Management Scheme for End-to-End Encryption. IEEE Trans. Comm. 37(10), 1014–1023 (1989)
2. Pierson, L.G., et al.: Key Management for Large Scale End-to-End Encryption. In: 28th Annual International Carnahan Conference on Security Technology, pp. 76–79. IEEE Computer Society Press, New York (October 1994)
3. Spanos, G.A., et al.: Performance Study of a Selective Encryption Scheme for the Security of Networked, Real-Time Video. In: 4th International Conference on Computer Communications and Networks, pp. 2–10 (September 1995)
4. Van Droogenbroeck, M., et al.: Techniques for a selective encryption of uncompressed and compressed images. In: Advanced Concepts for Intelligent Vision Systems (ACIVS) 2002, Ghent, Belgium, pp. 90–97 (September 2002)
5. Van Droogenbroeck, M.: Partial encryption of images for real-time applications. In: Fourth IEEE Signal Processing Symposium, Hilvarenbeek, The Netherlands, pp. 11–15 (April 2004)
6. Podesser, M., et al.: Selective bitplane encryption for secure transmission of image data in mobile environments. In: NORSIG 2002. 5th IEEE Nordic Signal Processing Symposium, Norway (October 2002)
7. Jungmaier, A., et al.: Performance evaluation of the Stream Control Transmission Protocol. In: High Performance Switching and Routing, Germany, pp. 141–148 (June 2000)
8. Ravier, T., et al.: Experimental studies of SCTP multi-homing. In: First Joint IEI/IEE Symposium on Telecommunications Systems Research, Dublin, Ireland (November 2001)
9. Islam, M., et al.: Throughput Analysis of SCTP over a Multi-homed Association. In: CIT 2006. Sixth IEEE International Conference on Computer and Information Technology, Seoul, Korea (September 2006)

10. Fu, S., et al.: Performance Modeling of SCTP Multihoming. IEEE Computer Society Press, St. Louis, MO (November 28-December 02, 2005)
11. Wei, G., et al.: simulation on NS. IEEE ICII 2001 2, 345–350 (2001)
12. Kashihara, S., et al.: Multi-path Transmission Algorithm for End-to-End Seamless Handover across Heterogeneous Wireless Access Networks. IEICE Transaction on Communication E87-B(3), 490–496 (2004)
13. Coene, L.: Stream Control Transmission Protocol Applicability Statement. RFC 3257, IETF (April 2002)
14. Postel, J.: Transmission Control Protocol. RFC 793, IETF (September 1981)
15. Caro Jr., A.L., et al.: SCTP: A Proposed Standard for Robust Internet Data Transport. IEEE Computer 36 (11), 56–63 (2003)
16. Schuba, C., et al.: Analysis of a denial of service attack on TCP. In: IEEE Symposium on Security and Privacy, pp. 208-223 (May 1997)
17. Stewart, R., et al.: Stream Control Transmission Protocol. RFC 2960, IETF (October 2000)
18. User Level Implementation of SCTP by University of Essen, http://www.sctp.de/sctp-download.html
19. Computer Emergency Response Team.: Ongoing Network Monitoring Attacks. CERT Advisory CA-94:01 (February 1994)
20. tcpdump, http://www.tcpdump.org
21. ethereal, http://www.ethereal.com
22. Snort, http://www.snort.org
23. Braden, R., et al.: Requirements for Internet hosts communication layers. RFC 1122, IETF (October 1989)
24. Jungmaier, A., et al.: On the use of SCTP in failover-scenarios. SCI 2002, USA, pp. 363-368 (July 2002)
25. Rosen, E., et al.: Multiprotocol label switching architecture. RFC 3031, IETF (January 2001)
26. Goldschlag, D., et al.: Hiding Routing Information. In: Anderson, R. (ed.) Information Hiding. LNCS, vol. 1174, pp. 137–150. Springer, Heidelberg (1996)
27. Reed, M., et al.: Anonymous Connections and Onion Routing. IEEE Journal on Selected Areas in Communications 16(4), 482–494 (1998)
28. Goldschlag, D., et al.: Onion Routing for Anonymous and Private Internet Connections. Communications of the ACM 42(2), 39–41 (1999)
29. Wikipedia. http://en.wikipedia.org
30. Li, Z., et al.: A New Multipath Routing Approach to Enhancing TCP Security in Ad Hoc Wireless Networks. In: ICPPW 2005. International Conference on Parallel Processing Workshops, pp. 372–379 (June 2005)
31. Bouam, S., et al.: Data Security in Ad Hoc Networks Using MultiPath Routing. In: 14th IEEE Personal, Indoor and Mobile Radio Communications (September 2003)
32. Papadimitratos, P., et al.: Secure Data Transmission in Mobile Ad Hoc Networks. In: WiSe 2003, ACM Press, New York (September 2003)
33. Lou, W., et al.: SPREAD: Enhancing Data Confidentiality in Mobile Ad Hoc Networks. In: IEEE INFOCOM 2004 (March 2004)
34. Burmester, M., et al.: Secure Multipath Communication in Mobile Ad hoc Networks. In: ITCC 2004. International Conference on Information Technology: Coding and Computing, vol. 2, p. 405 (April 2004)

Improving the Performance of Read-Only Transactions Through Speculation

T. Ragunathan and P. Krishna Reddy

International Institute of Information Technology, Hyderabad (A.P), India
ragunathan@students.iiit.ac.in, pkreddy@iiit.ac.in

Abstract. A read-only transaction (ROT) does not modify any data. The main issues regarding processing ROTs are correctness, data currency and performance. Two-phase Locking (2PL) protocol is widely used for concurrency control with serializabilty as correctness criteria. Even though 2PL processes ROTs correctly with no data currency related issues, the performance deteriorates as data contention increases. To improve the performance over 2PL, snapshot isolation (SI)-based protocols have been proposed. SI-based protocols process ROTs by reading from a snapshot of the committed data and ignoring the modifications produced by the concurrent active transactions. Even though SI-based algorithms improve the performance of ROTs, both data currency of ROTs and correctness (serializability) are compromised. In this paper, we propose an approach to improve the performance of ROTs using speculation without compromising data currency of transactions and correctness. The proposed approach improves the performance of ROTs by trading extra computing resources without violating serializability as correctness criteria. The simulation results show that with the proposed protocol the throughput performance is improved significantly over 2PL and SI-based approaches with manageable extra resources.

Keywords: Speculation, Transaction processing, Read-Only Transactions, Serializability.

1 Introduction

In the emerging web databases and e-commerce scenario, information systems should meet intensive information requirements from a large number of users. The information systems frequently execute read-only transactions (ROTs) or queries. To meet the demands, efforts are being made in the literature to investigate improved approaches to process ROTs in an efficient and correct manner. Two-phase locking protocol (2PL) [1] [2] which is widely used for concurrency control, lack the power of meeting this growing throughput demand. The performance of ROTs under 2PL degrades with data contention as the ROTs have to wait for the commitment of conflicting update transactions (UTs) to ensure serializability criteria.

A new isolation level called "Snapshot Isolation (SI)" [3] was proposed for improving the performance of ROTs. In SI-based protocols, ROTs are processed by reading only the committed data object values which are available at the time of

S. Bhalla (Ed.): DNIS 2007, LNCS 4777, pp. 203–221, 2007.

submission. The ROTs ignore the data object values produced by concurrent active transactions. Even though SI-based protocols improve the performance of ROTs, both correctness of transaction processing (serializability) and data currency of ROTs are compromised [3][4]. The term "data currency" [5] can be defined as *the time elapsed (t) between the time point the value of a data object returned to the transaction that requested the data object and the time point the most recent change performed to the data object*. If "t" is less/more, it means that transactions are provided with high/low data currency. It can be noted that 2PL processes transactions at low performance with serializability as correctness criteria, whereas SI-based protocols process transactions at high performance, however by compromising correctness. Also, 2PL provides high data currency and SI-based protocols provide low data currency for ROTs.

Due to fast improvements in hardware technology, high speed computer systems are available at affordable cost. Research is going on to improve the performance by identifying application parallelism to exploit low cost computing power. In the literature, an effort has been made to improve the transaction processing performance by using the notion of speculation [6]. By exploiting the notion of speculation, it has been shown that it is possible to improve the transaction processing performance by trading extra computing resources.

In speculative processing, a transaction carries out multiple executions by accessing the uncommitted values produced by the preceding transactions. It was observed that UTs create more uncommitted object versions under speculation. As a result, the waiting transactions have to carry out increased number of speculative executions. By restricting speculation only to ROTs, it is possible to improve the performance by processing ROTs with few speculative executions. Based on this observation, in this paper, we have proposed a protocol to improve the performance of ROTs with speculation. The simulation results show that the proposed protocol improves the performance significantly over 2PL and SI-based methods with manageable extra processing resources.

1.1 System Model

A database is a collection of data objects. Users interact with the database by invoking transactions. Transactions are represented with T_i, T_j,.... A transaction is a sequence of read and writes operations that are executed atomically on the data objects. A transaction can read a set of data objects from the database which forms the read set (RS) of the transaction and modify the values of another set of data objects which forms the write set (WS) of the transaction. An ROT contains only read operations. A UT consists of both read and writes operations. The transactions T_i and T_j are said to have a conflict, if $RS(T_i) \cap WS(T_j) \neq \emptyset$, or $WS(T_i) \cap RS(T_j) \neq \emptyset$ or $WS(T_i) \cap WS(T_j) \neq \emptyset$. The execution of a transaction must be atomic [1]; i.e., a transaction either commits or aborts. The commit of a transaction results in all of its changes being applied to the database, whereas the abort results in the changes being discarded. A commonly accepted correctness criterion in database systems is to ensure that interleaved executions of concurrent transactions are serializable [1]. A concurrency control protocol is employed in database management systems to ensure that all transaction executions are serializable.

The database management system consists of modules like transaction manager and a data manager [7]. Processing of transactions is managed by the transaction manager component of database management systems, while database is managed by the data manager.

Notations. Data objects are denoted with 'x','y', ... For the data object 'x', 'x_i' (i = 0 to n) represents i^{th} version of "x". The notation $r_i[x_j]$ indicates that read operation is executed by T_i on 'x_j' and $w_i[x_j]$ denotes that write operation is executed by T_i on a particular version of 'x' and 'x_j' is produced. The notations 's', 'c', and 'a' depict the start, commit and abort of transactions. T_{ij} indicates j^{th} speculative execution of T_i.

1.2 Paper Organization

The rest of the paper is organized as follows. In the next section, we discuss the related work. In section 3, we explain 2PL, SI-based protocols and speculation-based protocols. In section 4, we explain the proposed protocol and the proof of correctness. In section 5, we present the simulation results. In section 6, we discuss the advantages and implementation issues of the proposed approach. The last section contains summary and conclusions.

2 Related Work

In this section, we review the approaches proposed in the literature for improving the performance of ROTs. We also discuss the approaches based on speculation.

Regarding correctness, four isolation levels are specified by ANSI/ISO SQL-92 standard [8] for the processing of transactions. These isolation levels are read uncommitted, read committed, repeatable read, and serializable. Serializabilty is the accepted correctness criteria for transaction processing. In [9], Atul Adya et. al. proposed generalized definitions for the ANSI-SQL isolation levels by covering optimistic and multi-version schemes. It has been proposed that the performance can be improved by processing ROTs at lower isolation levels, other than serializable isolation level. However, both the data currency and correctness are compromised by processing ROTs at lower isolation levels.

An approach has been proposed in [10] for distributed environment in which read-only queries are processed with a special algorithm that is different from the algorithm used for update transactions. The ROTs are executed with specific currency requirements (strong or weak consistency) and can read the updates produced by the preceding committed UTs by examining transaction log in reverse chronological order until the desired data are reconstructed.

A protocol is proposed in [11] for managing data in a replicated multi-version environment. In this approach, the ROTs are processed independently of the underlying concurrency control and replica control mechanisms. As a result, the data availability for ROTs increases significantly since they can be executed as long as any one of the object is available in the system. In [12], an approach has been discussed by maintaining multiple versions of data objects in which ROTs read the particular versions of the data objects based on version period in which they arrived. It avoids undesirable interferences between ROTs and UTs, but the ROTs cannot see the modifications performed by the other active UTs.

A dual copy method has been proposed in [13], which separates the processing of ROTs from UTs to improve ROTs performance. In dual copy method, two copies of data are managed for each data object; a master and a slave. Master copy is used by UTs and ROTs use slave copy. Multiple versions of the slave copy are   maintained and these copies are synchronized by the master copy at appropriate times. In this approach, the data currency of ROTs depends on the update frequency of slave copies.

An approach has been proposed in [14] for processing ROTs in mobile environment by considering the data consistency and currency related issues. The data currency requirements of transactions are divided into three categories: transactions with strong, firm and weak requirements. In that paper only "firm currency" is discussed, which means the data objects read by an ROT must be at least as recent with reference to ROT's starting time.

To improve performance of ROTs, a new isolation level called "Snapshot Isolation (SI)" was proposed in [3]. (Pease refer Section 3.2 for details). Note that ROTs processed at SI violate serializability criteria and receives low data currency.

In [4], a theory is discussed to convert non serializable executions under SI into serializable executions by modifying the program logic of the applications. However, this approach requires programmers to detect the static dependencies between the application programs and to modify the program which will lead to a semantically equivalent application program that can be executed correctly without violating serializability criteria.

Regarding speculation, speculation has been extended in [15] to optimistic protocol for improving the deadline performance in centralized real-time environments. In [6], speculation has been extended to improve the performance of distributed database systems (please refer section 3.3 for details) by considering transactions which contain both read and writes operations.

The approaches proposed so far (other than speculation approaches), improve the performance of ROTs by compromising data currency. In this paper we have proposed an approach to improve both performance and data currency of ROTs by extending the notion of speculation.

3   Two-Phase Locking, Snapshot Isolation-Based and Speculative Locking Protocols

3.1   Two-Phase Locking Protocol

Under 2PL [7], a transaction obtains "read (R) lock" to read an object and a "write (W) lock" to write/update the data object. In 2PL, a transaction should obtain all the required locks before performing any unlock operation. We have considered a variation of 2PL called "strict two-phase locking protocol" [7]. The strict 2PL scheduler releases all of a transaction's locks together, when the transaction terminates.

The lock compatibility matrix for 2PL is shown in Figure 1. The terms *yes* and *no* in the matrix means that the corresponding lock requests are compatible and incompatible respectively.

The processing of ROTs under 2PL is depicted in Figure 2. Both T_1 and T_3 are UTs and T_2 is an ROT. It can be observed that T_2 has to wait for a lock on the object 'x' until T_1 commits. Similarly T_3, has to wait for a lock on 'y'. (The space between the last operation and 'c' notation in the transaction diagram depicts the time required to carry out logging and commit operations.)

Lock Request by T_i	Lock Held by T_j	
	R	W
R	yes	no
W	no	no

Fig. 1. Lock compatibility matrix for 2PL

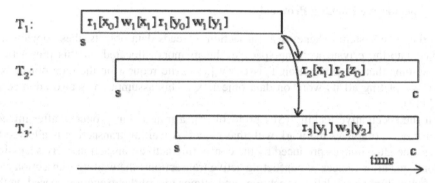

Fig. 2. Depiction of transaction processing for 2PL

3.2 Snapshot Isolation-Based Protocol

A new isolation level called snapshot isolation (SI) is proposed in [3]. The SI-level lies between READ COMMIITTED and REPEATABLE READ isolation levels [3]. In SI-based techniques, an ROT reads data from the snapshot of the (committed) data available when the transaction has started or generated the first read operation. The modifications performed by other concurrent UTs which have started their execution after the executing ROT (T_i), are unavailable to T_i. A variation of SI-based protocol called "First Committer Wins Rule (FCWR)" works as follows. Let T_i and T_j be UTs. T_i will successfully commit if and only if no concurrent T_j has already committed writes of data objects that T_i intends to write. SI-based protocols are not serializable [4].

The processing of ROTs using FCWR is depicted in Figure 3. Both T_1 and T_3 are UTs, and T_2 is an ROT. It can be observed that T_2 reads the currently available values 'y_0' and 'z_0' and proceeds with the execution. Simultaneously T_3 also reads and issues update operation to 'x'. As T_1 commits, T_3 has to be aborted as per the FCWR. However, as per FCWR, T_2 commits with the old values and it has not accessed the updates produced by T_1 even though T_1 commits before its completion. It can be observed that T_2 misses the updates produced by T_1 and therefore violates the serializability criteria.

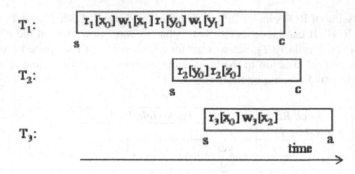

Fig. 3. Depiction of transaction processing for FCWR

3.3 Speculative Locking Protocol

In a database system, whenever a transaction T_i reads data objects, these objects are copied into the private working space in the memory allocated for this transaction. We assume that the transaction T_i issues $w_i[x]$ (write request on the data object 'x') after completing all its work on data object 'x'. This assumption is also adopted in [16] [17].

In the speculative locking (SL) protocol [6], a transaction produces after-images whenever it completes the work with that object. A waiting transaction is allowed to access the after-images produced by the conflicting active transactions. By accessing before- and after-images of conflicting active transactions, the waiting transaction carries out multiple speculative executions and retains one of the executions based on the termination status of preceding UTs. The requesting transaction commits only after the termination of preceding transactions with which it has formed commit dependencies. When a transaction T_i forms commit dependency with T_j, T_i commits only after the termination of T_j.

Figure 4 depicts the processing of transactions with SL. T_{ij} indicates j^{th} ($j > 0$) speculative execution of T_i. It can be observed that T_2 starts speculative executions T_{21} and T_{22}, once T_1 produces the after-image 'x_1'. T_2 accesses both 'x_0' and 'x_1' and starts speculative executions. Here T_2 forms commit dependency with T_1. If T_1 commits, T_2 commits by retaining the execution T_{22}. Otherwise, if T_1 aborts, T_2 commits by retaining T_{21}.

Lock compatibility matrix of SL protocol is shown in Figure 5. Here the W-lock is partitioned into two locks: executive write (EW)-lock and speculative write (SPW)-lock. Transactions request R-lock for read and EW-lock for write. When a transaction produces after-image for a data object, the EW-lock is converted into SPW-lock. Under SL, only one transaction holds an EW-lock on a data object at any time. However, note that multiple transactions can hold the R- and SPW-locks on a data object at the same time. The entry "sp_yes" indicates that the requesting transaction carries out speculative executions and forms commit dependency with the preceding transactions that hold SPW-locks.

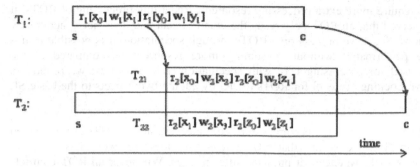

Fig. 4. Depiction of speculative transaction processing

Lock Re-quest by T_i	Lock Held by T_j		
	R	EW	SPW
R	yes	no	sp_yes
EW	sp_yes	no	sp_yes

Fig. 5. Lock compatibility matrix for SL

In SL, at a time, a data object may have multiple versions which are organized using tree data structure with one committed value (at the root) and uncommitted values at other nodes. Whenever a transaction reads a particular version ('x_i') and produces new versions by executing a write operation, new object versions are created and are added as children to the corresponding node ('x_i').

A family of SL protocols, SL(n), SL(1), and SL(2) are proposed in [6]. Through simulation experiments, it has been shown that SL improves the performance significantly over 2PL by trading extra resources. Also, SL protocol produces serializable executions.

4 Proposed Speculation-Based Protocol for ROTs

We first explain the basic idea. Next, we present the proposed protocol. Subsequently, we prove the correctness.

4.1 Basic Idea

The SL protocol [6] was proposed to process UTs; i.e., the transactions that contain both read and write operations. In that protocol, at a time, a data object may have multiple versions which are organized using tree data structure. Whenever a transaction executes a write operation, new uncommitted object versions are created and are added to the corresponding object trees. It can be observed that write operations are the cause for the generation of new uncommitted versions. As a result, the number of speculative executions that are to be carried out by waiting transactions and the number of versions stored in the trees explode with the increase in data contention. As a result, we require more extra processing resources. Regarding processing of ROTs, it can be observed that an ROT only reads the existing data and does not generate any new versions. So, if we process only ROTs through speculation, it is possible to improve the performance without consuming more resources as compared to the resources used for processing UTs with speculation. Based on this, we propose a Speculative Locking protocol for ROTs (SLR) by adding two aspects to the basic SL protocol.

a) In SLR, only ROTs are processed with speculation. The UTs are processed with 2PL. We assume that a UT releases the locks (converts EW-lock into SPW lock) whenever it produces after-images. Whenever an ROT conflicts with a UT, it carries out speculative executions by accessing both before- and after-images of the preceding UTs.

b) The other aspect is regarding commitment of ROTs. In the SL [6], a waiting transaction carries out speculative executions and waits for the commitment of preceding transactions. Whereas, in SLR, whenever ROT completes execution, it commits by retaining appropriate execution. In SLR, an ROT does not wait for the termination of conflicting active transactions. However, it can be noted that, a UT waits for the termination of preceding UTs and ROTs.

Lock Re-quest by T_i	Lock Held by T_j			
	RR	RU	EW	SPW
RR	yes	yes	no	sp_yes
RU	yes	yes	no	no
EW	no	no	no	no

Fig. 6. Lock Compatibility Matrix for SLR

The lock compatibility matrix of SLR is shown in Figure 6. Similar to the case of speculative locking, W-lock is divided into EW-lock and SPW-lock. UTs request EW-lock for writing the data object. The EW-lock is converted into the SPW-lock after the work on the data object is completed. Separate read-locks are employed for UTs and ROTs. A UT requests RU-lock (read lock for UT) for reading a data object and an ROT requests RR-lock (read lock for ROT) for reading a data object. The entry "sp_yes" indicates that the requesting transaction carries out speculative executions and forms commit dependency with the lock holding transaction.

It can be noted that commit dependency in SLR is different from SL. Let T_i be an ROT and T_j be a UT. Suppose T_i forms commit dependency with T_j. In SL, T_i commits only after the termination of T_j. Whereas in SLR, whenever T_i completes, it can commit by retaining one of the speculative executions without waiting for T_j to terminate.

Figure 7 depicts processing under SLR. Here, T_2 is an ROT and T_1 and T_3 are UTs. Whenever T_1 produces after-image 'x_1', T_2 accesses both 'x_0' and 'x_1' and carries out two executions T_{21} and T_{22}, respectively. After T_2's completion, T_{21} is retained even though T_1 is not yet committed. Note that, being a UT, T_3 waits for T_1 for the release of the lock on 'x' as per 2PL rule.

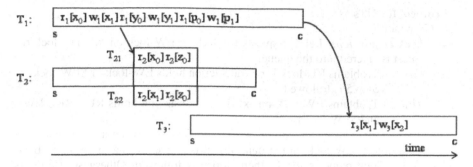

Fig. 7. Depiction of transaction processing with SLR

4.2 The SLR Protocol

In SLR, a "dependset" data structure is maintained for each transaction which is the set of transactions with which the executing transaction has formed the commit dependencies. For each data object, a FIFO queue is maintained to store the pending lock requests. In SLR, a UT requests for RU-lock to read and EW-lock to write. An ROT always requests for RR-lock to read. The protocols for ROTs and UTs are as follows.

Protocol for ROTs

Let T_i be an ROT

1. *Lock acquisition*: Let T_i requests for RR-lock on 'x'. The lock request is entered into the queue.
   (a) If no transaction holds EW- or SPW-locks, the RR-lock is allocated to T_i. Step 2(a) is followed.
   (b) If the preceding transaction is holding SPW-lock, lock is granted in speculative mode (sp_yes). The identifiers of preceding transactions which hold lock on 'x' are included in the T_i's dependset. Step 2(b) is followed.
2. *Execution*:
   (a) T_i continues with the current executions by accessing 'x'. If no further lock requests for T_i, then step 3 is followed. Otherwise, step 1 is followed.
   (b) Each execution of T_i is split into two executions one with the before-image and the other one with the after-image. If no further lock requests for T_i, then step 3 is followed. Otherwise, step 1 is followed.
3. *Commit/Abort Rule:* Once the transaction T_i is completed, one of the speculative executions of T_i is chosen as follows: Suppose T_i has completed at time 't'. T_i retains that speculative execution which contains the effect of all committed transactions which have committed before 't'. If T_i is aborted, then all its speculative executions are aborted. Also the locks allocated to T_i are released. The dependset of T_i is deleted.

Protocol for UTs

Let T_j be a UT.

4. *Lock acquisition:* Let T_j requests for RU- or EW-lock on 'x'. The lock request is entered into the queue.

    (*a*) T_j obtains RU-lock if no transaction holds EW-lock or SPW-lock. Step 5 is followed.

    (*b*) T_j obtains EW-lock on 'x', if no transaction holds RU-, RR-, EW-, and SPW-locks.

5. *Execution:* During execution, whenever T_j produces the after-image for a data object, EW-lock on the data object is converted into SPW-lock. If no further lock requests for T_j, then step 6 is followed. Otherwise, step 4 is followed.

6. *Commit/Abort Rule:* Whenever T_j commits, the speculative executions of ROTs which have been carried out with before-images of T_j are terminated. Whenever T_j aborts, the speculative executions of ROTs carried out with after-images of T_j are terminated. The information regarding T_j is deleted from the dependsets of ROTs. All the related lock entries are deleted.

4.3 Correctness

The terms, transaction, history over a set of transactions, and serializability theorem are formally defined in [7].

A transaction is defined as a partial ordering of operations (read, write, commit and abort). A history H, over a set of transactions indicates the order in which the operations of the transactions were executed relative to each other and it is formally defined as a partial order. Committed projection of history H denoted as C(H) is the history obtained from H by deleting all operations that do not belong to transactions committed in H.

A history is serial if there is no interleaving, i.e. once a transaction starts executing it finishes without any other transaction executing some operations in-between. A history H is serializable, if its committed projection C(H) is equivalent to a serial history for the same set of transactions

A graph derived from the history H is called as serialization graph SG(H). SG(H) is a directed graph whose nodes are the transactions which are committed in H and whose edges are all $T_i \rightarrow T_j$ ($i \neq j$) such that one of T_i's operations precedes and conflicts with one of T_j's operations in H. To prove that a history is correct, we must prove that the serialization graph formed from the history is acyclic.

Three types of conflicts occur while transactions are processed under 2PL: R-W, W-W and W-R conflicts. However, in SLR, the following conflicts occur: RR-EW, RR-SPW, RU-EW, RU-SPW, EW-RR, EW-RU, EW-EW, EW-SPW conflicts. (Note that in SLR, the W-lock is split into EW-lock and SPW-lock. Also, UTs request RU-lock for read and EW-lock for write. The UTs convert EW-lock into SPW-lock after completing the work on the data object. ROTs request RR-lock for read.) The lock compatibility matrix is given in Figure 6.

Let $p_i[x]$ denote an operation (RR, RU, EW) requested by transaction manager (TM) for T_i and $pl_i[x]$ denotes a lock (RR-, RU-, EW- and SPW-locks). The notation '«' indicates partial order. The notation '$T_i « T_j$' indicates T_i precedes T_j in the history. Also we use the notations 'o' to denote an operation (RR, RU, EW), 'l' to denote locking and 'u' to denote unlocking operations. The SLR protocol manages locks using the following rules.

1) When scheduler receives an operation $p_i[x]$ from the TM, the scheduler tests the conflict between $pl_i[x]$ with some $ql_j[x]$ that is already set.

    i.   If the conflict is of "no", it delays $p_i[x]$ by forcing T_i to wait until it can set the lock it needs. This is equivalent to $ql_j[x] \ll pl_i[x]$ (or $T_j \ll T_i$).

    ii. If the conflict of "sp_yes", T_i forms commit dependency with T_j and it carries out speculative executions. After its completion, T_i retains one of the speculative executions based on the T_j's termination status. If T_j has already committed, it will retain that execution which is being carried out by reading the after-images produced by T_j. This is equivalent to the order $ql_j[x] \ll pl_i[x]$ (or $T_j \ll T_i$). Otherwise T_i retains that execution which is carried out by reading the before-images of T_j. This is equivalent to the order $pl_i[x] \ll ql_j[x]$ (or $T_i \ll T_j$).

2) Once the scheduler has set a lock for T_i, say $pl_i[x]$, it may not release that lock at least until T_i is committed or aborted.

3) Once a transaction completes the work on a data object, the scheduler converts the EW-lock into the SPW-lock for that object in an atomic manner.

4) Once the scheduler has released a lock for a transaction, it may not subsequently obtain any more locks for that transaction (on any data object).

Based on SLR rules we propose the following propositions.

Proposition 1. Let H be a history produced by an SLR scheduler. If $o_i[x]$ is in C(H), then $ol_i[x]$ and $ou_i[x]$ are in C(H), and $ol_i[x] \ll o_i[x] \ll ou_i[x]$.

There are three kinds of operations: read by ROTs, read by UTs and write by UTs. Whenever TM requests read operation on behalf of an ROT or UT, the operation is executed after obtaining the corresponding lock. When a UT requests a write operation, the operation is executed after obtaining the EW-lock. After completion of work on the data object, the EW-lock is converted into SPW-lock atomically. All the locks are released after the commit/abort of transactions. So, lock is obtained for every operation and released after the completion of the operation.

Proposition 2. Let H be a history produced by an SLR scheduler. If $p_i[x]$ and $q_j[x]$ ($i \neq j$) are conflicting operations in C(H), then either $p_i[x] \ll q_j[x]$ or $q_j[x] \ll p_i[x]$.

Suppose we have two operations $p_i[x]$ and $q_j[x]$ that are in conflict. Then, the corresponding locks are also in conflict. The order of the operations is equivalent to either $ql_j[x] \ll pl_i[x]$ or $pl_i[x] \ll ql_j[x]$, (or $T_j \ll T_i$ or $T_i \ll T_j$) according to rule (1).

Proposition 3. Let H be a complete history produced by an SLR scheduler. If $p_i[x]$ and $q_i[y]$ are in C(H), then $pl_i[x] \ll qu_i[y]$.

As per the rule (2), a transaction cannot obtain any lock after releasing any other lock. It means every locking operation is executed before any unlocking operation. (However, it can be noted that we are converting EW-lock into SPW-lock which is not equivalent to the unlocking operation.)

Theorem 1. Let H be history of the committed transactions under SLR. Then, H is serializable.

Proof: To prove H is serializable, we have to prove that SG (H) is acyclic.

Suppose an edge $T_i \rightarrow T_j$ is in SG (H). As per the propositions 1 and 2, T_j might have waited for T_i or formed a commit dependency with T_i. Suppose $T_i \rightarrow T_j \rightarrow T_k$ is in SG (H). This means that T_k might have waited for the completion of T_j or formed a commit dependency with T_j. By transitivity, T_k has waited for T_i or formed commit dependency with T_i. By induction, this argument extends to long paths. For any long path $T_1 \rightarrow T_2 \rightarrow \ldots \rightarrow T_n$, T_n has waited for T_1 or formed commit dependency with T_1.

Suppose SG (H) had a cycle $T_1 \rightarrow T_2 \rightarrow \ldots \rightarrow T_n \rightarrow T_1$. This means T_1 has waited for T_n or formed commit dependency with T_n. T_n in turn waits for T_1 or forms commit dependency with T_1. By transitivity, T_1 waits for T_1 or forms commit dependency with T_1. This is a contradiction as per the propositions 2 and 3. Thus SG (H) has no cycles and therefore H is serializable. .

5 Simulation Results

In this section, we first explain the simulation model. Next, we present experimental results.

5.1 Simulation Model

We have developed a discrete event simulator based on a closed-queuing model. We have a pool of CPU servers, all having identical capabilities and are serving one global queue of transactions. Each CPU manages two I/O servers. A CPU server serves the requests placed in the CPU queue in FCFS order. The I/O model is a probabilistic model of a database that is spread out across all the disks. A separate queue is maintained for each I/O server. Whenever a transaction needs service, it randomly (uniform) chooses a disk and waits in the I/O queue of the selected I/O server. For the I/O queue also we follow FCFS order [18].

The description of parameters with values is shown in Table 1. The database size is assumed to be "dbSize". The parameters "cpuTime" and "ioTime" are amounts of I/O and CPU time associated with reading and writing an object (equivalent to an operating system page). Regarding transaction size, we have chosen different parameter values for ROTs and UTs by considering the load character in modern information systems [19]. The parameters "rotMaxTranSize" and "rotMinTranSize" are the maximum and minimum number of objects in ROT respectively. The maximum and minimum number of objects in UT is represented by the parameters "utMaxTranSize" and "utMinTranSize" respectively. Each resource unit (RU) constitutes 1 CPU and 2 I/O servers by considering that one CPU can drive two I/O servers. The parameter "noResUnits" represents the number of resource units. The parameter "MPL" denotes the number of active transactions existing in the system. The parameter "% of UTs" means the percentage of UTs currently active in the system. Let "p" indicates the "% of UTs", which means that at any point of time, there are "p" percent UTs and (100-p) percent ROTs are active in the system..

Table 1. Simulation Parameters, Meaning and Values

Parameter	Meaning	Values
dbSize	Number of objects in the database	1000
cpuTime	Time to carry out CPU request	5 ms
ioTime	Time to carry out I/O request	10 ms
rotMaxTranSize	Size of largest ROT transaction	20 objects
rotMinTranSize	Size of smallest ROT transaction	15 objects
utMaxTranSize	Size of largest UT transaction	15 objects
utMinTranSize	Size of smallest UT transaction	5 objects
noResUnits	Number of RUs(1 CPU, 2 I/O)	8
MPL	Multiprogramming Level(10 – 100)	Simulation Variable
% of UTs	Percentage of UTs (10 - 90)	Simulation Variable

The value for "dbSize" is chosen as 1000 data objects [18]. The value for "cpuTime" is chosen as 5 ms by considering the speed of modern processors [20]. The value for "ioTime" is fixed as 10 ms by considering the speed of recent hard disk drives [21]. The values for "rotMaxTranSize" and "rotMinTranSize" are fixed at 20 and 15 respectively and the values for "utMaxTranSize" and "utMinTranSize" are 15 and 5 objects respectively [14]. The size of a ROT is a random number between 15 and 20 (both inclusive) and UT is a random number between 5 and 15 (both inclusive). We conducted the experiments by varying "percentageOfUts" from 10 to 90 and "mpl" from 10 to 100.

Performance Metrics. The main performance metric is "throughput" which is the number of transactions completed per second. We also use the metric "percentage of transaction aborts", which is the ratio of the number of aborted transactions to the number of committed transactions.

We have compared SLR with 2PL and FCWR. We have already explained SLR, 2PL, and FCWR in other sections. In both 2PL and SLR, transactions request locks in a dynamic manner, one by one. For SLR, we have assumed that all the speculative executions of a transaction are carried out in parallel. In FCWR, the conflicts between UTs are managed by aborting the transactions. Aborted transactions are resubmitted after the time duration equals to average response time.

We assume the cost of performing concurrency control operations is negligible compared to the cost of accessing objects. Also we have not taken into account the cost of deadlock detection as it is same for all locking-based protocols.

In the experiments, the graphs show the mean results of 20 experiments; each experiment was carried out for 10,000 transactions. The results were plotted with a mean of 95 percent confidence intervals. These confidence intervals are omitted from the graphs.

5.2 Performance Results

We have evaluated the performance by analyzing the overall throughput, percentage of transaction aborts and extra resource requirement. At first, we explain the experimental results carried out by assuming unlimited resources. Next, the details of additional resources consumed are analyzed.

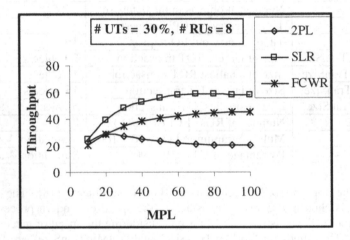

Fig. 8. MPL versus Throughput

Figure 8 shows how throughput performance for 2PL, FCWR and SLR vary with MPL. It can be observed that performance of SLR is significantly higher than 2PL and FCWR. 2PL performs poorly because, the waiting time of the transactions increases with data contention. In FCWR, ROTs are processed without any waiting. However, the performance of FCWR decreases with data contention, as more number of UTs gets aborted because of first committer wins rule which is explained in section 3.2. SLR improves performance over 2PL due to reduced waiting and over FCWR due to the reduced number of aborts. The performance of SLR is improved as the transaction waiting time is reduced due to speculation. Under SLR, an ROT is able to read uncommitted values produced by preceding transactions and able to start the execution early. So SLR performs significantly better than 2PL and FCWR protocols.

Figure 9 shows how abort performance of 2PL, SLR and FCWR protocols with the increase in number of UTs. It can be observed that the number of transaction aborts under FCWR increases with the increase in data contention. However, the number of transaction aborts under 2PL and SLR protocols is very less in comparison with FCWR.

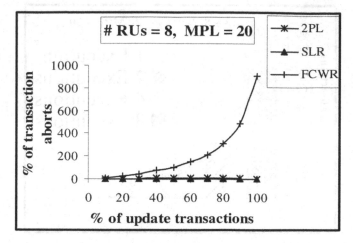

Fig. 9. % of UTs versus % of Transaction Aborts

Figure 10 shows how throughput performance for 2PL, FCWR and SLR vary with percentage of UTs. It can be observed that SLR exhibits high performance over 2PL and FCWR. The performance of FCWR falls sharply as the data contention increases due to more number of transaction aborts. The performance of 2PL is also significantly less than SLR as transactions spend more time in waiting. It can be observed that the performance of SLR is close with 2PL at higher 'p' value. This is because at higher 'p' value, all transactions are UTs. So these transactions are processed with 2PL only. Overall SLR performs better than 2PL and FCWR.

We now discuss the details regarding the additional resource consumption. In Figure 11, the details regarding percentage of transactions which consumed 1, 2, 4 and 8 speculative executions are shown. We conducted this experiment by fixing

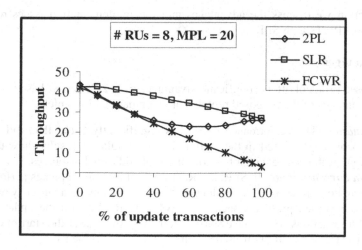

Fig. 10. % of UTs versus Throughput

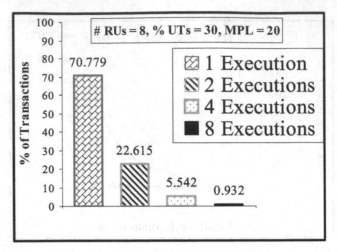

Fig. 11. Details of Speculative Executions

percentage of UTs as 30 (percentage of ROTs is 70), which is the normal values in online transaction processing systems [22]. It was observed that around 70% of the transactions require only one execution, 22% of transactions require 2 speculative executions, 6% of transactions require 4 speculative executions and mere 1% of transactions require 8 speculative executions. The average number of executions comes to 1.4. This indicates that it is possible to get an improved performance with a fraction of (0.40) additional resources. However, the detailed experiments will be carried out as a part of future work.

6 Advantages and Implementation Issues

In this section we discuss the advantages and the implementation issues regarding processing of ROTs with SLR.

6.1 Advantages

The proposed protocol offers significant advantages over 2PL and SI-based protocols. The advantages of the proposed protocol are summarized in Table 2.

(i) Performance. The performance of SLR is significantly better than 2PL and SI-based protocols as discussed in the section 5. The simulation results show that SLR achieves high performance with a fraction (0.40) of additional resources.

(ii) No data currency issue. In SLR, the ROTs do not miss any updates performed by the preceding committed UTs. So, it does not suffer from any data currency related issues. In the SI-based protocols such as FCWR, the ROTs miss the updates of the committed transactions which have started their executions after the start of currently executing ROTs. So, it suffers from data currency related problems.

(iii) **No correctness issues.** SLR does not suffer from correctness issues as the histories generated by SLR scheduler are serializable. It can be noted that SI-based based protocols are not serializable.

(iv) **No additional memory to maintain data object versions.** During execution, a UT maintains the after-images of the data objects in its address space. The memory space required to keep these after-images can be shared by ROTs using the shared memory technique. So, An ROT can read the before-images from the database and the after-images from the shared memory. Note that for each data object only two versions (before-image and after-image) are needed. So, no extra memory is required for before-and after-images.

(v) **No additional disk I/O.** SLR maintains the after-images produced by UTs in the main memory. So, additional disk I/O is not required for reading these images.

Table 2. Comaprison of SLR, 2PL and FCWR protocols

Parameter	SLR	2PL	FCWR
Throughput	high	low	medium
Data Currency	no problem	no problem	suffering from data currency problem
Correctness	yes	yes	no

6.2 Implementation Issues

(i) Pre-compiling. In this paper, we assume that a UT releases the lock whenever it produces the after-image. To implement this operation, a software module is required which can put a lock conversion marker for each data object after the last write operation specified in the transaction program for that object, by scanning the entire program. During execution, when the lock conversion marker is encountered, the EW-lock on the data object is converted into the SPW-lock. Since the transactions are stored procedures, we believe that it is possible to put the lock conversion markers by analyzing the stored procedures.

(ii) Speculative executions. We have assumed that speculative executions of a transaction are carried out in parallel by considering multi-processor environment. It can be noted that additional memory can be added to the system at lesser cost. Since CPU speed is high in the orders of magnitude than the disk I/O, even in a single processor environment, the CPU time can be used productively to improve the performance of ROTs. However, the detailed study will be carried out as a part of future work.

7 Conclusions and Future Work

We have proposed a protocol by extending speculation to improve the performance of ROTs. The proposed protocol does not suffer from any data currency and correctness related issues. Through simulation experiments, it has been shown that the proposed protocol improves the performance significantly over 2PL and SI-based protocols with manageable extra processing resources.

As a part of future work, we are planning to carry out simulation experiments under limited resource environments. Also, in SLR, all speculative executions of a transaction are carried out synchronously. We are planning to investigate the performance by allowing each speculative execution of a transaction to proceed in an asynchronous manner.

Over the years, costs of both CPU and memory are falling down. Multi-processor systems with huge main memory would be affordable soon. Also, data currency of ROTs is a crucial factor in several environments like stock marketing, airline operating systems and other crucial web services. Speculation-based ROT algorithm provides the scope for improving the performance and data currency of ROTs in such environments without violating the correctness criteria.

References

1. Eswaran, K., Gray, J., Lorie, R., Traiger, I.: The notions of consistency and predicate locks in database systems. Communications of the ACM 19(11), 624–633 (1976)
2. Gray, J., Reuter, A.: Transaction Processing: Concepts and techniques. Morgan Kaufmann, San Francisco (1993)
3. Berenson, H., Bernstein, P., Gray, J., Melton, J., O'Neil, E., O'Neil, P.: A Crtique of ANSI SQL Isolation Levels. ACM SIGMOD (1995)
4. Fekete, A., Liarokapis, D., O'neil, E., O'neil, P., Shasha, D.: Making Snapshot Isolation Serializable. ACM Transactions on Database Systems 30(2), 492–528 (2005)
5. Theodoratos, D., Bouzeghoub, M.: Data Currency Quality Factors in Data Warehouse Design. In: Proceedings of the International Workshop on Design and Management of Data Warehouses, Germany, pp. 1–15 (June 1999)
6. Krishna Reddy, P., Kitusuregawa, M.: Speculative Locking Protocols to Improve Performance for Distributed Database Systems. IEEE Transactions on Knowledge and Data Engineering 16(2), 154–169 (2004)
7. Bernstein, P.A., Hadzilacos, V., Goodman, N.: Concurrency Control and Recovery in Database Systems. Addison-Wesley, Reading (1987)
8. ANSI X3.135-1992, American National Standard for Information Systems- Database Language-SQL (November 1992)
9. Adya, A., Liskov, B., O'Neil, P.: Generalized Isolation Level Definitions. In: Proceedings of the IEEE International Conference on Data Engineering, IEEE Computer Society Press, Los Alamitos (March 2000)
10. Garcia-Molina, H., Wiederhold, G.: Read-Only Transactions in a Distributed Database. In: ACM Transactions on Database Systems, pp. 209–234. ACM Press, New York (June 1982)
11. Satyanarayanan, O.T., Agrawal, D.: Efficient Execution of Read-only Transactions in Replicated Multiversion Databases. IEEE transactions on Knowledge and Data Engineering 5(5), 859–871 (1993)
12. Mohan, C., Pirahesh, H., Lorie, R.: Efficient and Flexible Methods for Transient Versioning of Records to Avoid Locking by Read-Only Transactions. In: ACM SIGMOD (1992)
13. Lu, B., Zou, Q., Perrizo, W.: A Dual Copy method for Transaction Separation with Multiversion Control for Read-only Transactions. In: Proceedings of the ACM Symposium on Applied Computing, pp. 290–294 (2001)

14. Seifert, A., Scholl, M.H.: Processing Read-Only Transactions in Hybrid Data Delivery Environments with Consistency and Currency Guarantees. In: Mobile Networks and Applications, vol. 8, pp. 327–342. Kluwer Academic Publishers, Dordrecht (2003)

15. Bestavros, A., Braoudakis, S.: Value-Cognizant Speculative Concurrency Control Protocol. In: Proceedings of 21st Very Large Databases Conference, pp. 122–133 (1995)

16. Agrawal, D., El Abbadi, A., Lang, A.E.: The Performance of Protocols Based on Locks with Ordered Sharing. IEEE Transactions on Knowledge and Data Engineering 6(5), 805–818 (1994)

17. Salem, K., Garcimolina, H., Shands, J.: Altruist Locking. ACM Transactions Data base Systems 19(1), 117–165 (1994)

18. Agrawal, R., Carey, M.J., Livny, M.: Concurrency Control Performance Modeling: Alternatives and Implications. ACM Transactions on Database Systems 12(4), 609–654 (1987)

19. Kwok-Wa, L., Son, S.H., Lee, V.C.S., Sheung-Lun, H.: Using Separate Algorithms to Process Read-Only Transactions in Real-Time Systems. In: Proceedings of the IEEE Real-Time Systems Symposium, pp. 50–59. IEEE Computer Society Press, Los Alamitos (1998)

20. Kam-Yiu, L., Rei-Wei, K., Kao, B., Lee, T.S.H., Cheng, R.: Evaluation of Concurrency Control Strategies for Mixed Soft Real-Time Database Systems. Information Systems Journal 27(2), 123–149 (2002)

21. Barracuda E.S.: The highest-capacity drives for the enterprise, (November 2006), http://www.seagate.com/docs/pdf/marketing/po_barracuda_es.pdf

22. Badrinath, B.R., Ramamirtham, K.: Performance Evaluation of Semantics-based Multi-level Concurrency Control Protocols. In: Proceedings of the ACM SIGMOD Conference, pp. 163–172. ACM Press, New York (1990)

LEASE: An Economic Approach to Leasing Data Items in Mobile-P2P Networks to Improve Data Availability

Anirban Mondal[1], Sanjay Kumar Madria[2], and Masaru Kitsuregawa[1]

[1] Institute of Industrial Science
University of Tokyo, Japan
{anirban,kitsure}@tkl.iis.u-tokyo.ac.jp
[2] Department of Computer Science
University of Missouri-Rolla, USA
madrias@umr.edu

Abstract. This work proposes LEASE, a novel Mobile-P2P lease-based economic incentive model, in which data requestors need to pay the price (in virtual currency) of their requested data items to data-providers. In LEASE, data-providing mobile peers lease data items to free-riders, who do not have any data items to provide, in lieu of a lease payment. Thus, LEASE not only combats free-riding, but also entices free-riders to host data items, thereby improving network connectivity due to higher peer participation. In essence, LEASE facilitates the collaborative harnessing of limited mobile peer resources for improving data availability. Our performance study shows that LEASE indeed improves query response times and data availability in Mobile-P2P networks.

1 Introduction

In a Mobile Ad-hoc Peer-to-Peer (M-P2P) network, mobile peers (MPs) interact with each other in a peer-to-peer (P2P) fashion. Proliferation of mobile devices (e.g., laptops, PDAs, mobile phones) coupled with the ever-increasing popularity of the P2P paradigm (e.g., Kazaa [11], Gnutella [6]) strongly motivate M-P2P network applications. Mobile devices with support for wireless device-to-device P2P communication are beginning to be deployed such as Microsoft's Zune [9].

M-P2P applications facilitate mobile users in sharing information with each other *on-the-fly* in a P2P manner. A car user could request other car users for information e.g., locations of nearby parking slots and restaurants, and traffic reports a few miles ahead. A pedestrian could request an available taxi nearby his current location. Customers in a shopping mall could share information about the cheapest 'Levis' jeans or swap shopping catalogues. Mobile users could exchange songs or video-clips (as in a future mobile eBay market). Such P2P interactions among mobile users are generally not freely supported by existing wireless communication infrastructures. Our target applications mainly concern slow-moving objects e.g., cars on busy streets, people moving in a market-place or students in a campus.

Data availability in M-P2P networks is typically lower than in fixed networks due to frequent network partitioning arising from user movement and users switching 'on'/'off' their mobile devices. Moreover, a large percentage of MPs typically do not have any

S. Bhalla (Ed.): DNIS 2007, LNCS 4777, pp. 222–231, 2007.

data to share with other MPs i.e., they are free-riders [10]. To exacerbate the problem, MPs generally have limited bandwidth, hence a data-providing MP can make available only few of its data items to be shared (i.e., the **shared data items**) based on the amount of bandwidth that it would like to share, but it has additional data items (i.e., the **unshared data items**) in the memory. Given the ephemeral nature of M-P2P environments, unshared data items may *expire* before they can be made available to M-P2P users, which further decreases data availability.

M-P2P data availability could be significantly improved if free-riders could be enticed to pool in their bandwidth resources by hosting unshared data items. Hence, we propose LEASE, a novel lease-based economic incentive model for effective collaborative data sharing among MPs with limited resources. In LEASE, data-providing MPs *lease* data items to those who do not have any data items to provide. A data item d (originally owned by MP P) is said to be **leased** by P to MP H when P provides d to H for a pre-specified lease period τ, in lieu of a lease payment (in *virtual currency*). During the period τ, H hosts d, and after τ expires, H deletes the copy of d at itself. Notably, P may lease d simultaneously to multiple MPs. In case any updates are required to the data (e.g., traffic reports in transportation application scenarios), P sends the updates to H. We shall henceforth refer to a data-providing MP P as a **provide-MP**, and the host MP H as a **host-MP**.

Each data item has a *price* (in *virtual currency*). Data item price depends on access frequency, data quality [13] (e.g., image resolution, audio quality) and the estimated response time for accessing the data item. A query issuing MP pays the *price* of the queried data item to the query-serving MP. Thus, LEASE provides an incentive for free-riding MPs to act as host-MPs so that they can earn **revenue** for issuing their own requests. **Revenue** of an MP is defined as the difference between the amount of virtual currency that it earns (by providing data) and the amount that it spends (by requesting data). Virtual currency is suitable for P2P environments due to high transaction costs of micro-payments in real currency [17]. Secure virtual currency payments have been discussed in [4].

Leasing benefits both provide-MPs and host-MPs. It facilitates a provide-MP in earning revenue from its unshared data items even without hosting them, especially since unshared data items may expire. It helps a host-MP in earning revenue using other MPs' data items. In the absence of a lease model, MPs without any data to provide cannot earn any revenue, thereby decreasing the overall MP participation. In M-P2P networks, leasing is better than *buying* (permanent ownership transfer) since data items have expiry times, hence their value depreciates significantly over time. Moreover, host-MPs wish to host as many 'hot' data items as possible to maximize their revenues.

The main contributions of LEASE follow:

1. Its lease model entices even those users, who have no data to provide, to host data items, thereby improving data availability and MP revenues.
2. Its economic model discourages free-riding, which improves connectivity due to higher peer participation.

Higher peer participation leads to better data availability due to higher available bandwidth and better connectivity. Existing M-P2P replication schemes [8,18] do not combat free-riding, while M-P2P incentive schemes [19,20] do not entice free-riders, which have no data, to provide service.

Our performance study indicates that LEASE indeed improves query response times and data availability in M-P2P networks. To our knowledge, this is the first work to propose a lease-based economic model for M-P2P networks.

2   Related Work

Economic models have been discussed in [5,7,12] primarily for resource allocation in distributed systems. These works do not address unique M-P2P issues such as frequent network partitioning, mobile resource constraints, free-riding and incentives for peer participation. Incentive mechanisms for static P2P networks have been proposed in [10,15]. However, pre-defined data access structures (e.g., distributed hash tables [16]) used in static P2P networks assume peers' availability and fixed topology, which makes incentive schemes for static P2P networks too static to be deployed in mobile ad-hoc networks (MANETs). Furthermore, the proposals in [10,15] do not consider economic models to combat free-riding.

Incentive mechanisms have also been investigated for MANETs [3,21], the main objective being to encourage an MP in forwarding information to other MPs. However, these works do not consider economic issues and M-P2P architecture. The **E-DCG+** replica allocation approach [8] for MANETs does not consider lease models, incentives and prices of data items. Interestingly, economic ideas for M-P2P networks have been discussed in [19,20]. However, these works propose opportunistic dissemination of data in M-P2P networks with the aim of reaching as many peers as possible, while we address on-demand data dissemination. The work in [1] proposes an barter-based economic model, but it does not consider M-P2P issues.

In our earlier work [14], we proposed an economic model for data replication based on the price of data items. However, in contrast with this work, the proposal in [14] does not consider a lease-based approach. Moreover, in [14], each peer behaves autonomously without any co-ordination among themselves, while this work considers peer collaboration for improving data availability in M-P2P networks.

3   The LEASE Economic Model

Each provide-MP maintains recent read-write logs (including timestamps) of its own data items as well as details (e.g., lease duration) of the data items that it leases. This information helps provide-MPs to select their respective shared and unshared data items. Each host-MP maintains recent access information of data items based on queries that pass through itself. Such information facilitates host-MPs in selecting data items that they want to host. Available memory space of MPs, bandwidth and data item sizes may vary. We define the **load** of an MP as its job queue length normalized w.r.t. bandwidth.

Table 1 summarizes the notations used in this paper. Using the notations in Table 1, price μ of a data item is computed as follows:

$$\mu = \int_{t_1}^{t_2} \int_0^{\delta} (\eta \, dt \times (1/\delta^2) \, d\delta \times DQ \times BA_{M_S}) / J_{M_S, t_j} \tag{1}$$

Table 1. Summary of Notations

Symbol	Significance
d	A given data item
η	Recent Access frequency of d
DQ	Data quality of d
$size$	Size of d
Ex	Time to Expiry time of d
BA_{M_S}	Bandwidth of the query-serving MP for d
J_{M_S,t_j}	Job-queue length of the query-serving MP at time t_j

where $[t_2 - t_1]$ represents a given time period and δ is the *Euclidean distance* between the query issuing MP M_I and the query serving MP M_S during the time of query issue. For unshared data items, the access frequency η refers to the number of access failures. DQ reflects the quality of data (e.g., image resolution, audio quality) provided by M_S for queries on d. The value of DQ is determined as in our previous work in [13], where we considered three discrete levels of DQ i.e., *high*, *medium* and *low*, their values being 1, 0.5 and 0.25 respectively. As BA_{M_S} increases and δ decreases, μ increases due to faster query response time. As J_{M_S,t_j} increases, μ decreases since M_S's response time for queries on d increases due to higher load.

The revenue earned by an MP M equals ($\sum_{i=1}^{p} (\mu_i \times accs_i)$, where p is the number of data items made available by M, and μ_i and $accs_i$ are the price and access frequency of the ith data item respectively. Similarly, the revenue spent by M equals ($\sum_{i=1}^{q} (\mu_i \times accr_i)$, where q is the number of items queried by M, and μ_i and $accr_i$ are the price and access frequency of M for the ith item respectively.

Role of the Provide-MPs and Host-MPs

A provide-MP P makes available at itself (i.e., *shares*) data items, with higher revenue-earning potential γ for maximizing its revenue, while leasing out some of its *unshared* data items. Given that $\mu_{i,P}$ is the price of a data item i at P and $acc_{i,P}$ is the recent access frequency of i, $\gamma = \mu_{i,P} \times acc_{i,P}$. (For data items that P is currently not making available, $acc_{i,P}$ is the number of times a query failed to obtain the data item at P.) P avoids leasing frequently updated data items due to the high communication overhead (e.g., energy, bandwidth) required for maintaining the consistency of such items. Periodically, P broadcasts its list of unshared data items, which have low write frequencies, for finding prospective host-MPs to host these items.

P selects host-MPs by accepting bids for its given data item d based on the quality of service and connectivity of the MPs. P leases d to higher-bidding MPs since MPs with better resources for providing good service would bid higher since they can earn more revenue from d. We define the connectivity of an MP as the number of its one-hop neighbours. P prefers to lease d to MPs with higher connectivity to facilitate it in sharing its data items with as many MPs as possible, thereby enabling it to earn more revenue.

Given an unshared data item d, P decides the number of copies of d to be leased based on the revenue λ that it wishes to obtain from leasing d. P computes λ as follows:

$$\lambda = 0.5 \int_{t_1}^{t_2} (\eta_d \times \mu_d) \qquad (2)$$

where $[t_2 - t_1]$ is a given time period, η_d is the number of failed queries on d, and μ_d is the price of d. In Equation 2, the term $(\eta_d \times \mu_d)$ reflects P's estimated lost revenue due to not making d available. Observe that λ is 50% of P's estimated lost revenues. Thus, the estimated revenue from leased data items is shared *equally* between the provide-MP and the host-MPs to ensure fairness. Furthermore, allowing the host-MPs to earn 50% of the revenues from d provides adequate incentive for them to host d since they also incur energy and bandwidth-related costs due to downloads of d. Hence, P essentially sums up the bids for d starting from the highest bid until the total value of the bids is greater than or equal to λ. Then, P leases d to the corresponding MPs that made these bids. Notably, unlike existing works, we determine the number of copies based on revenue.

Host-MPs decide which data items to bid for as well as their bid values based on the queries for these items that pass through themselves. A host-MP H bids for data items with higher revenue-earning potential γ for maximizing its revenue. The number of data items for which H bids depends upon its available bandwidth and memory space. Given a data item d, H bids the amount β of currency for d based on d's revenue-earning potential, which depends upon d's popularity, quality, size, estimated expiry time, amount of bandwidth that it would likely make available for d and its current job-queue length. (Recall that d's price at H depends upon H's bandwidth and job-queue length.) Using Table 1 (see Section 3), H computes β as follows:

$$\beta = \int_{t_1}^{t_2} (\eta \, dt \times DQ \times Ex \times BA_{M_S})/(size \times J_{M_S,t_j}) \qquad (3)$$

where $[t_2 - t_1]$ represents a given time period. The access frequency η is based on the queries for d that passed through H. A data item *expires* when its access frequency falls below a certain application-dependent threshold. Data items with higher time to expiry facilitate H in earning more revenue by hosting d. Higher bandwidth of H implies better response time for queries on d, while larger job-queue length signifies higher load on H, thereby increasing response time. Smaller-sized data items help H to maximize its revenue per unit of its limited memory space.

Data providers periodically broadcast the unique identifiers of host-MPs, to whom they have leased their data items. Thus, MPs can download *updated* copies of data items from the authorized lease-holders, thereby improving the quality of service. (Provide-MPs send updates only to authorized host-MPs.) In case a host-MP H illegitimately hosts a given data item d or if H continues to host d after its lease period of d has expired, other MPs (e.g., relay MPs through which messages for downloads of d would pass) would inform the corresponding provide-MP P, and P would blacklist H. Periodically, provide-MPs broadcast their list of blacklisted MPs. Blacklisted MPs have to pay double the lease payment the next time they want to lease data items from any provide-MP, which acts as a deterrent.

Host-MPs make the lease payments to provide-MPs at the time of expiry of the lease so that host-MPs can earn revenue from hosting data items before they pay for the lease. This facilitates seamless integration of newly joined MPs, which may initially be unable to make the lease payment. Host-MPs, which fail to make the lease payment at the end of the lease expiry period, are blacklisted, thereby deterring malicious MPs from abusing the leasing system.

Algorithm *LEASE_Provide_MP*

Spc: Its available memory space

(1) Sort all its data items in in descending order of their revenue-earning
    potential γ into a list L.

(2) for each data item d in L

    /* WF_d is d's write frequency, THW_F is the write frequency threshold */

(3)   if ($WF_d < THW_F$)

(4)   if ($size_d \leq$ Spc)   /* $size_d$ is the size of d */

(5)   Fill up its memory space with d

(6)   Spc = Spc - $size_d$

(7)   if (Spc = = 0)   **exit**

(8) Create set CL comprising its **unshared** data items

    /* CL is the set of candidate data items for lease */

(9) Broadcast the set CL to its n-hop neighbours

(10) for each data item d in CL

(11)   Receive bids from prospective host-MPs, which wish to host d

(12)   Arrange the bids in descending order of bid value

(13) $Bid_{Sum} = 0$

(14)   for each bid β from host-MP i

(15) $Bid_{Sum} = Bid_{Sum} + \beta$

(16)   if $Bid_{Sum} \leq \lambda$

(17)   Add i to set $Host_d$

(18)   if set $Host_d$ is non-empty

(19)   Lease d to the MPs in set $Host_d$ with bid values as lease payment

(20)   Initialize set $Host_d$ by making it a NULL set

end

Fig. 1. LEASE algorithm for provide-MP

4 Algorithms in LEASE

Figure 1 depicts the algorithm for a provide-MP P. In line 3, write frequency WF_d of a data item d is computed as (nw_d / τ), where nw_d is the number of writes on d and τ is the lease period. Write frequency threshold THW_F is computed as the average write frequency of all the shared and unshared items in P. In Line 9 of Figure 1, n= 3 or n =4 were found to be reasonable values for our application scenarios (as indicated by preliminary experimental results). In Line 9, P's broadcast message contains the

Algorithm *LEASE_Host_MP*

CL_i: Candidate data items for lease from provide-MP i

Spc: Its available memory space

(1) for each provide-MP i
(2)   Receive broadcast message from i containing items for lease
(3)   Add all data items in CL_i to a set $bigCL$
(4) Sort all data items in $bigCL$ in descending order of γ
(5) for each data item d in $bigCL$
(6)   /* $size_d$ is the size of d */
(7)   if ($size_d \leq Spc$)
(8)   Add d to a set BID
(9)   Spc = Spc - $size_d$
(10)   if (Spc = = 0)   **exit**
(11) for each data item d in set BID
(12)   Send the bid of β_d to the corresponding provide-MP
(13)   if bid is successful
(14)   Obtain d from corresponding provide-MP with β_d as lease payment
end

Fig. 2. LEASE algorithm for host-MP

unshared data items and their prices to help prospective host-MPs to determine their bid values. In Lines 14-16, the values of λ and β are computed by Equations 2 and 3 respectively.

Figure 2 depicts the algorithm executed by a host-MP H to facilitate it in *simulating* the choice of data items that it should bid for. H may not necessarily be able to obtain a lease for all the data items that it bids for since other MPs may outbid H, hence it is a *simulation*. Thus, H greedily *simulates* the filling up of its memory space by data items with higher value of γ. (γ is computed in Section 3). In Lines 12-14, the value of β_d is computed by Equation 3.

5   Performance Evaluation

MPs move according to the *Random Waypoint Model* [2] within a region of area 1000 metres $\times$ 1000 metres. The *Random Waypoint Model* is appropriate for our application scenarios, which involve random movement of users. A total of 100 MPs comprise 30 data-providers and 70 free-riders (which provide no data). Each data-provider owns 8 data items comprising 4 *shared* items and 4 *unshared items*. Each query is a request for a single data item. 20 queries/second are issued in the network, the number of queries directed to each MP being determined by a highly skewed Zipf distribution with Zipf factor of 0.9. Communication range of all MPs is a circle of 100 metre radius. Table 2 summarizes our performance study parameters.

Performance metrics are **average response time (ART)** of a query, **data availability (DA)** and **average querying traffic (QTR)**. ART equals $((1/N_Q) \sum_{i=1}^{N_Q} (T_f - T_i))$,

Table 2. Performance Study Parameters

Parameter	Default value	Variations
No. of MPs (N_{MP})	100	20,40,60,80
Zipf factor (ZF)	0.9	
Queries/second	20	
Bandwidth between MPs	28 Kbps to 100 Kbps	
Probability of MP availability	50% to 85%	
Size of a data item	50 Kb to 350 Kb	
Memory space of each MP	1 MB to 1.5 MB	
Speed of an MP	1 metre/s to 10 metres/s	
Size of message headers	220 bytes	

where T_i is the query issuing time, T_f is the time of the query result reaching the query issuing MP, and N_Q is the total number of queries. ART includes data download time, and is computed only for successful queries. DA equals $((N_S/N_Q) \times 100)$, where N_S is the number of successful queries and N_Q is the total number of queries. Queries can fail due to MPs being switched 'off' or due to network partitioning. QTR is the average number of hops per query.

As reference, we adapt a non-economic model **NL (No-Lease)** since existing M-P2P proposals do not address economic lease-based models. In NL, leasing is not performed and querying is broadcast-based. As NL does not provide incentives for free-riders to become host-MPs, only a single copy of any given data item d exists at the owner of d.

Performance of LEASE: Figure 3 depicts the performance of LEASE using default values of the parameters in Table 2. Leasing procedures are initiated only after the first 4000 queries, hence both LEASE and NL initially show comparable performance. The ART of both LEASE and NL increases with time due to the skewed workload ($ZF = 0.9$), which overloads some of the MPs that store 'hot' data items, thereby forcing queries to incur high waiting times and consequently high ART. However, over time, the economic incentives of LEASE entice more MPs to host data items, thereby

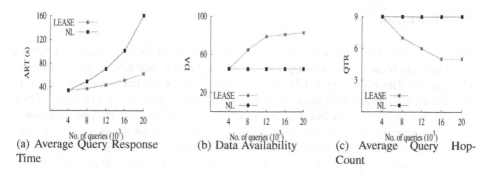

(a) Average Query Response Time   (b) Data Availability   (c) Average Query Hop-Count

Fig. 3. Performance of LEASE

increasing the resources (e.g., bandwidth, memory space) in the network for creating multiple (leased) copies for the same data item to facilitate load-balancing as well as reduction of QTR. Moreover, LEASE considers the connectivity of host-MPs, which further decreases its QTR, thereby decreasing ART. In Figure 3b, DA eventually plateaus for LEASE due to network partitioning and unavailability of some of the MPs.

In contrast, the non-economic nature of NL does not entice the free-riders to host data items via leasing, thus the ART of NL keeps increasing due to overloading of MPs storing 'hot' data items. For NL, DA remains relatively constant since it depends only on the probability of availability of the MPs. The QTR for NL remains relatively constant as only one copy of any given data item d exists in the network.

Effect of variations in the number of MPs: To test LEASE's scalability, we varied the number N_{MP} of MPs, while keeping the number of queries proportional to N_{MP}. In each case, 30% of the MPs were data-providers, the rest being free-riders. As the results in Figure 4 indicate, ART increases for both approaches with increasing N_{MP} due to larger network size. At higher values of N_{MP}, LEASE outperforms NL due to the reasons explained for Figure 3.As N_{MP} decreases, the performance gap decreases due to limited leasing opportunities, which results in lesser number of copies for leased data items, thereby making the effect of leasing less prominent.

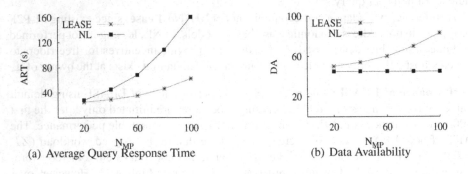

(a) Average Query Response Time   (b) Data Availability

Fig. 4. Effect of varying the number of MPs

6   Conclusion

We have proposed LEASE, a novel Mobile-P2P lease-based economic incentive model, in which data requestors need to pay the price (in virtual currency) of their requested data items to data-providers. In LEASE, data-providing mobile peers lease data items to free-riders, who do not have any data items to provide, in lieu of a lease payment. Thus, LEASE not only combats free-riding, but also entices free-riders to host data items, thereby improving network connectivity due to higher peer participation. In essence, LEASE facilitates the collaborative harnessing of limited mobile peer resources for improving data availability. Our performance study shows that LEASE indeed improves query response times and data availability in M-P2P networks.

References

1. Anagnostakis, K.G., Greenwald, M.B.: Exchange-based incentive mechanisms for peer-to-peer file sharing. In: Proc. ICDCS (2004)
2. Broch, J., Maltz, D.A., Johnson, D.B., Hu, Y.C., Jetcheva, J.: A performance comparison of multi-hop wireless ad hoc network routing protocol. In: Proc. MOBICOM (1998)
3. Buttyan, L., Hubaux, J.P.: Stimulating cooperation in self-organizing mobile ad hoc networks. Proc. ACM/Kluwer Mobile Networks and Applications 8(5) (2003)
4. Elrufaie, E., Turner, D.: Bidding in P2P content distribution networks using the lightweight currency paradigm. In: Proc. ITCC (2004)
5. Ferguson, D.F., Yemini, Y., Nikolaou, C.: Microeconomic algorithms for load balancing in distributed computer systems. In: Proc. ICDCS, pp. 491–499 (1988)
6. Gnutella. http://www.gnutella.com/
7. Grothoff, C.: An excess-based economic model for resource allocation in peer-to-peer networks. In: Proc. Wirtschaftsinformatik (2003)
8. Hara, T., Madria, S.K.: Data replication for improving data accessibility in ad hoc networks. IEEE Transactions on Mobile Computing (2006)
9. http://www.microsoft.com/presspass/presskits/zune/default.mspx
10. Kamvar, S., Schlosser, M., Garcia-Molina, H.: Incentives for combatting free-riding on P2P networks. In: Proc. Euro-Par (2003)
11. Kazaa. http://www.kazaa.com/
12. Kurose, J.F., Simha, R.: A microeconomic approach to optimal resource allocation in distributed computer systems. Proc. IEEE Trans. Computers 38(5), 705–717 (1989)
13. Mondal, A., Madria, S.K., Kitsuregawa, M.: CADRE: A collaborative replica allocation and deallocation approach for Mobile-P2P networks. In: Proc. IDEAS (2006)
14. Mondal, A., Madria, S.K., Kitsuregawa, M.: EcoRep: An economic model for efficient dynamic replication in Mobile-P2P networks. In: Proc. COMAD (2006)
15. First Workshop on the Economics of P2P Systems (2003), http://www.sims.berkeley.edu/research/conferences/p2pecon
16. Stoica, I., Morris, R., Karger, D., Kaashoek, M.F., Balakrishnan, H.: Chord: A scalable peer-to-peer lookup service for internet applications. In: Proc. ACM SIGCOMM (2001)
17. Turner, D.A., Ross, K.W.: A lightweight currency paradigm for the P2P resource market. In: Proc. Electronic Commerce Research (2004)
18. Wolfson, O., Jajodia, S., Huang, Y.: An adaptive data replication algorithm. Proc. ACM TODS 22(4), 255–314 (1997)
19. Wolfson, O., Xu, B., Sistla, A.P.: An economic model for resource exchange in mobile Peer-to-Peer networks. In: Proc. SSDBM (2004)
20. Xu, B., Wolfson, O., Rishe, N.: Benefit and pricing of spatio-temporal information in Mobile Peer-to-Peer networks. In: Proc. HICSS-39 (2006)
21. Zhong, S., Chen, J., Yang, Y.R.: Sprite: A simple, cheat-proof, credit-based system for mobile ad-hoc networks. In: Proc. IEEE INFOCOM, IEEE Computer Society Press, Los Alamitos (2003)

Real-Time Event Handling in an RFID Middleware System

Kaushik Dutta[1], Krithi Ramamritham[2], B. Karthik[2], and Kamlesh Laddhad[2]

[1] College of Business, Florida International University, Miami, FL
[2] Indian Institute of Technology Bombay, Mumbai, India

Abstract. Radio Frequency IDentification (RFID) tags have emerged as a key technology for real-time asset tracking. Wide application of RFID leads to huge amounts of data being generated from scan of each of these RFID tags on individual items, for example, the RFID system of a moderate size retail chain will generate 300 million RFID scans per day. Extracting meaningful information out of this huge amount of scan data is a challenging task. Moreover CIOs are looking for real time business decision from this RFID scan data. In this paper we show how to add value to an RFID middleware system by enabling it to handle a large number of RFID scan data and execute business rules in real-time. Experimentally we demonstrate that our proposed approach is very time efficient compare to a similar implementation with existing technologies. Lastly we also propose an architecture for a distributed RFID middleware system to handle raw RFID scan data.

1 Introduction

RFID is an automated identification technology that allows for non-contact reading of data [1], making it attractive in verticals such as manufacturing, warehousing, retail [2, 3], logistics, pharmaceutical [4], health care [5] and security. RFID systems are foreseen as replacement to the legacy bar code system of identifying an item. One of the major advantages of RFIDs over bar codes is that it is a non-line-of-sight technology - thus every item need not be handled manually for reading. In addition, RFID readers can read tags even when they are hidden. However, wide application of RFIDs leads to huge amounts of data being generated from the scan of these RFID tags on individual items; extracting meaningful information out of this huge amount of scan data is a challenging task. Moreover, today's CIOs are looking for real time business decisions based on RFID scan data. For example, in the retail case, when the sale of a product within an hour crosses a threshold, to avoid out of stock situations, a store manager may want more products be ordered for the inventory. In a hospital, when the drug to be given to a patient is considered along with the particulars of an 'RFID tagged' patient, the doctor or the nurse can be informed of drug interactions with the drugs the patient is currently taking. Once RFID tags are scanned and subsequently analyzed at the back-end, such decisions are definitely possible in today's technology. However with RFID technology, users want real-time information to take a decision on the spot, e.g. users of an online e-commerce site may want to know, in real-time, details of what stage of production an on-demand produced item is.

S. Bhalla (Ed.): DNIS 2007, LNCS 4777, pp. 232–251, 2007.

In this paper we develop and present the architecture of a system that will allow businesses to take real-time decisions on the basis of huge amounts of data generated by the RFID scans.

To further motivate our work in the next section we briefly describe a case study on the deployment of RFID technology in supply chain scenarios from Harvard Business Case studies [6] and explain how our research can help in this scenario.

1.1 A Motivating Scenario

Consider a Supermarket; *Metro Group (MG)*. Let us assume supply to *MG* follows this route: Manufacturers make pallets containing cases of same products and send them to *distribution centers* (DC). At the DC, these pallets are reassembled into mixed pallets and then sent to different stores of *MG*. Pallets arrive at stores in accordance with the need specified by that store. Similarly, the pallets arrive at the DC in accordance with the requirement from the *DC*. Using bar codes, mistakes happen while counting the number of pallets loaded on to a truck at the manufacturer's site or the kinds of cases transferred to the mixed pallet at the DC. All these result in slowing down the process, revenue loss or out of stock position at the stores [6].

To overcome such problems, pallet-level and case-level RFID tagging is used. The pallets leaving the manufacturer's site are tagged. They are loaded on to the truck and the RFID reader counts the number of pallets being loaded. At the DC, pallets are reassembled into mixed pallets. This can be automated with the help of RFID tags. Thus, mistakes in packing the cases into the right pallets, will be reduced. Reassembled pallets are now sent to the stores. These pallets are received at the backroom of the store and are opened and verified for correct configuration using RFID readers.

The cases are kept at the backroom before replenishing the shelf when the store closes. When the number of items at the shelf for a certain product falls below a threshold, it is refilled from the backroom. This must be done in an automated way. The employees may not always remember the number of items remaining on each shelf. Thus, an indicator might inform the backroom to refill a certain item as its number falls below the threshold.

Also, keeping excess stocks of all items in the backroom is not advisable. Thus, ordering for new stocks should happen if the number of cases of that product in the backroom is below a threshold. All these processes could not be efficiently handled by human beings. At the shelf, sensors can be monitoring the conditions (e.g., temperature for food items). Alarms can be raised if conditions worsen for a certain item. Items which are about to expire can be identified for an extra discount. These can be automated as the RFID tags may contain the item's expiry date.

RFID Enabled Tasks. In this paper we develop an event based RFID system that will enable us to observe and detect the state of the system and take actions accordingly in real-time. Our proposed approach will enable the RFID middleware system of *MG* to address the following business rules.

1. If the number of items on a shelf goes below a threshold, send an alert to the backroom.
2. If the number of items in the backroom is below a threshold, send an alert to the store manager, or ask DC to send the item.

3. If the DC is out of stock for an item, a request for a new supply for the item should be sent to the manufacturer.
4. If the DC has already dispatched the item in the last consignment to the store, recognize this and alert the store manager.
5. If the temperature of a food item is being monitored by a sensor and it observes abnormal temperature variations such that the item placed in that shelf can not be kept in such situations, alert the concerned authority about this problem.
6. If a certain food item kept on a shelf is about to expire, make visual announcements that the cost has been reduced.

Challenges in Exploiting RFID. Full utilization of the RFID system deployed in *MG* will occur only when these events (e.g., number of items on a shelf going below a threshold) can be detected and respective action (e.g., send an alert to the backroom) be taken in the least possible time. Some of the key challenges in developing such a system are,

(i) *The number of RFID events is huge.* Every scan by RFID scanner generates an event, leading to large number of such events from which identifying a particular event (e.g., number of items on a shelf goes below a threshold) is a challenging task. Following the example given in [7] (which is also very much applicable in *MG* case), suppose the retailer with 3,000 stores sells 10,000 items a day per store. Assume that we record each item movement with a tuple of the form: (EPC, location, time), where EPC is an Electronic Product Code which uniquely identifies each item. If each item leaves only 10 traces before leaving the store by going through different locations, this application will generate at least 300 million tuples per day. Extracting meaningful information from this 300 million tuples is the challenge.

(ii) The event should be detected in the least time possible for real time business decisions to be taken. The time difference between a shelf becoming almost empty and the time by which the shelf is replenished from the backroom should be as small as possible. So, nightly or periodic batch processing of such huge number of data will not serve the purpose. The problem is more complex when the RFID system is distributed, where both generation of events and detection of events happen in various systems distributed geographically at various locations.

In the next section we describe the related work. In Section 3, we develop the event based model for RFID system. In Section 4, we describe the detailed architecture and approach. In Section 5 we describe an implementation of RFID event handling system using relational database. We experimentally demonstrate the performance of our system in Section 6. In Section 7 we explain how our proposed approach can be extended in a distributed system. Lastly in Section 8 we conclude the paper.

2   Related Work

Recently, a number of RFID middleware systems have attracted industry attention [10, 11, 12]. The event handling mechanism in these RFID systems is very rudimentary in nature, based on Java's event handling mechanism. In general they can handle very simple basic events e.g., raw RFID scans. They do not have any infrastructure to handle large number of complex events including events that combine RFID scans along with other environmental conditions such as temperature. In real life business scenarios, however, it is very natural to look for complex events. Developing an infrastructure to answer the questions mentioned in Section 1.1 will require elaborate system development effort in these systems and will not be efficient.

Research on active database systems and event-condition-action model seems relevant to large extent. In this, some of the important work to mention are [13, 8, 14]. Because the nature of events in active databases and RFID systems is different, we can not directly apply any of the existing research, however we have borrowed several concepts from these areas to apply in our system, e.g., we borrowed the idea of indexing events on the basis of parameter values as proposed in [15]. The key differences between an RFID system and active database systems are (i) the number of events to be monitored is huge in RFID systems compared to typical active database systems (ii) the events in RFID systems are more complex in nature linking several events (including non-RFID events) together to form composite events and (iii) to achieve the desired business goal real-time response is required in RFID event handling systems. These differences make the handling of RFID events a challenging goal in it own right.

With respect to research related to real-world event handling, from [9], we borrowed the idea of representing complex RFID events with the help of state diagrams.

In [16, 17, 18], authors proposed pre-processing of raw RFID scan data for cleaning such as identifying missing data and detecting outliers. We show how the data-cleaning as proposed in these can be integrated with our proposed system. In [19], authors proposed a security mechanism for RFID data, which is orthogonal to the research of this paper.

In essence, though we borrow some of the concepts from existing research work, so far there has not been an end-to-end solution proposed in either academics or in industry to handle large number of RFID events generated from RFID scans and hence this research.

3 Events in RFID Systems

According to Wikipedia, an event is something that takes place at a particular place and time. For software systems, an event is something that needs to be monitored and may trigger a specific action. Specifying an event is therefore providing a description of the happening. Following [8], each RFID event can be described with some set of *dimensions* which includes *source* of event, event *granularity*, *location* of event, *time* at which the event occurred and a possible set of *operations* for combining events. An Event can be of *primitive* type or *composite* type. A *primitive event* occurs at a particular place and time. A *composite event* is a combination of a number of such primitive events linked by predefined operators (e.g., AND, OR, NOT etc.) [9]. In addition to this classification, in an RFID system, we define two types of events - **Basic Events** and **Events of Interest**.

3.1 Basic Events

A *basic event* (b_e) is an event generated by a source, e.g. an individual scan of RFID tag affixed to an object in the system. Following [9] and [7] a basic event(b_e) can be defined as a tuple (L, S, T) where L is the label dimension containing details of the event, S is the location dimension of the event occurrence and T is the time dimension at which the event occurs. As for example, going back to our example of *MG*'s supply chain system, consider the following situation: An object o is being loaded on a truck in a warehouse at location s at a certain point of time t. This object gets scanned by different readers. Scan of this object by a reader attached to truck at time $T = t$ at location $S = s$ is a basic event generating a tuple (l, s, t). The label l will contain details of the object o such as the scanned RFID tag number of manufacturer details.

In an RFID system a basic event (b_e) can be generated in four ways.

Object Scan: A RFID reader scans a RFID tagged object and generates the basic event, which is termed as RAW RFID event [7]. The RFID scan generates the label L and the scanner id. The location S can be derived from the RFID database [20] that contains the information about the location of the scanner corresponding to the scanner id and the time when the scan happened. The time (T) will contain when the actual event occurs.

Clock: A clock event is raised at some point in time independent of objects and other state of the system. The clock time can be absolute (e.g. 15th of August at 7:55 AM), relative (the next day after the match), or periodic (every day at 11:30 PM). The label of a clock event will just identify it as "CLOCK" event. The location (S) will remain empty in case of Clock event.

External: An external event is raised by a happening outside the system, which includes environmental conditions (e.g., the temperature of the hall/room goes above 30 degrees Celsius). Typically such external events will be generated by various sensors deployed in a system e.g., temperature sensor. The label of the external event will contain sensor data e.g., temperature value in case of temperature sensor. The location will identify the location of the sensor which can be derived from a separate sensor database. The time will identify, when the sensor generated the external event.

Internal: An internal event is related to internal state change of the system at time t. The internal event can be the effect of cascading action of some basic event which got fired by one of previously defined three ways. These are mostly program generated events. The label(L) of an internal event will identify the details of the event such as "Number of Items > 100" or "Average Price of item sold > 40". The location (S) of an internal event will identify the generator i.e., the particular systems or applications that is generating this internal event. The time(T) will denote the time at which such an event is generated.

3.2 Events of Interest (EI)

Events of Interest (EI) are the events which need to be monitored. EI can be of two types (i) *primitive EI* and (ii) *composite EI*.

Primitive EI. A *primitive event*(p_e) can be defined as a tuple (L, S, T) where L is the label containing details of the event, S is the location of the event occurrence and T is the time at which the event occurs. However, unlike basic events, in case of primitive EIs each of this may or may not be pointedly specified.

The *label of a primitive EI* is an indication of range of products, items or objects for which this event has been defined. This may include a particular supplier, a particular product from a particular supplier, etc. Since the EPC is a hierarchical representation of entities in supply-chain, we can use bits of EPC code to define label for primitive EI in case RFID tag contains the EPC code.

The *time of a primitive EI* is an indication of a range of time e.g. morning 8AM-10AM or today or month September. As a specific case the time of a primitive EI may be a specific time at the granularity at which it is defined in the basic event, e.g., 7.00 AM 29 Sept 2006.

The *location of a primitive EI* is a region that may contain one or more locations at the granularity of basic events, e.g., a basic event may occur at a scanner located at the distribution center in Miami, whereas an EI may be specified as a scan in a distribution center in South Florida region which contains the distribution centers both in Miami and Tampa.

For example, in the following p_e^1 is a primitive EI.

$$p_e^1 = (L = \{product\_type = \#54567\},$$
$$S = Shelf\#583, T =' Morning')$$

Here the primitive EI p_e^1 is looking for product with "product_type" as "#54567" defined in the label (L). The location (S) dimension of EI p_e^1 is defined by the shelf number "#583". The time (T) is morning.

Composite EI. A *composite EI*(o_e) is a combination of multiple primitive EIs (p_e) or multiple composite EIs linked by operators drawn from the following set [13].

AND ($\wedge$): Conjunction of two events E_1 and E_2, denoted as $E_1 \wedge E_2$, occurs when both E_1 and E_2 occur (the order of occurrence of E_1 and E_2 is irrelevant).

OR ($\vee$): Disjunction of two events E_1 and E_2, denoted as $E_1 \vee E_2$, occurs when either E_1 or E_2 occurs.

SEQ ($\Rightarrow$): Sequence of two events E_1 and E_2, denoted by $E_1 \Rightarrow E_2$, is when E_2 occurs provided E_1 has already occurred. This implies that the time of occurrence of E_1 is guaranteed to be less than the time of occurrence of E_2.

NOT (!): The NOT operator, denoted by !(E_1, E_2, E_3), detects the non-occurrence of the event E_2 in the closed interval formed by E_1 and E_3. It is rather similar to the SEQ operator except that E_2 should not occur between E_1 and E_3 [13].

Relative periodic (R_p): An event E_2 which occurs periodically with specified frequency after event E_1 has occurred till event E_3 occurs. This can be indicated as $P(E_1, E_2, E_3)$.

Composite EIs are reactive. Some action is associated with such events and every time such events occur, the system identifies them and executes these actions. Following ECA model [8], a composite EI has three parts (i) event definition (ii) condition and (iii) action to be executed.

Event Definition: The composition of EIs to create a composite EI is defined here, e.g., $o_e^1 = p_e^1 \wedge p_e^2$ is the event definition of composite EI o_e^1 which links two primitive EIs p_e^1 and p_e^2 by 'AND' operators. The derivation of L, S and T dimensions of composite EI o_e^1 from dimensions of EI p_e^1 and p_e^2 is done following [9]. Note that, the event definition of a *composite EI* can be expressed as a regular expression of multiple EIs. Thus the event definition of a composite EI can be represented as a state graph (DFA for regular expressions) as shown in [9]. We will use this *event state graph* of an EI in section 4.1.

Condition: For a given composite EI, condition is a side-effect free boolean computation or set of boolean computations on dimensions (L, S and T) of *two or more* primitive EIs, which when evaluated as true may trigger an action associated with the EI. The condition is not a mandatory specification. The result of condition is presumed to be true if no condition is specified. Formally condition will be boolean combination of multiple conditional elements of the form $x \otimes y$ where,

$$\otimes \in \{>, <, \geq, \leq, =, \neq\}$$
$$x \to q_e^i.d.v$$
$$y \to q_e^j.d.v \tag{1}$$

Where q_e^i and q_e^j are EIs (primitive or composite), d is one of the dimensions *label(L)*, *location(S)* and *time(T)* of EIs q_e^i and q_e^j, and v is some attribute value of this dimension, e.g., *company* is one attribute of dimension *label(L)*. Here v is an optional item. If there is no v specified, the default value of the dimension is used, e.g., the default value for the dimension label, L, corresponding to an RFID event on a product will be the complete EPC code of the product. Note that, join conditions like $p_e^1.L.company = p_e^2.L.company$ is a valid condition for composite EI $o_e^1 = p_e^1 \wedge p_e^2$, whereas $p_e^1.L.company = $ "HP" is not a valid condition for composite EI o_e^1, because this is defined based on single primitive event and should have been represented in the primitive EI p_e^1 itself.

Action: An action is arbitrary sequence of predefined operations which are executed when the corresponding event gets fired on evaluations of associated conditions. Actions depend on the type of business where the system is being deployed.

Composite EI as a composition of composite EI: A composite EI may be composed of two more composite events, e.g.

$$o_e^3 = o_e^1 \wedge o_e^2$$
$$o_e^1 = p_e^1$$
$$o_e^2 = p_e^2 \wedge p_e^3$$

o_e^3 is a composite EI composed of two more composite EIs o_e^1 and o_e^2. The event definition of o_e^3 will be composition of o_e^1 and o_e^2. The condition of o_e^3 can be separately defined based on dimensions of o_e^1 and o_e^2, whereas the individual condition of o_e^1 and o_e^2 will be intact which will be evaluated before o_e^1 and o_e^2 triggers. Similarly the action of o_e^3 will be separately defined whereas the individual action of o_e^1 and o_e^2 will remain intact.

3.3   Example of ECA in *MG*

Here we present some examples of how the business rules regarding *MG*'s supply chain system can be expressed as EI using ECA form.

Rule 1: When the number of items of product "54567" in the shelf "583" falls below a threshold (let us say 5), then the backroom needs to be alerted.
This EI (o_e^1) can be expressed in ECA form as follows.

Event:

$$o_e^1 = p_e^1 = (L = \{product\_type = \#54567\}, S = \#583, T = t_1)$$

Here the composite EI o_e^1 contains single primitive EI p_e^1. The primitive EI p_e^1 is looking for product with "product_type" as "#54567" defined in the label (L). The location (S) dimension of EI p_e^1 is defined by the shelf number "#583". The time (T) is t_1, where t_1 is anytime when the system will look for primitive EI p_e^1.

Condition:

$$p_e^1.L.count < 5$$

Here, we assume RFID reader has an attribute count. The value of count holds the number of items the reader would read for a given product-type.

Action: *Notify backroom.*

Rule 2: The store manager wants the system to alert him when the temperature sensor on a shelf "124" finds the temperature to be unsuitable for a certain item "54567".

Event:

$$p_e^1 = (L = TEMP, S = \#124, T = t_1),$$
$$p_e^2 = (L = \{product\_type = \#54567\}, S = \#124, T = t_2)$$
$$o_e^1 = p_e^1 \wedge p_e^2$$

Here, p_e^1 and p_e^2 are primitive EIs with dimensions: (L, S, T), o_e^1 is the composite EI denoting the event corresponding to rule 2. In this example, time denotes the time at which the event occurs. Here we wait for only those events which satisfy the label and location as specified above. The EI p_e^1 has label temperature from an external source (temperature sensor) with location identifier indicating shelf #124, and the EI p_e^2 has label RFID tag with source as RFID scanner located at shelf #124. When both of these events occur the following condition is evaluated.

Condition:

$$(p_e^1.L.temperature > 70 \vee p_e^1.L.temperature < 50)\wedge$$
$$(p_e^1.T < p_e^2.T + timethreshold \wedge p_e^1.T > p_e^2.T - timethreshold)$$

In the condition we check if the temperature is within a certain range and the time difference between the two EIs does not exceed a threshold.

Action: *Notify backroom*

4   Architecture

Figure 1 describes the overall architecture of our proposed system. The key component in this architecture is the *RFID Event Handler* (EH) and the *Events of Interest database*(EIDB). Basic events are generated by various RFID sources (RFID reader) and non-RFID sources (Clock, External and Internal). These basic events are reported to EH. EIDB contains all EIs of the system. Based on incoming basic events, generated out of RAW RFID scans by RFID readers, EH identifies matching EIs in the EIDB. EH passes the identified EIs and the related basic events to the rest of the IT system. In an RFID system as the number of basic events generated from RFID readers is huge, the scalability of the EH is the biggest concern addressed by our design. In its simplest form, EH is a centralized system. In complex systems, this EH may be a distributed system. In this paper we primarily concentrate on the centralized EH and briefly describe distributed RFID event handling mechanism in section 7.

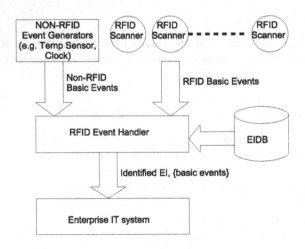

Fig. 1. RFID Event Handler Architecture

4.1   EI Database

The EI database contains all EIs along with their respective details. In EIDB, EIs are first broken into primitive events of interest (primitive EIs). Primitive EIs are stored and maintained in EIDB in a multi-dimensional R-tree [21] structure. In each of the three dimensions (L, S and T) of EI, the tree is formed by the semantic hierarchy similar to R-tree. So we have three R-trees in the structure, e.g., if the label contains EPC code, the hierarchy in the R-tree of label is defined by the EPC hierarchy. In the label dimension, a primitive EI with label "Compaq" will lie above a primitive EI with label "Compaq-laptop". Similarly in the R-tree of source dimension, the hierarchy is defined by geographical hierarchy. A primitive EI with source (i.e. location of event) "Mumbai" will stay above a primitive EI with source "Powai" (a place in Mumbai). Such a hierarchy indicates, if a primitive EI p_e^1 is identified for "Powai", another primitive EI p_e^2 with source "Mumbai" and location & time same as that of p_e^1 should also be identified. In the R-tree of time(T) dimension, the hierarchy is defined by natural hierarchy of time, e.g. the EIs related to a particular time (e.g. 15th Oct 06, 8.00 PM) will reside below the EIs related to a particular day (e.g. 15th Oct 06). Also against each primitive EIs m in the EI tree, we maintain a list (Q_m) of composite EIs that are composed of the primitive EI m (for simplicity this list is not shown in the Figure 2).

Consider the following example of a set of primitive EIs.

$p_e^1 = \{\ L$=Compaq-laptop, S= Mumbai, , T= Jan, '06 $\}$

$p_e^2 = \{\ L$= HP-Printer 3650, S=Powai, T=15th Oct, '06 $\}$

$p_e^3 = \{L$=HP-Printer , S=San Francisco, T=15th Oct, '06, 8:00 PM $\}$

$p_e^4 = \{L$=Compaq, S=IIT Bombay, T=*$\}$

$p_e^5 = \{\ L$=*, S=IIT Bombay Convocation Hall, T=Every morning-Jan, '06$\}$.

These events are kept in the EIDB in a multi-dimensional R-tree as shown in Figure 2. Whenever a basic event (b_e) arrives at EH, EH looks into the EIDB and does three tree traversal one on each of the event dimensions - label, source and time. The traversal on each of these dimensions results in three sets S_L, S_S and S_T. Where S_L is the set of all primitive EIs that match with b_e on the label (L) dimension, S_S is the set of all primitive EIs that match with b_e on the location (S) dimension, and S_T is the set of all primitive EIs that match with b_e on the time (t) dimension. Next we compute the set $P = S_L \cap S_S \cap S_T$, where each primitive EI $p_e \in P$ matches with the basic event b_e in all three dimensions L, S and T. Next Q_{p_e}, the list of composite EIs, that are dependent on primitive EI $p_e \in P$ is computed.

Handling Composite EIs. The semantic of composite EIs can be captured by a state transition diagram. The current state of a composite EI is maintained in the EIDB. The state graph of a composite EI is give by $G = (V, \mathcal{E})$ where V is set of all the possible states for given composite EI and $\mathcal{E}$ is set of edges indicating occurrence of primitive EIs which leads to state transition. The final states of the state graph G will denote the occurrence of corresponding composite EI.

For example, a typical composite EI can be a combination of two or more primitive events such as $o_e^1 = p_e^1 \wedge p_e^2$. This specifies that when p_e^1 and p_e^2 both occur, the occurrence of composite EI o_e^1 will be identified. However, this does not specify what will

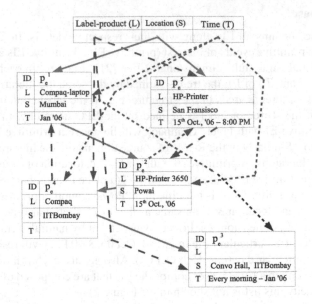

Fig. 2. R-Tree Structure of Primitive EIs

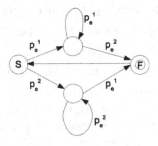

Fig. 3. State Diagram for Composite EI $o_e^1 = p_e^1 \wedge p_e^2$

happen if more than one p_e^1 occurs before occurrence of a p_e^2. Both the Figure 3 and Figure 4 represents the composite EI o_e^1. In Figure 3, once event p_e^1 has occurred all subsequent occurrence of p_e^1 will be ignored unless a p_e^2 occurs. Whereas in Figure 4, once p_e^1 has occurred, subsequent occurrence of event p_e^1 will lead to a state where consecutive two occurrences of event p_e^2 will result in the identification of composite EI o_e^1 twice. Thus the event state graph of a composite EI will also help us to clarify these details of the composite EI.

Handling Conditions in RFID ECA. As explained in section 3.2, the condition part of an EI is described based on dimensions L, S and T of the EI. Conditions are typically join conditions on multiple primitive EIs. Such conditions can result in monitoring the same EI with different parameters. E.g., consider an EI $o_e^1 = p_e^1 \wedge p_e^2$, with condition $p_e^1.L.company = p_e^2.L.company$, which identifies two primitive events

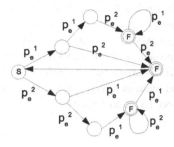

Fig. 4. State Diagram for Composite EI $o_e^1 = p_e^1 \wedge p_e^2$

on the product of the same company. The composite EI o_e^1 will be identified if two events having the same company on the product label occur. Once primitive EI p_e^1 with $p_e^1.L.company = company_1$ occurs how to store it efficiently so that occurrence of p_e^2 with $p_e^2.L.company = company_1$ at some later time can be linker together very efficiently.

This is an example of multiple instances of the same composite EI (o_e^1) occurring simultaneously for different companies. So we need to maintain multiple instances of the state graph as well. Here we borrow from the implementation of "scalable trigger" in [15]. We maintain single state graph of a composite EI. In an indexed structure similar to the one proposed in [15] we keep each instance of the composite EI (o_e) along with following information - (i) $\mathcal{B}_{o_e}$ a list of all basic events related to a particular instance of the composite EI o_e, e.g. basic events related to a particular company (e.g., $L.company$), (ii) the current state in the event state graph for that instance of the composite EI and (iii) values of fields (e.g., $L.company$) in condition clause (e.g., $p_e^1.L.company = p_e^2.L.company$) in each state transition. The structure is indexed on the field values of condition clause e.g., in the example it will be indexed on $l.company$. A sample of this structure is shown in table 1.

Table 1. Multiple Instances of Composite EI

Instances of EI	Basic Events	Current State	Field1	Field2	...
Instance 1 of o_e^1	$b_e^i, b_e^j, ...$	S_a	$p_e^i.L.company=$"HP"	$p_e^j.S.city = $"Mumbai"	...
Instance 2 of o_e^1	$b_e^m, b_e^n, ...$	S_b	$p_e^m.L.company=$"Compaq"	$p_e^n.S.city = $"Miami"	...

5   A Solution Using RDBs

We are not aware of any RFID system that can identify complex events from input stream of basic events. So to compare the efficacy of our approach we developed an alternate approach of detecting EIs from input of basic events. In this section we describe this "RDB" approach which primarily exploits and uses existing relational database technology. In Section 6 we compare the performance of our approach with this RDB approach. The details of the experiment can be found in section 6 where we compare the performance of our approach with this RDB approach.

In the RDB approach all EIs are stored in a relational database and whenever a new basic event occurs, this database is queried using standard SQL to get matching EIs. The database schema to store EIs is as follows.

Composite event table:

$CET = (o_e^{id},$ Composite EI expression, conditions, actions).

Primitive event table:

$PET = (p_e^{id},$ location, time, label)

Primitive-Composite event map:

$PCM = (o_e^{id}, p_e^{id})$

The CET table maintains definitions of all composite EIs. PET maintains all primitive EIs that are part of these composite EIs and the table PCM maintains which primitive EI is being used by which composite EI. We also maintain three tables namely $Time$, $Label$ and $Space$, each of which keep the dimension related abstractions mapped to actual values of corresponding dimension i.e., each of them has schema as ($abstraction$, $actual\_value$) e.g. $Time$ table will tell us 'morning' is '6:00 AM to 11:59 AM', $Space$ table will tell us 'Powai' is part of 'Mumbai', and so on. Along with these, we maintain a datastructure, an adjacency list (AL) of active composite EIs - all composite EIs that have been partially fulfilled based on already occurred basic events. So AL maintains the detail of all composite EIs waiting for one or more primitive EIs to occur. AL will be empty when the system is initialized.

Having described our approach for real-time RFID event handling and a RDB approach, in the next section we compare these two approaches to experimentally demonstrate the efficacy of our approach.

6   Experimental Results

In this section we experimentally demonstrate the efficacy of our approach of event handling in an RFID middleware system. We study the experimental result from two dimensions (i) performance of our approach (denoted as "R" approach) compared to the RDB approach (ii) how the performance and scalability of our approach varies on various parameters.

We developed the RFID EH system using JDK 1.5. We ran the system on a Windows XP machine with pentium 2.8 GHz processor and 1 GB RAM. The EIDB database is a main-memory data-structure of multi-dimensional R-tree. The input basic events are simulated using a Java program communicating with the RFID EH using shared memory. The basic events are randomly generated by event simulator. The set of primitive EIs in EIDB are randomly generated based on a predefined hierarchy of label, location and time. The composite EIs are randomly generated by combining multiple primitive EIs with operators defined in section 3. In RDB system, the database tables as defined in Section 5 are created and populated with EIs in Oracle 10g Express Edition(XE) [22] on the Windows XP machine. The Oracle 10g XE was set to use memory to store tables

Table 2. Experimental Parameters

Parameter	Values	Base Value
Number of Composite EIs	2000 4000 6000 8000 10000	10000
Rate of incoming basic events (number/sec)	50 70 90 110 130	50
Number of Primitive EIs per Composite EIs	4 8 12 16 20	4

and indices required for all experiments. This ensures that no disk access was required in Oracle for the RDB approach. To ensure that same set of events are generated by simulator in both RDB and our approach (denoted as "R" approach), same seed was used for generation of random numbers in both the approaches.

The parameters of our experimental analysis are chosen as shown in table 2. During the experiment we measure (i) the time it takes to identify primitive EIs, we denote this time as "Time to identify Primitive EI" (TP) (ii) the time it takes to identify composite EIs, we denote this time as "Time to identify Composite EI" (TC). The total time required to process an input basic event will be the sum of TP and TC.

To compare the performance of our "R" approach with the "RDB" approach, we first initialize the EIDB with 2000 composite EIs and we programmatically simulate the basic event generation and send basic events to RFID EH at a fixed rate (50 basic events per seconds). We continue this for 5 minutes. We note the TP and TC for each basic event. We compute the average value of all TP and TC during the run for 5 minutes. We repeat this process for each value of number of composite EIs in the EIDB as given in Table 2. In all cases, we keep the rate of incoming basic events constant at the base value 50 per seconds. We complete this experiment both for "R" and "RDB" case.

In Figure 5 and Figure 6, we plot the TP and TC respectively as it varies with the number of composite EIs in EIDB. As can be seen, at any number of composite EIs our approach provides much lower value of TP and TC compare to RDB approach. The more number of composite EIs will require more search time in the R-tree in "R" case and in the database in "RDB" case. Thus as number of composite EIs increases, the TP increases both for "R" and "RDB" case. The improved performance of "R" approach can be explained due to mainly two reasons (i) R Tree - The R-Tree approach of finding primitive EIs is taking $\frac{1}{3}$rd time of that being taken by the SQL query in Oracle in-memory database (ii) State Diagram - Representing composite EIs as state diagram means maximum one operation per basic event to determine whether the composite EI has occurred or not, whereas in "RDB" approach this requires computing the boolean function based on matching primitive EI. As a result TC in "R" approach is $\frac{1}{20}$th of that of "RDB" approach.

To demonstrate the scalability of our approach we kept the number of composite EIs in EIDB constant at its base value 10000. We varied the rate of incoming basic events in our system. For each value of rate of incoming basic events, we measured the total time (TP + TC) required to process each basic event and average it over 5 minutes. We plot the total processing time with the rate of incoming basic events in Figure 7. The y-axis

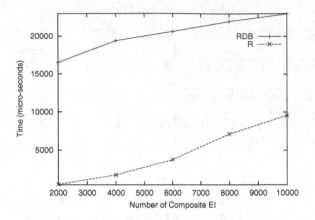

Fig. 5. Variation of TP with # EIs

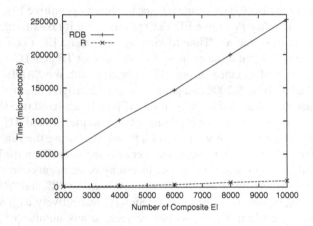

Fig. 6. Variation of TC with # EIs

denoting "Average Processing Time" (i.e., TP + TC) starts at 18900. As the incoming rate of basic events increases the average processing time for each basic event increases gradually, however one should note that the increase in processing time is very minimal, from 19000 μseconds at 50 basic events per seconds to 19500 μseconds at 130 basic events per seconds.

The complexity of composite EI should also affect the scalability of our system. One way to measure the complexity of a composite EI is the number of primitive EI it depends on as described in Section 3. We kept the total number of composite EIs in EIDB constant at the base value of 10000, the rate of incoming basic events to 50 per seconds and vary the average number of primitive EIs per composite EIs as described in Table 2. For each value of average number of primitive EIs per composite EI, we compute the average processing time of each basic event (TP + TC) over 5 minutes and report it in Figure 8. The y-axis denoting the "Average Processing Time"

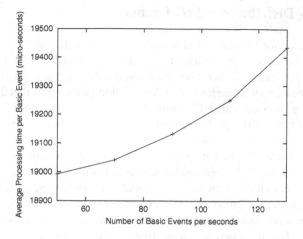

Fig. 7. Total Performance

(i.e., TP + TC) starts at 18900 and the x-axis denoting the "Average Number of Primitive EIs per Composite EI" starts at 4. It is obvious that as the complexity of the composite EIs increases the system is taking more time to process each basic event and identify corresponding composite EIs, however the increase in processing time is very minimal compared to the increase in the complexity of composite EI - from 19000 μseconds at 4 primitive EI per composite EI to 19900 μseconds at 20 primitive EI per composite EI.

Thus we can conclude that not only our approach of handling incoming RFID basic events provides much improved performance compared to "RDB" approach based on latest technologies, our approach also scales very well with increased rate of incoming basic events and the complexity of composite EIs giving *real-time performance* (few milliseconds) in all scenarios.

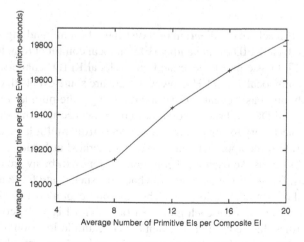

Fig. 8. Performance with Complexity of Composite EI

7   Handling Distributed RFID Events

So far in the paper we have assumed a single instance of RFID EH where all RFID basic events are being gathered for identifying and handling EIs. If the number of event sources (i.e. RFID scanners) increases, both the number of basic events and EIs will increase. This may result in performance degradation of the centralized event handler system. Moreover in a centralized system, there is a potential risk of system failing even with single failure. The performance of the system, scalability and reliability can be improved by having a distributed event handling system. Moreover in real life situations raw RFID scans happen at distributed locations e.g. at various DCs or various stores in case of *MG*'s RFID deployment scenario. Each DC and store will have their own respective computer system to handle local data. Following this, in this section we briefly describe a distributed RFID event handler system that processes local basic events locally in local computer system and cooperate with other locations to address global business rules, that spans across multiple DC or multiple stores.

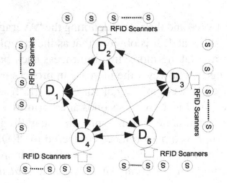

Fig. 9. Distributed Event Handler

Figure 9 depicts a schematic diagram of a distributed event handling system, where $D_1, D_2,...,D_5$ are local RFID event handler (EH) in local computer (e.g. local computer in DC). Local RFID basic events generated out of local RFID scanners are processed by the corresponding local RFID EH. However there are some events that require monitoring and combining basic events at two locations e.g. "the number of a pallets of a particular color in all DCs in Powai should not go below a certain threshold". To identify such EIs it is necessary to consolidate basic events from multiple locations. For this in a simplistic brute force approach, basic events generated at location will be broadcast to all other locations. So each local computer system will be aware of basic events in other locations. However the obvious drawback for such brute force approach is increased overhead of communication cost. The communication cost can be reduced by designating primary handler for each global EI (a global EI is an EI corresponding to global business rules that require basic events from multiple locations) and keeping a distributed directory of global EI along with its primary handler. A number of variations

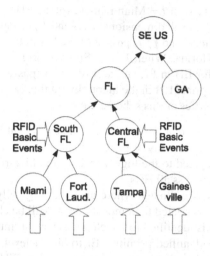

Fig. 10. Location based Hierarchy in Distributed Event Handling

of such approach can be borrowed from distributed cache architecture and invalidation literature [23].

Typically, geographically distributed enterprise systems follow organizational hierarchy, e.g., in *MG* case, store specific systems communicate with the DC responsible for a region. Computer systems at regional DC communicate with state-wide DC and so on. In developing an efficient distributed RFID handler system we exploit such a hierarchy in the enterprise distributed system. Even if such hierarchy does not exist in other distributed systems of an enterprise, based on location we can create such hierarchy in the distributed RFID EH system of the enterprise. We group the EHs located at geographically distributed locations based on "location" parameter, which coincides with the location dimension (L) of an RFID basic event. An example hierarchy is shown in Figure 10. Here, we group city based store EHs (Miami and Fort Lauderdale) on the basis of the distribution centers (South Florida region) serving them.

RFID basic events may arrive at any EHs in the hierarchy. The EIs are also specified at any EH in the hierarchy. An EI defined for a particular EH can have label dimension specified at the same level in the hierarchy or below it, e.g., at the label dimension (L) of any EI for EH at South Florida can have one of following values South Florida, Miami or Fort Lauderdale. When an EI is defined at an EH, corresponding primitive EIs are appropriately propagated to all respective EHs, e.g., when an EI $o_e^i = p_e^1 \wedge p_e^2$ with label dimensions $p_e^1.L=$'South Florida' and $p_e^2.L=$'Central Florida' is specified at the EH in FL, the primitive EI p_e^1 is passed to EH in South Florida and subsequently to EHs in Miami and Fort Lauderdale. Similarly the primitive EI p_e^2 is passed to EH in Central Florida and subsequently to EHs in Tampa and Gainesville. At each level primitive events are stored along with the EH where the EI has been originally specified, in this case EH in FL for both p_e^1 and p_e^2.

Assume a basic event b_e with L='Miami' arrives at the EH in Miami. If EH identifies a match between b_e with p_e^1 in all dimensions (L, S and T) following the R-tree structure described in section 4, the b_e and p_e^1 is passed to the EH in the immediate higher level, here the EH in South Florida. When EH in South Florida receives this message, it identifies that the p_e^1 is for EH in FL and accordingly it passes this to the higher level until it reaches the appropriate EH in the hierarchy (in this case the EH in FL).

Few noteworthy properties of this scheme are -

1. Basic events are screened at the level where a basic event is first reported. Thus if a basic event does not meet any of the EIs in the system the basic event will not be processed and propagated to the higher level in the hierarchy. This distributes the identification of EIs across all EHs.
2. Conditions of an EI is evaluated where an EI is originally specified, e.g., in our example any condition related to composite EI o_e^i will be checked in the EH at FL.
3. Once a primitive EI is identified to match a basic event, intermediate nodes passes the basic event and identified primitive EI to higher level. The identification of a primitive EI for a basic event happens only once at the EH where basic event first arrives in the system.
4. A single basic event may identify multiple primitive EIs and accordingly multiple message may need to be passed to higher level.
5. Number of basic events processed at each EH is limited by the number of basic events being first reported in the system at that EH, making it a very efficient scheme for EI identification.

8 Conclusions

In this paper we described the architecture of a system that can handle large number of incoming RFID events and identify events of interest in real-time. We developed an event based model for RFID system using the ECA framework. We demonstrated that our approach in identifying RFID events of business interest can perform significantly better than an implementation using latest technologies. Lastly we described how our approach can be extended in a distributed scenario. In future we intend to implement the distributed architecture for RFID event handling mechanism and show the performance of our proposed distributed architecture.

References

1. Glover, B., Bhatt, H.: Rfid essentials - theory in practice (2005)
2. IDTechEx: Rfid progress at wal-mart (2005)
3. TechWeb: Albertsons launches rfid initiative (2005)
4. RFID Journal: Rfids in pharmaceuticals (2005)
5. RFID Journal: Rfids in healthcare (2005)
6. Ton, Z., Dessain, V., Stachowiak-Joulain, M.: Rfids at the metro group (2005)
7. Gonzalez, H., Han, J., Li, X., Klabjan, D.: Warehousing and analyzing massive rfid data sets. In: ICDE 2006. Proceedings of the 22nd International Conference on Data Engineering, p. 85 (2006)

8. Paton, N.W., Díaz, O.: Active database systems. ACM Comput. Surv. 31(1), 63–103 (1999)
9. Nagargadde, A., Varadarajan, S., Ramamritham, K.: Semantic characterization of real world events. In: Zhou, L.-z., Ooi, B.-C., Meng, X. (eds.) DASFAA 2005. LNCS, vol. 3453, pp. 675–687. Springer, Heidelberg (2005)
10. Hoag, J., Thompson, C.: Architecting rfid middleware. IEEE Internet Computing 10(5), 88–92 (2006)
11. Sun Microsystems: Software solutions: Epc and rfid (2006)
12. IBM Inc.: Integrate your enterprise application with ibm websphere rfid middleware (2006)
13. Chakravarthy, S.: Sentinel: an object-oriented dbms with event-based rules. In: SIGMOD 1997. Proceedings of the 1997 ACM SIGMOD international conference on Management of data, pp. 572–575. ACM Press, New York, NY, USA (1997)
14. Chakravarthy, S., Le, R., Dasari, R.: Eca rule processing in distributed and heterogeneous environments. In: Proceedings of the International Symposium on Distributed Objects and Applications, pp. 330–339 (1999)
15. Hanson, E., Carnes, C., Huang, L., Konyala, M., Noronha, L., Parthasarathy, S., Park, J., Vernon, A.: Scalable trigger processing. In: Proceedings. of 15th International Conference on Data Engineering, pp. 266–275 (1999)
16. Rao, J., Doraiswamy, S., Thakkar, H., Colby, L.S.: A deferred cleansing method for rfid data analytics. In: Proceeding of VLDB Conference, pp. 175–186 (2006)
17. Jeffery, S.R., Garofalakis, M.N., Franklin, M.J.: Adaptive cleaning for rfid data streams. In: Proceeding of VLDB Conference, pp. 163–174 (2006)
18. Subramaniam, S., Palpanas, T., Papadopoulos, D., Kalogeraki, V., Gunopulos, D.: Online outlier detection in sensor data using non-parametric models. In: Proceeding of VLDB Conference, pp. 187–198 (2006)
19. Song, J., Kim, H.: The rfid middleware system supporting context-aware access control service. In: ICACT 2006. Proceedings of The 8th International Conference Advanced Communication Technology, 2006, p. 4 (2006)
20. Hoag, J., Thompson, C.: Architecting rfid middleware. IEEE Internet Computing 10(5), 88 – 92 (2006)
21. Guttman, A.: R-trees: a dynamic index structure for spatial searching. In: SIGMOD 1984. Proceedings of the 1984 ACM SIGMOD international conference on Management of data, pp. 47–57. ACM Press, New York, NY, USA (1984)
22. Oracle Inc.: Oracle database 10g express edition (2006)
23. Pong, F., Dubois, M.: Verification techniques for cache coherence protocols. ACM Comput. Surv. 29(1), 82–126 (1997)

Sustainable Creation of Conversational Content Using Conversation Quantization

Hidekazu Kubota[1,2]

[1] Japan Society for the Promotion of Science (JSPS) Research Fellow
[2] Graduate School of Informatics, Kyoto University,
Yoshida-Honmachi, Sakyo-ku, Kyoto 606-8501, Japan
kubota@ii.ist.i.kyoto-u.ac.jp
http://www.ii.ist.i.kyoto-u.ac.jp/~kubota/

Abstract. The purpose of this paper is to support the sustainable creation of conversation content. My approach is from the viewpoint of conversational records as materials for new conversations. I introduce the concepts of conversation quanta and quantization spiral, the latter being a feasible framework for constructing conversational content. A conversation quantum is a reusable conversation material that is applicable to conversational systems. I survey the systems using conversation quanta and their evaluations.

Keywords: Conversation, Content creation, Content visualization, Agents.

1 Introduction

People exchange a wide variety of knowledge during their daily activities by means of conversations. A sustainable conversation is indispensable for constructive knowledge creation. In a conversation, sustainability implies the production of conversation materials for the next conversation. A record of conversations is a source for new conversations. For example, consider a regular meeting. People generally write a transcript or prepare notes of a meeting to build up a store of knowledge. People cannot continue constructive discussions without the transcripts of previous meetings. There are many other examples of conversation materials such as the proceedings of an annual conference, notes of an educational dialogue, OHP or PowerPoint slides, handouts, and so on.

The objective of this paper is to support the sustainable creation of conversational content by means of computer-reinforced conversation materials. I assume that knowledge creation is facilitated by generating new conversations from conversational records by a computational method. I would like to discuss the end-user applications of conversation generation in greater detail. Content is the final output that is tailored to the end-users. Knowledge should be generated such that the end-users understand it well. Abstract knowledge is characterized by specific expressions so that it can be used in practical situations. The persuasiveness of knowledge depends on some rhetoric and direction. Content is not so much a formal and reductive form of knowledge as it is an informal representation such as voices,

S. Bhalla (Ed.): DNIS 2007, LNCS 4777, pp. 252–271, 2007.

images, and movies. I would like to investigate the feasibility of computational conversation generation in the practical activities of people. This is not sufficiently clear because it is difficult to develop a durable conversational system by using the current reductive framework of natural language processing.

I will approach this topic from the viewpoint of conversational records as material for new conversations. Therefore, I will discuss reductive knowledge and natural language syntax and semantics very briefly. I will discuss conversation quanta, which are larger-grained informal content as compared to formalized knowledge, in greater detail. A conversation quantum represents a reusable conversation material that is an individually functional, interactive, and synthesizable conversation block, similar to a LEGO® toy block. Conversation quantization is performed by processing a large number of conversation quanta using a computer. This is a feasible framework for constructing conversational content.

Conversation quantization was first proposed by Nishida [1], and this concept has been developed in many applications. CoMeMo-community [2] and EgoChat II [3] are early systems in which avatars exchanged messages in a conversational fashion. In these systems, keywords or short sentences are used as a quantum. A detailed description of EgoChat II is provided in Section 3. Conversation quanta have been developed from sentences to paragraphs (Section 4), to utterance pairs (Section 5), and to video clips (Section 6).

Conversation quantization has applications in several areas. The following areas are mentioned in the subsequent sections: an avatar of an actual person (Sections 3 and 5), community broadcasting service (Section 4), pedagogical agent (Section 5), and proceedings of meetings (Sections 6 and 7).

2 Conversation Quantization

Conceptually, the framework of conversation quantization is generalized to a quantization spiral that comprises (1) quantization, (2) construction, (3) tailoring, and (4) re-quantization (Fig. 1). I consider a scenario wherein people conduct conversations in many real-world situations supported by computer supported collaborative work (CSCW) systems. Situations (A) and (B) are different in time and place. People talk about topical real-world objects or topical electronic contents. Conversation quanta are extracted from the conversation in situation (A) and then utilized in situation (B) in the quantization spiral.

(1) Quantization

Conversation quanta are quantized by identifying and encoding the segments of interactions in a conversational situation. These segments are identified by recognizing intentional and unintentional clues provided by people. People often explicitly record a certain point of a conversation in a notebook. Such a process is regarded as a rudimentary method of conversation quantization. I will discuss handwritten quanta in Sections 3, 4, and 5. In order to make such handwritten quanta more sophisticated, it is essential to utilize a medium that is richer than text and to employ methods that are more casual than handwriting. The interaction corpus project [4] is an approach for automatically capturing people's activities by using ubiquitous

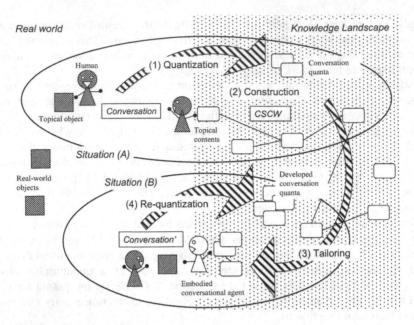

Fig. 1. Generalized quantization spiral

and wearable sensor technologies. The discussion mining project [5] is another approach that has not yet been fully automated; it attaches more importance to a person's intentional behavior. A point of conversation is annotated by the manipulation of colored discussion tags such as "comment," "question," and "answer." These tags are manipulated up and down by people to indicate their intentions in a formal discussion.

The problem with quantization is that the conversation models and their interpretations are not entirely clear in several day-to-day conversational situations. I will consider knowledge-oriented conversation in a room, the goal of which is to acquire knowledge, in Sections 6 and 7. The restrictions of such a conversation are more informal than that of a decision-making meeting. A sample situation is a meeting room where people share and create new knowledge. An exhibition is another situation where people acquire knowledge from various presenters.

The encoding of conversation quanta into knowledge representation is another problem. Although the encoded quanta should not lack conversational nuances, they should be computationally retrievable. The use of video with annotations is currently a popular encoding approach. Several types of annotations have been proposed in previous studies, such as time and place information, computationally recognized speech, identified speaker, interaction primitives, intention of speaker, and any other context such as manipulations of the equipment. A knowledge card that uses rich media with natural language annotations is another remarkable approach [6]. Rich media such as photo images or video clips can include extracts of conversational situations. It appears that reusing knowledge cards can be highly economical, although the text annotations will have to be written by hand. Text annotations are more likely to be retrieved as compared to rich media because of the recent evolution

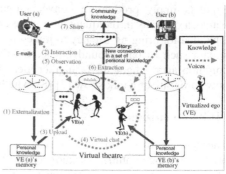

Fig. 2. Screenshot of EgoChat II   **Fig. 3.** Quantization spiral in EgoChat II

of text processing technologies. Text annotation also has several good conversational applications such as embodied conversational agents or communication robots. I will discuss knowledge cards in Sections 4, 5, and 6.

(2) Construction

CSCW and archive systems are required to build up the store of conversational knowledge. The amount of daily conversation quanta is expected to be large. The CSCW system enables us to collaboratively arrange large content.

A knowledge landscape is a virtual content space where people can store and manage conversation quanta. The use of a spatial world is a general approach for supporting group work and managing large content. People can look over content if they are arranged spatially. People can also grasp content locations by using several spatial clues such as direction, distance, and arrangement.

Long-term support requires the construction of conversational knowledge because people cannot talk constructively unless they have past knowledge. The knowledge landscape facilitates spatial and long-term quanta management. A zoomable globe of the knowledge landscape enables the accumulation of large amounts of quanta. Its panoramic content presentation enables people to edit and discuss the quanta according to conversational situations. The details of knowledge landscapes are presented in Section 6.

(3) Tailoring

The constructed quanta are reused and tailored to people's requirements by using conversational systems such as embodied conversational agents, communication robots, or conversation support systems. Conversation quanta are designed such that they can be applied to conversational situations that differ from the original situation from where they were quantized. The knowledge in a previous conversation is shared and reused by people collaborating with these conversational systems. A unique issue that concerns the use of conversation quanta is the coherency of quantized context. Rich media such as video includes several contexts, for example, the scenery of the conversation location, clothes of the speakers, camera angles, and so on. Various contexts imply productive conversational knowledge. The problem is that gaps between contexts in the quanta are large and it is quite difficult to modify rich media. One idea for bringing fluency to joined quanta is to combine them in another coherent stream such as background music. Combining them in a coherent talk by using a common agent character is another idea. This implies a TV style program such as a news program or a talk show [7] [8], which is fully discussed in Section 4.

(4) Re-quantization

The re-quantization of new conversations in the applied situations is an essential process for the evolution of knowledge. People's conversations are a developed form of past conversations that are derived by collaborating with conversational systems. Conversational knowledge grows with such a quantization spiral. Re-quantization involves the design of a system for the smooth coordination of conversational applications and the quantizing process. In fact, the spiral development of knowledge is generally observed in knowledge processes such as a conversational knowledge process [1] or the SECI model [9]. The front-end of the knowledge spiral is realized in the form of several conversational systems [2] [3] [6] [7] [9], while the back-end is realized in the form of knowledge landscape systems [10] [11]. The spiral reuse of conversation quanta by using knowledge landscape systems is empirically discussed in Section 7.

3 Virtualized Egos that Expand Conversation in a Community

An embodied conversational agent is a desirable interface for knowledge media because it can provide an automated method for knowledge circulation by employing a conversational style that is familiar to humans. An agent-mediated conversation system that uses a virtualized ego (VE) [2] [3] is the first implementation of the quantization spiral. The agent called VE is a person's other self; it works independent of the real self and can talk on the person's behalf. By using their VEs, people separated by a large spatiotemporal distance can converse with each other. An early study of VEs was carried out through EgoChat II, which is a virtual environment for conversation among humans and VEs [3]. With EgoChat II, people can exchange their experiences with each other even if one of them is absent because VEs can talk on behalf of the absentee.

A VE is one of the major application areas of a sustainable conversation system. In EgoChat II, the VE stores and uses microcontent that consists of one or two sentences. The microcontent is a small amount of content that is almost self-sufficient like an e-mail or a blog entry. In EgoChat II, a summary of e-mails is stored and used for generating a new conversation.

A screenshot of EgoChat II is shown in Fig. 2. A user is talking with three VEs, each of which has a 3-D head of another user. The quantization spiral in EgoChat II is shown in Fig. 3. The personal memory of a VE is represented by microcontents, that is, a set of summarized sentences from previous e-mails of the original person. One e-mail is summarized into one or two sentences that relate to a particular topic category and is labeled by a keyword ((1) Externalization).

When a user desires to retrieve the knowledge of other people, he/she communicates his/her interest to the corresponding VEs. The user's voice messages are segmented into topic words by a commercial speech recognition system, and a VE then answers the sentences related to the same topic as that of the VE ((2) Interaction and (3) Upload). The VEs attempt to chat by maintaining conversation-like coherence ((4) Virtual chat). The coherence of their chat is judged by using two knowledge representations. Topic-and-summaries representation is a set of topics related to summaries. The VEs chat on one topic for a certain period of time. Flow-of-topics representation is the associative knowledge of the original person. The VEs sometimes change a topic by using this associative knowledge. By observing a virtualized chat among the VEs ((5) Observation), users can find and share knowledge ((7) Share) that is extracted from the newly connected personal knowledge ((6) Extraction).

The personal memories of four VEs were generated experimentally from a chat about liquor on a mailing list, in which participants exchanged ideas on how to market brandy. When "brandy" is the topic of interest for a user, the conversation among VE-(a), VE-(b), VE-(c), and VE-(d) progresses as follows[1]:

1: VE-(a): Speaking of brandy, generally, people don't particularly like brandy.
2: VE-(b): I dislike the alcohol that was tasted.
3: VE-(a): I suppose you dislike drinking just a little at a time.
4: VE-(c): Brandy tastes bitter.
5: VE-(d): I used to drink V.S.O.P.
6: VE-(b): The ordinary way of drinking brandy is straight, isn't it?
7: VE-(d): Wow, is straight the ordinary way?
. . .

Each part of the message is connected to the subsequent part by a keyword, if possible. VE-(a) generates line 3 after line 2 because the sentence, "I suppose you dislike drinking just a little at a time," is labeled by the same keyword "dislike" as that of line 2. Both lines 6 and 7, which are labeled by "straight," are generated in the same way. EgoChat II suggested that exchanging parts of past conversations may be a kind of new conversation.

The conversation generation on EgoChat II is certainly not powerful enough to provide concise responses to users. On the other hand, EgoChat II can easily store and process a significant number of messages because the representation of the microcontent and the methods of ordering messages are very simple.

4 Generation of Conversational TV Representation from the Text Contents

A sustainable conversation implies that people often simply listen to a conversation in the long-term view. People may sometimes desire to have an interactive conversation with questions and answers, while they may sometimes choose to passively listen to others' conversations. In order to continuously provide content, certain techniques for tailoring conversations to passive users are required. Unlike questions and answers for

---

[1] Utterances were originally written in Japanese.

a few minutes, a POC caster [7] [8] aims to generate a long-term conversation that does not bore people. TV presentations are used in a POC caster for tailoring microcontents to passive users. TV is an easy conversational presentation that is tailored to couch potatoes. The microcontent described in Section 3 is text-styled content and is not suited to couch potatoes; however, I believe that I can change it into conversation-styled content by dividing sentences and adding a comment.

A POC caster is a broadcasting agent system that uses certain methods to transform a text form into a conversation form. The POC caster is based on the public opinion channel (POC), which is an interactive broadcasting system that supports community knowledge creation [12]. In a POC, the community members evolve their knowledge by exchanging POC cards. A POC card comprises one title line and one paragraph of sentences, similar to a general BBS. One or two sentences in the microcontent are considered too short to describe a detailed presentation. Therefore, the POC card contains more information.

The POC caster plays the role of a broadcaster in the POC that introduces POC cards through a conversation. The process used for conversation generation divides the original text into two parts by using a period and adds a comment to a text. The POC caster comprises a main caster agent and an announcer agent. They provide the following information: (1) the main caster reads out the title, and this allows users to obtain a summary of the topic, (2) the announcer introduces the first few parts of the original text, (3) the main caster comments on this text based on certain rules, and (4) the announcer proceeds to the latter parts of the original text. Three generation rules that are applied to the original text are listed in Table 1. Rule (1) is applied when the original text can be divided into the former introductory part and the latter detailed part by using a period. Rule (1) inserts a comment between the former and the latter sentences that make a listener expect a detailed continuation by presenting contextual information. First, (1) the announcer agent (Agent I) speaks the former introductory part, (2) the main caster (Agent II) randomly inserts a comment from the list in Table 1, and (3) the announcer agent speaks the latter detailed part (as shown by the processes in Fig. 4). Similarly, Rule (2) makes the listener pay attention to the topic by repeating questions from the former parts of the original sentences. Rule (3) is applied when the original text comprises more than three sentences. Rule (3) divides the original long text into two short parts by using a period. Rule (3) gives a listener time to understand long text.

Two experiments were conducted to evaluate the transformation rules of the POC caster [8]. The results of the validity of the rules are summarized as follows. The conversation generated by the transformation rules promotes a more effective comprehension of the content in the case of longer sentences than in the case of shorter ones; inserting words having rich context information has a stronger effect on the participants' comprehension than simple responses. In addition, the results of the relationship between a participant's knowledge level and conversational form are summarized as follows. (1) The conversational form had a beneficial effect when the participants were knowledgeable about the topic; however, it had no effect when the participants had little knowledge of the topic. (2) In the case of the conversational form, participants with no knowledge of the topic were more likely to find the sentence easier to understand as compared to the knowledgeable participants. (3) The response time was shorter in the conversational form when the participants had little knowledge of the topic. These results indicate that the effect of the conversational form depends on the user's relevant knowledge.

Table 1. The POC caster transformation rules[2] [8]

Rule 1: Presentation of the context	**Objective:** To facilitate the listener's understanding of the context by inserting words **End of sentence:** Words providing context **Processing:** Inserting words after the sentence to provide information **Inserted words:** "What is that?" or "Give me more details"
Rule 2: Repetition of question	**Objective:** To promote the listener's understanding by repeating previously asked questions **End of sentence:** Question **Processing:** Inserting words after the sentence represents asking a question **Inserted words:** "What do you think about that?"
Rule 3: Simple response	**Objective:** Allowing the listener time to understand by inserting a simple response **End of sentence:** No use **Processing:** Inserting simple responsive words after the second sentence **Inserted words:** "Yes" or "Uh-huh"

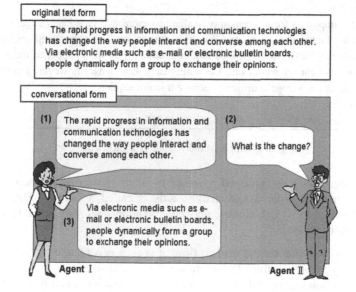

original text form

The rapid progress in information and communication technologies has changed the way people interact and converse among each other. Via electronic media such as e-mail or electronic bulletin boards, people dynamically form a group to exchange their opinions.

conversational form

(1) The rapid progress in information and communication technologies has changed the way people interact and converse among each other.

(2) What is the change?

(3) Via electronic media such as e-mail or electronic bulletin boards, people dynamically form a group to exchange their opinions.

Agent I   Agent II

Fig. 4. Transformation from a text form to a conversation

---

[2] Utterances are originally written in Japanese.

5 Conversational Agent Using Knowledge Card and Utterance Pair

In order to facilitate the processes in the conversation spiral by using VEs, I have expanded the microcontent of EgoChat II (Section 3) and the POC card of the POC caster (Section 4) to a knowledge card. A VE can tell a longer story than EgoChat II by using one story that comprises multiple knowledge cards (the original concept of such a story is proposed in [12]). A VE can also speak with people in a question and answer style by using an utterance pair that comprises a set of knowledge cards. The length of the sentences on a knowledge card is not rigorously defined; rather, it is determined by the amount the author chooses to semantically bind. Based on experience, I might expect this amount, which is equivalent to that required to briefly express a single topic, to be roughly equivalent to a paragraph of text, an OHP slide, or an e-mail message.

Fig. 5 shows an example of the format of a knowledge card that deals with sharing tacit community knowledge. The card comprises the following three elements. The <title> element is a line of text describing the topic of the card. The <body> element consists of a few sentences that present the text information. The <image> element indicates the location of the picture related to the card.

```
<opinion>
<title>Sharing tacit community knowledge</title>
<body>
    It is important for a community to share its members' tacit knowledge, which is
    one's subjective view and intuition based on one's personal experience.
    Informal messages exchanged in a community, i.e., what happened today, what
    one thought, or whom one met imply tacit knowledge; however, sharing this
    type of knowledge is difficult because it is not explicit.
</body>
<image> http://xxx.yyy.zzz/community.jpg</image>
```

Fig. 5. Example of the format of a knowledge card

Knowledge cards are grouped together by their creator and ordered within the group. A series of ordered knowledge cards is called a "story." Three classes of stories—basic stories, QA stories, and conversation logs—are used to express the following information.

- Basic story: expresses the general flow of speech generated by a VE
- QA story: expresses the VE's utterance pair knowledge
- Conversation log: records a conversation between a VE and a user

The knowledge described by the utterance pair appears to be a minimal form of conversational knowledge. An utterance pair implies not only a question and its correct answer but also a dialogue that includes suggestive, characteristic, and comedic content that may not always be correct.

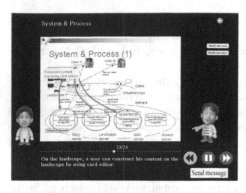

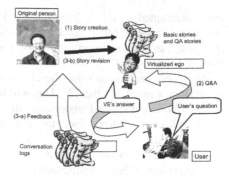

Fig. 6. Screenshot of EgoChat III   **Fig. 7.** Lifecycle of knowledge cards

EgoChat III is a system that supports a quantization spiral by using knowledge cards [6]. Fig. 6 illustrates the screenshot of EgoChat III and Fig. 7 illustrates the life cycle of knowledge cards in a conversation mediated by VEs. EgoChat III facilitates the life cycle of knowledge cards by using an utterance pair as follows:

Step 1: Story creation
The creator of a VE uses the knowledge-card editor to create a basic story that relates to his interests, which is stored as conversational content for the VE. A basic story comprises N knowledge cards in order; a conversation begins with card 1 and proceeds linearly toward card N. In cases where the VE's creator wants to use the VE as a stand-in to respond to questions, the creator generates a QA story to store the conversational content. In a QA story, one knowledge card provides a question (the question card), followed by any number of responding knowledge cards (answer cards). The answer cards are ordered in the same manner as basic story cards and taken together they constitute the answer to the question.

Step 2: Q&A
A user can interrupt an avatar in midstory to ask a question. The chat client sends the question to the QA module and awaits a response. The QA module follows the method proposed by Kiyota et al. [13] to process the questions and answers precisely; it parses the user's question to find the question card with the nearest syntax tree and then looks that up from the collection of QA stories stored in the VE. Syntax-tree similarity is calculated based on the agreement of unique words and syntax-tree dependency relationships. The QA module then sends the retrieved question card and its related answer cards to the chat client. The chat client displays the questioner-agent (shown on the left of Fig. 6 as a female figure) and reads the question text aloud. After reading the question, the VE reads the answer cards and returns to the original basic story. When a single basic story is completed, the VE begins reading the next basic story in the list, although the user can select a different basic story at any point.

Step 3: Feedback and story revision
The VE records conversation logs between itself and users in a form that only the VE's creator can view. Conversation logs are stories consisting of the story that the

user interrupted to ask a question, the knowledge card corresponding to the user's question, and the VE's response cards. The conversation log inserts the user's question card at the interruption point in the basic story, followed by the related answer cards.

The VE's creator can use the chat client to replay the conversation between the questioner's agent and the VE. This will help the creator grasp the context and content of the Q&A session and understand the questioner's reaction to the basic story and Q&A session. Further, since the conversation logs can be directly edited in the knowledge-card editor, the editor can add new QA stories for questions that could not be answered previously. If the creator's expectation of the user's level of understanding differs significantly, the same Q&A session may be repeated many times, and the conversation log will clarify these explanatory shortcomings in the basic story.

I predicted that the example-based approach, which uses knowledge cards to generate conversation content for VEs, would result in more predictable conversations between VEs and users as compared to the rule-based methods and would be easier to create. Over a three-month period from May 7, 2002 to August 9, 2002, a professor created a VE as part of his "Knowledge Communication" coursework in Information Engineering at the University of Tokyo Graduate School. The VE included a total of 274 knowledge cards organized into 27 basic stories on the subject of knowledge communication, which he presented to his students during the semester. The other students prepared a report on agents and workflow models; 58 students created approximately 1000 knowledge cards organized into 140 basic stories, which were submitted as reports. When a card was submitted, it was published on Web for all the class members, and each class member could refer to the cards created by others when adding to or editing their own reports so that a comparative study of one's own report with those of the others could help produce an even more insightful report. This example demonstrates that creating knowledge cards is simple and that they can be used for cross-referencing knowledge.

The VEs discussed in this paper are not capable of advanced conversations; however, the inclusion of the conversation log, which makes Q&A sessions between the VE and user highly reusable, can assist the VE creator. Since the conversation log can bring specific user questions to the attention of the creator, the creator can write stories that address those questions more easily. Since the creator is prompted to write cards, QA stories will accumulate, and the example-based QA module will yield good search results.

6 Sustainable Management of Conversation Quanta

I have discussed the concept of landscape in a quantization spiral. When people sustainably store conversational content, they require an archiving system for managing the content, arranging the content, reducing the cost of creating content, increasing efficiency by cooperating with CSCW systems, and searching for and displaying contents casually in a conversational situation. Utilizing space is generally a good idea for managing a large amount of content. I therefore propose a spatial system called sustainable knowledge globe (SKG) [10] [11].

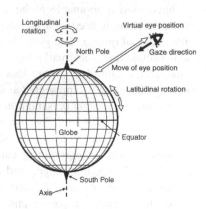

Fig. 8. Screenshot of SKG

Fig. 9. Zoomable globe

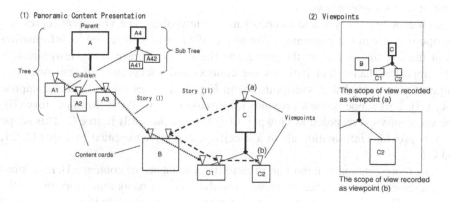

Fig. 10. Panoramic content presentation

6.1   Sustainable Knowledge Globe

A sustainable knowledge globe (SKG) is a system for managing conversational content. Fig. 8 shows a screenshot of SKG. The shape of SKG's landscape is similar to that of a terrestrial globe. Each content item is represented as a content card and laid on the global surface. Here, a user can explore and construct his/her content anywhere on the surface by rotating and zooming the globe.

The content card is an extension of the knowledge card. A knowledge card allows only a picture file, whereas a content card allows a document file to be more expressive. The simplest method for creating a new content card is to use a card editor. It is a built-in tool for writing and drawing, and it saves an embedded file in the PNG or RTF format. In addition, an SKG provides some methods for importing existing pictures, movies, and documents on a PC to content cards.

I have used a zoomable globe to manage a large archive; in my approach, the zoomable globe is used as a memory space to facilitate an arbitrary arrangement of conversational quanta. The globe is a sphere such as a terrestrial globe with latitude and longitude lines, landmarks for the north and south poles, and the equator (Fig. 9). It also has a zoomable feature that allows multiscale representation like electronic atlases. A quantum is represented as a rectangle on the globe.

A panoramic content presentation is a topological model of content arrangement in the knowledge landscape. The panorama includes tree-structured content cards and stories, each of which consists of a sequence of viewpoints (Fig. 10). A tree structure is a standard way of representing categories. By categorizing the content cards, a set of cards can be easily arranged and retrieved. Cards A1, A2, A3, and A4 in Fig. 10 are the children of card A, and A4 represents a subcategory of card A. The relationship between a parent and a child is drawn by an edge. The knowledge landscape allows multiple trees on a virtual globe because the arrangement is limited if it allows only one tree. An independent card that does not belong to a tree is also allowed for the same reason.

Each content card is bound to one or more viewpoints. The viewpoint represents a scope of view in the panorama. The scope of view is represented as information about the eye distance from the globe and the center of gaze. The viewpoint is a very important factor that includes the context and background of the card. The scope of view recorded as viewpoint (a) in Fig. 10 focuses the cards that compose story (II). This scope of view provides information about all the flows of story (II). The scope of view recorded as viewpoint (b) focuses the cards in tree C. This scope of view provides information about a specific category represented by cards C, C1, and C2.

Story representation is important to describe a sequence of contents. Here, a story is represented by a sequence of viewpoints that show a panoramic storyline of the contents. The relationship between a previous viewpoint and the subsequent viewpoint is drawn by an arrow that cuts across the trees. In Fig. 10, the sequence from A1 to E constitutes story (I). The sequence from C to E constitutes story (II). A content card is allowed to belong to multiple stories. The story is so orderly and cross-boundary that people can easily create cross-contextual contents.

Contour map representation [14] has also been proposed as an extension of panoramic content presentation. This is a method of visualizing tree data structures arranged on a plane. The graphical representation of a spatially arranged tree becomes more complex as the number of nodes increases. The tree nodes get entangled in a web of trees where several branches cross each other. Contour map representation simplifies such complex landscapes by using contour lines. The contour lines show the arrangement of the peaks; moreover, the enclosures can be regarded as hierarchical structures. For people, a contour map has a familiar design that can be memorized. The space efficiency of contour map representation is relatively high because the shapes of the contour lines are freer than the shapes of primitives like circles or rectangles. A 3D view of the contour map representation of SKG is shown in Fig. 11.

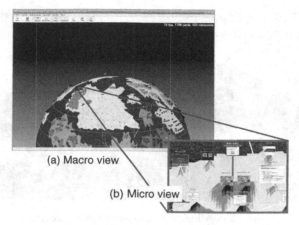

(a) Macro view

(b) Micro view

Fig. 11. Contour map representation

6.2 Experiment on Creating Proceedings

The meeting proceedings of a working group were experimentally created by using a SKG. The participants discussed the study of an intelligent room. The virtual landscape of SKG was projected on a large screen during the meetings. One operator imported handouts, noted down the speeches, and located the transcripts on the surface of the globe. Speeches on the same topic were grouped into one or two cards; these cards were then grouped into a tree-like structure that was sorted on the basis of their dates. The operator also manipulated the globe according to the requests from participants to focus on specific content.

Ten meetings were held from August to November in 2004. The total duration of the meetings was 20 h. The average number of participants was six. From the experiment, we have acquired 151 contents that include 12 trees and 3 stories.

A summary story of 10 meetings was created. The viewpoints in the story show the group of monthly proceedings and significant topics. The storyline is shown in Fig. 12 (a). Viewpoints 1 to 9 are connected, and the synopsis of the viewpoints and annotations is as follows:

Viewpoint	Scope of view	Annotation
1	A close view of the card that describes the title of the story	I would like to talk to you about a meeting supported by SKG.
2	A distant view of the entire proceedings	151 contents have been created in 3 months on SKG.
3	A distant view of the proceedings in August	These are the proceedings in August.
4	A close view of the cards that describe the main topics in the proceedings in August	We discussed the methods for evaluating knowledge productivity in a meeting.
⋮	⋮	⋮

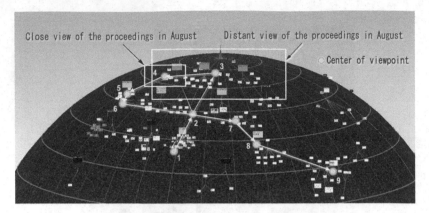

Close view of the proceedings in August   Distant view of the proceedings in August

Center of viewpoint

(a)   A storyline about a summary of proceedings
(the number shows the order of the story)

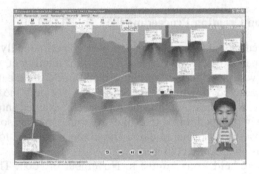

(b) A presentation agent who talks about a scope of view on SKG

Fig. 12. Example of the story

The walking path of the presentation agent is shown in Fig. 12 (a). The walk from viewpoints 1 to 2 indicates a transition from a focused view to an overview. The walk from viewpoints 2 to 3 indicates a transition from the overview to a focused view. The same transition is observed from viewpoints 3 to 4. These fluctuating transitions lend a panoramic aspect to the entire presentation. The speaking manner of the presentation agent is shown in Fig. 12 (b). The total playtime of the presentation is approximately 2 min. The presentation agent can summarize the presentation in order to provide a concise overview.

The participants were interviewed about the effectiveness of SKG and positive comments were received with regard to the good overview and the reminder of the meetings. One participant commented that it is difficult for the speaker to operate SKG by using a mouse. The user should not be occupied with operating the mouse in conversational situations because it impedes communication using natural gestures. Nomura and Kubota et al. [11] are now developing a novel immersive browser that can improve the operativity of SKG by using physical interfaces such as a motion capturing system in the surrounding information space.

7 Reusing Video Conversation Quanta

The implementation of conversation quantization depends on the data structure used for representing conversation quanta. Plain video clips can be used for representation, but the efficiency of retrieving and processing would be rather limited and large expenses would be involved in retrieving, editing, and applying the conversation quanta. Alternatively, a deep semantic representation using logical formulas or case frames would not be ideal due to the expenses involved and the limited capability of representing nonverbal information. The use of annotated videos and images to represent a conversation quantum appears to be an acceptable implementation.

Saito and Kubota et al. [15] targeted a conversation using slides. They simulated conversation quantization by hand to investigate the nature of conversation quanta in a real situation. They gave shape to the concept of conversation quanta as follows:

1. By setting up a practical conversational situation
2. By capturing conversation using a video camera
3. By extracting conversation quanta from the video stream by hand

The nature of conversation quanta was empirically analyzed by extracting the conversation quanta from captured videos and creating new conversational contents using these conversation quanta. The video consists of 3 meetings between subject A and subject B. The subjects were aware that their conversations would later be utilized as conversation quanta. Each meeting was held in a different place and at different times. Each subject interacted using PowerPoint slides on a mobile PC (with a web camera and a microphone) to capture the voices, faces, and context. As a result of these meetings, three-and-a-half-hour-long videos of subjects A and B were obtained. In their conversations, the presentation style and discussion style were both given equal importance, and the topics were related to conversation quantization—its history, problems, approaches, systems, and so on.

In this study, a VE is assumed to be created by quantizing the video and mixing the quanta. To confirm this assumption, an experiment was conducted in which the conversation quanta were extracted arbitrarily from these videos and a new presentation video in which the VEs of the participants talk about their study was created.

The first approximate model has been proposed for extracting conversation quanta. First, the video is divided at the point of transition of the slides (Division 1 in Fig. 13) because the speeches seem to be almost coherent and united in a slide.

The second division point is the start of a dialogue (Division 2 in Fig. 13). A set of dialogues and a single speech unit are distinguished because conversation quantum is rather a holistic unit than a reductionistic unit, typical examples of the latter being morphemes, words, or sentences..A set of dialogues is expected to include a unique conversational scene, for example, a question and answer with a rich context, comic dialogue, or any other inter and intrapersonal synchronization of speeches and gestures.

Using this model, 41 quanta for subject A and 66 quanta for subject B were extracted from Video A. Table 2 shows the number of conversation quanta in the archive of each subject. "Single speech" implies that the quanta include only the subject, while "Dialogue" implies that the quanta include two subjects. As a general

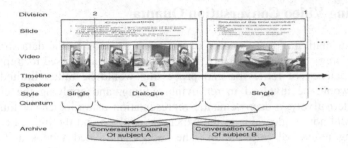

Fig. 13. The first approximate model for extracting conversation quanta [15]

Table 2. Conversation quanta from three-and-a-half-hour videos [15]

Subject	Single speech (total time)	Dialogue (total time)
A	24 quanta (16 min)	17 quanta (21 min)
B	49 quanta (35 min)	17 quanta (21 min)

rule, a conversation quantum is stored in the archive of only one speaker. It is stored in the archive of every speaker only when the quantum is in a dialogue style. The number of dialogue quanta of subject A is identical to that of subject B.

A VE system has been simulated based on these conversation quanta. Conversation quanta of subject A are arranged on the assumption that the system talks with a user on behalf of subject A. Fig. 14 provides an overview of the simulation. The user first stands in front of a system screen on which the face of subject A is displayed. The system then begins to talk on behalf of subject A when the user asks for his interest ("Greeting"). The system talks by arranging past conversation quanta that are related to the interest of the user ("Quantum1" and "Quantum2"). While the system is talking, the user can ask any question ("Question"). The system then answers the question by searching an answering conversation quantum ("Quantum 3"), and continues talking ("Quantum4").

I have obtained some valuable suggestions about conversation quanta from the simulation and analysis above. First, conversation quanta that depended on context are reusable in situations wherein a user is familiar with the original situations of the conversation quanta. In Fig. 14, Quantum1, Quantum2, and Quantum4 were acquired in different rooms. Thus, new conversational content can be created from the past conversation quanta that were obtained in different situations. When searching conversation quanta that are suitable for a user, thinking about the background knowledge of the user is very important. The conversation in Fig. 14 was fragmented; however, the user managed to continue the conversation because he is a colleague of the speakers on the screen. Second, a dialogue style quantum that contains a rhythmic conversation and jokes is interesting. They have good points of conversation such as conversational rhythms, presence, and dynamics. In addition, from the viewpoint of a VE, a quantum that contains individual experiences and know-how is also interesting.

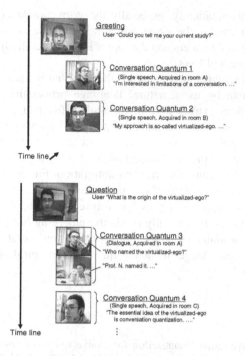

Fig. 14. Simulation of conversational video content using conversation quanta

8 Future Work

There are many interesting works on conversation quantization. Conversation quanta can be extracted from real-world conversations by expanding the ubiquitous sensor room proposed by Sumi et al [4]. A sustainable knowledge globe has been developed for the spatiotemporal management of conversation quanta [10] [11]. A virtualized ego (VE) [6] that can talk on behalf of an actual person (as mentioned in Sections 3 and 5) is a good utilization of conversation quanta. A video and sound collage system of one's experience [4] would also constitute another good application. Minoh and Kakusho [16] developed a robust method for automatically recognizing communicative events in a real classroom by integrating audio and visual information processing. Rutkowski [17] studied a method for monitoring and estimating the efficiency of interpersonal communication based on the recorded audio and video content by computing the correlation between the activities of the sender and responses of the receiver.

Studies on extraction of topics from a conversation are increasing. A greater understanding of discourse structure is indispensable to extract the essence of conversation automatically. Shibata et al. [18] studied the discourse analysis of a cooking video by using linguistic and visual information. In this research, I have not considered a cooking program controlled by a video director, but casual conversations that arise in any situations; therefore, it is extremely difficult to understand the discourse structure automatically, especially the correspondence structure, when an indication word is omitted in the video. My aim is to make conversation quantization feasible by helping people understand the topics in the video in the conversation spiral such as the Q&A system of EgoChat.

The annotation of conversation quanta is the text that is mainly used for retrieving and combining quanta by using natural language processing. The annotation can constitute the sentences extracted from the slides, the text generated by a speech recognition system, and notes annotated by people. I plan to expand the manner in which a VE in the EgoChat system uses the annotation to generate conversation. In this work, the VE answers the user's question by displaying a slide with a synthesized voice that is retrieved by searching the text annotation. Future systems will show the video, slide, and human voice instead of simply the slide.

There are many interesting topics for future research. Capturing and mining conversational experience in our daily life is the emerging hot topic that is related to lifelogs. A sustainable daily system of the quantization spiral is needed to acquire additional empirical knowledge about the quantization spiral and develop it into a more detailed theory.

References

1. Nishida, T.: Conversation Quantization for Conversational Knowledge Process, Special Invited Talk, S. In: Bhalla, S. (ed.) DNIS 2005. LNCS, vol. 3433, pp. 15–33. Springer, Heidelberg (2005)
2. Hirata, T., Kubota, H., Nishida, T.: Talking Virtualized Egos for Dynamic Knowledge Interaction. In: Dynamic Knowledge Interaction, 6th edn., pp. 183–222. CRC Press, Boca Raton, USA (2000)
3. Kubota, H., Nishida, T., Koda, T.: Exchanging Tacit Community Knowledge by Talking-virtualized-egos. In: Proceedings of the Fourth International Conference on Autonomous Agents (Agents 2000), pp. 285–292 (2000)
4. Sumi, Y., Mase, K., Mueller, C., Iwasawa, S., Ito, S., Takahashi, M., Kumagai, K., Otaka, Y.: Collage of Video and Sound for Raising the Awareness of Situated Conversations, In: Proceedings of International Workshop on Intelligent Media Technology for Communicative Intelligence (IMTCI (2004), 167–172 (2004)
5. Nagao, K., Kaji, K., Yamamoto, D., Tomobe, H.: Discussion Mining: Annotation-based Knowledge Discovery from Real World Activities, In: Proceedings of the Fifth Pacific-Rim Conference on Multimedia (PCM 2004) (2004)
6. Kubota, H., Kurohashi, S., Nishida, T.: Virtualized Egos using Knowledge Cards. Electronics and Communications in Japan (Part II: Electronics) 88(1), 32–39 (2005)
7. Kubota, H., Yamashita, K., Nishida, T.: Conversational Contents Making a Comment Automatically, In: Proceedings of KES 2002 (2002), pp: 1326–1330.
8. Yamashita, K., Kubota, H., Nishida, T.: Designing Conversational Agents: Effect of Conversational Form on our Comprehension, AI & Society, Vol.20, No.2, 125–137 (2006)
9. Kumagai, K., Sumi, Y., Mase, K., Nishida, T.: Detecting Microstructures of Conversations by Using Physical References: Case Study of Poster Presentations. In: Washio, T., Sakurai, A., Nakajima, K., Takeda, H., Tojo, S., Yokoo, M. (eds.) JSAI Workshop 2006. LNCS (LNAI), vol. 4012, pp. 377–388. Springer, Heidelberg (2006)

10. Kubota, H., Sumi, Y., Nishida, T.: Sustainable Knowledge Globe: A System for Supporting Content-oriented Conversation. In: Proceedings of AISB, Symposium Conversational Informatics for Supporting Social Intelligence & Interaction (2005), pp. 80–86 (2005)
11. Kubota, H., Nomura, S., Sumi, Y., Nishida, T.: Sustainable Memory System Using Global and Conical Spaces, Special Issue on Communicative Intelligence in Journal of Universal Computer Science, 135–148 (2007)
12. Azechi, S., Fujihara, N., Kaoru, S., Hirata, T., Yano, H., Nishida, T.: Public Opinion Channel: A Challenge for Interactive Community Broadcasting. In: Ishida, T., Isbister, K. (eds.) Digital Cities. LNCS, vol. 1765, pp. 427–441. Springer, Heidelberg (2000)
13. Kiyota, Y., Kurohashi, S., Kido, F.: Dialog Navigator: A Question-answering System based on a Large Text Knowledge Base, In: Proceedings of the 19th International Conference on Computational Linguistics (COLING 2002) (2002), pp. 460–466.
14. Kubota, H., Sumi, Y., Nishida, T.: Visualization of Contents Archive by Contour Map Representation. In: Washio, T., Satoh, K., Takeda, H., Inokuchi, A. (eds.) JSAI 2006. LNCS (LNAI), vol. 4384, Springer, Heidelberg (2007)
15. Saito, K., Kubota, H., Sumi, Y., Nishida, T.: Support for Content Creation Using Conversation Quanta. In: Washio, T., Sakurai, A., Nakajima, K., Takeda, H., Tojo, S., Yokoo, M. (eds.) New Frontiers in Artificial Intelligence. LNCS (LNAI), vol. 4012, pp. 29–40. Springer, Heidelberg (2006)
16. Minoh, M., Nishiguchi, S.: Environmental Media - In the Case of Lecture Archiving System. In: Palade, V., Howlett, R.J., Jain, L. (eds.) KES 2003. LNCS, vol. 2773, pp. 1070–1076. Springer, Heidelberg (2003)
17. Rutkowski, T.M., Seki, S., Yamakata, Y., Kakusho, K., Minoh, M.: Toward the Human Communication Efficiency Monitoring from Captured Audio and Video Media in Real Environment. In: Palade, V., Howlett, R.J., Jain, L. (eds.) KES 2003. LNCS, vol. 2773, pp. 1093–1100. Springer, Heidelberg (2003)
18. Shibata, T., Kawahara, D., Okamoto, M., Kurohashi, S., Nishida, T.: Structural Analysis of Instruction Utterances. In: Palade, V., Howlett, R.J., Jain, L. (eds.) KES 2003. LNCS, vol. 2774, pp. 1054–1061. Springer, Heidelberg (2003)

Understanding the Dynamics of Crop Problems by Analyzing Farm Advisory Data in eSagu^{TM}

R. Uday Kiran and P. Krishna Reddy

Media Lab Asia Project,
Center for Data Engineering,
International Institute of Information Technology (IIIT-H), Hyderabad, India
uday_rage@research.iiit.ac.in, pkreddy@iiit.ac.in

Abstract. Developing personalized information services for the effective delivery of functional information by exploiting latest advances in information and communication technologies is one of the active research area. At IIIT-H, Hyderabad, (India), efforts are being made to design a personalized agricultural advisory system called eSagu^{TM}. In eSagu system, an agricultural expert give an expert advice to the farms based on the crop photographs and other related information. During 2004-05, the eSagu system was operated for 1051 cotton farms. Every farm received the expert advice once in a week. As a result, the data set of about 20,000 such advice texts has been generated. In this paper, we have carried out the text analysis experiments on the data set to understand the dynamics of farm problems. The graph of cluster size versus number of clusters on the advices of a particular day resulted into an exponential curve. In addition, it was observed that the group of farms which have received the same advice in particular week were not receiving the same advice in the subsequent weeks. The analysis shows that farming community needs a personalized advisory service in a regular manner as crop problems are dynamic and influenced by multiple factors. At the same time, there is a scope for improving the performance of eSagu by exploiting common problems of crops.

Keywords: Information dissemination, personalized agricultural service, personal information service, personalization agricultural extension, recommendation systems, IT for Agriculture.

1 Introduction

Print media such as news papers, magazines, and books, and communication media like radio, television and telephone are being used widely to disseminate information or knowledge to masses. Normally, these methods disseminate information in an ad-hoc and generalized manner. Currently, research is going on to investigate improved information dissemination methods by exploiting the recent developments in information and communication technologies (ICTs). Efforts are being made to disseminate functional information through search engines, web sites/portals, question/answering systems, publish/subscribe systems, topic directories, discussion forums, information blogs and call centers. In the literature,

S. Bhalla (Ed.): DNIS 2007, LNCS 4777, pp. 272–284, 2007.

it was proposed that developing information systems to deliver personalized information service to each individual is one of the problem areas [1]. Progress in database, data warehousing, data mining [3], mobile, and internet technologies are enabling mass customization and personalized information services [2] .

We are making an effort to design a personalized agricultural advisory system to improve the utilization and performance of agricultural technology to improve the productivity of Indian farmers[1]. In the field of agriculture, agriculture extension wing deals with the dissemination of both advanced agriculture technologies and experts' advices to the farming community. It is often claimed that, the knowledge delivered through agricultural extension is the cheapest input in bringing a noticeable increase in agricultural output through judicious use of inputs, cost minimization and sustainability. Efforts are being made to reach farmers through gatherings, news papers, magazines, journals, seminars, broadcast media and Web sites. However, such methods are not meeting the expectations of the farmers due to the lack of coverage, accountability, timeliness and personalization [4]. The traditional system does not consider the cases at the individual farmer's level as each farmer needs a distinct guidance for each crop which he/she cultivates.

By extending the developments in information and communication technologies to agriculture, we are making an effort to develop a personalized agricultural advisory system called $eSagu^{TM}$ [2](The word "Sagu" means cultivation in Telugu language.) [5] [6] [7] . The eSagu system aims at providing agricultural expert advices to the farmers in a timely and personalized manner. In eSagu, the agricultural experts generate the advice by using the latest information about the crop situation received in the form of both digital photographs[3] and text. The expert advice is delivered to each farm[4] on a regular basis (typically once in a week/two weeks depending on the type of crop) from sowing stage to the harvesting stage.

During 2004-05, the eSagu prototype was designed and the expert advices were delivered to 1051 cotton farms which belongs to the same area. Each farm received the expert advice once in a week. As a result, about 20,000 such advice texts were gathered. In this paper we have analyzed the data set through the text analysis methods. The clusters obtained on advice data set has resulted into an exponential curve. It was also observed that the group of farms which received the same advice in one week are not receiving the same advice in the subsequent weeks. The results indicate that the crop problems are dynamic and influenced by multiple factors.

Traditionally, information is disseminated to farming community in an ad-hoc and generalized manner by using print and communication media technologies.

---

[1] The proposed system has been developed by considering the agriculture situation in the state of Andhra Pradesh, India. Typically the farm size is about 1.5 hectares.
[2] eSagu is a trademark of IIIT-H, Hyderabad and Media Lab Asia.
[3] Here, the photograph is taken by visiting the farmer's field.
[4] The word "farm" means a piece of land in which a crop is cultivated. In this paper the words "farm" and "crop" are used interchangeably.

The eSagu system differs from other methods, as it delivered personalized agricultural expert advice to each farm once in a week. So the results show that the farm problems are influenced by several factors and it is necessary to complement the traditional methods with the personalized agricultural advice service in a regular manner for an efficient farming.

In the next section, we briefly explain about eSagu. In section 3, we describe the data set and explain the preprocessing steps. In section 4, we present the experimental results. The last section consists of summary and conclusions.

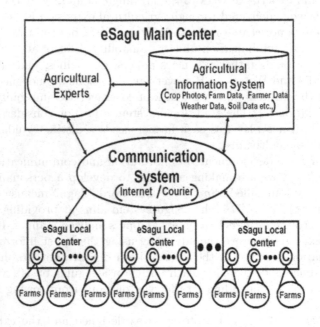

Fig. 1. The Architecture of eSagu System. The letter 'C' in the eSagu Local Center indicates Coordinator.

2   Overview of eSagu Prototype

In this section, we briefly explain about eSagu system, its operation, prototype implementation details and impact results.

2.1   The eSagu System

In eSagu, rather than visiting the crop in person, the agricultural scientist delivers the expert advice by getting the crop status in the form of digital photographs and other related information. The following are the parts of eSagu (Figure 1): Farms, eSagu local center, coordinators, eSagu main center, agricultural experts, agriculturalinformation system and communication system. Farms are owned by farmers, who are the end users of the systems.[5] One eSagu local center is established for about 10 to 20 villages. It is equipped with few computers, printer and dial-up internet connection and managed by a computer operator. The coordinator are literate having farming experience. In eSagu main center, a team of agricultural experts interact with agriculture information system to deliver the expert advice. Agricultural experts possess a university degree in agriculture and are qualified to provide the expert advice. Agricultural information system is a computer information system which contains the crop observation photographs and the related text. It also contains farmer registration data, farm registration data and weather data. Communication System is a mechanism to transmit the farm observation data (photographs and text) to the agricultural experts and the corresponding expert advice from the eSagu main center to the eSagu local centers. Transmission of digital photographs from the field to the eSagu main center requires a considerable bandwidth. If enough bandwidth is unavailable, information can be written onto compact disks and sent by a courier service. However, the expert advice (which is a text) is transmitted from the eSagu main center to the eSagu local center through a dial-up internet facility.

2.2   Operation of eSagu

The operation of eSagu is as follows. A team of agriculture experts [6] work at the eSagu main center (normally in a city or a university) supported by agricultural information system. One eSagu local center (few computers and one computer operator) is established for about ten villages. Educated and experienced farmers (local residents) are selected as coordinators. Depending on the crop, each coordinator is assigned with a fixed number of farms. The coordinator collects the registration details of the farms under him including soil data, water resources, and capital availability and sends the information to the eSagu main center. Once in a week, the coordinator visits each farm and takes four to five digital photographs depending on each farm condition. Later the accumulated data concerned to each farm condition and other information like weather details etc, are burnt on to a compact disk and delivered to the eSagu main center by a regular courier service. At the eSagu main center, the agricultural experts will analyze the crop situation with respect to soil, weather and other agronomic practices and prepares a farm specific advice, which will be stored back in agricultural information system. At the eSagu local center, the advice is downloaded electronically through a dial-up internet connection. The coordinator delivers the printed advices to the concerned farmer. In this way each farm gets the proactive advice at regular intervals starting from pre-sowing operations to post-harvest precautions.

---

[5] Farmers own the farms and are the end-users of the system.

[6] The Agricultural experts are from diverse backgrounds in agriculture like agronomy, entomology, pathology etc.

Table 1. Details of eSagu prototype

Variable	Value
Duration	June 2004 to March 2005
Location of Main lab	Hyderabad, India
Number of villages covered	3
Location of villages	Warangal district, Andhra Pradesh, India.
Distance between the eSagu main lab and villages	About 200KM.
Name of the crop	Cotton
Number of farmers	984
Number of farms	1,051
Number of agricultural experts	5
Number of coordinators	14
Number of farm observations	20,035
Number of advices delivered	20,035
Number of Photographs	1,11,515

2.3   Implementation Details of eSagu Prototype

The eSagu prototype system was implemented from June 2004 to March 2005 for 1051 cotton farms of 984 farmers (some farmers have multiple farms) belonging to three nearby villages. The implementation details of the eSagu prototype are shown in Table 1. The locations of eSagu main center and eSagu local center are shown in Figure 2. In one of the three villages, eSagu local center was established. The distance between the eSagu main center and the eSagu local center is about 200 kilometers. In eSagu main center, five agricultural experts have delivered agro-advisories. At the eSagu local center, fourteen coordinators were identified to take farm observation photographs. Each coordinator was given a digital camera and have about 80 to 100 farms. Each farm observation contains about five photographs. In total 20,035 farm observations consisting of 1,11,515 digital photographs were received and the respective advices were delivered.

2.4   The Impact of eSagu Prototype

The main results of eSagu implementation are summarized as follows.

Through eSagu it has been demonstrated that it is possible for the agricultural expert to provide the expert advice based on the crop photographs and other information available in the agricultural information system.

It has also been found that the expert advice helped the farmers to improve the efficiency by encouraging integrated pest management methods, judicious use of pesticides and fertilizers by avoiding their indiscriminate usage.

The impact study shows that the farmers have realized considerable monitory benefits by reducing the fertilizers and pesticide sprays, and getting the additional yield [8].

Fig. 2. The Geo-Location of eSagu Implementation in 2004-2005

3   Data-Set and Preprocessing

In this section, we explain the details of data-set generated as a result of eSagu prototype. Next, we explain the preprocessing steps applied on advisory data set.

3.1   Details of the Data Set

Different types of data have been collected and generated during the operation of eSagu prototype. The details are given below:

- **Location data:** It contains the location details of villages, eSagu local center and the eSagu main center.
- **Personnel data:** The profiles and background of agricultural scientists, coordinators, farmers.
- **Farm data:** It contains the details of soil, irrigation and sowing.
- **Farm observations:** Farm observation data contains crop photographs. About five photographs are collected for each farm once in a week.
- **Weather details:** The details of temperature, humidity and rainfall.
- **Advice data:** The advice data is a collection of advice texts prepared by agricultural experts. The agriculture experts prepares advice based on the crop photographs. Each advice is a piece of English text which contains the list of steps that the farmer should take to improve the efficiency. The number of sentences varies with the crop problem. It contains pest names, latest

Table 2. Sample of agricultural advices

Advice ID	Advice
100	All the coordinators are requested to strictly implement following advices immediately 1. Steam Application: Do stem application with Monocrotophos:water = 1:4, Imidacloprid(CONFIDOR):water = 1:20. When to do is at 20, 40 and 60 days after sowing. 2. Advice all farmers to plant trap crops like castor and marigold (BANTHI) at the rate of 100 plants per acre at random through out the field. 3. Border crops: Advice the farmers to grow border crops like jowar (JONNA) or maize (MOKKA JONNA) around the field in 3 to 4 rows.
5043	Helicoverpa (PACCHA PURUGU) infestation was identifide, follow the following practices. 1. Setup pheromone traps (4 traps/acre) for pest intensity identification as well as to trap the male months. 2. Setup light traps (1 light trap/5 acre) to know the range of pest incidence as well to kill moth population. 3. Arrange atleast 10 bird perches/acre. 4. In this stage there may be eggs and larvae, so it is better to spray ovicidal (GUDDU MEEDA PANICHESEDI) and larvicidal (purugu meeda (panichesedi) chemical i.e. Thiodicarb (Larvin) @ 10 grams or 15lit. OR Endosulfon+Sesamum oil in 2:1 ratio.
10509	1. Magnesium deficiency is observed (Drying of leaves) and it is very severe so suggest the farmer to spray Magnesium sulphate @ 100 g/tank. Repeat the spray agter one week. 2. Immediately put traps of Hel. and pInk boll worm. 3. It is better to spray 5% NSKE solution every week to kill the eggs of boll worms and to control sucking pests like Whitefly, Mealy bug etc.

cultural practises and/or integrated pest management methods to be followed, fertilizers and pesticides to be applied and so on. The sample advices are shown in Table 2.

3.2   Data Preprocessing

We have selected only advice data set for the text analysis experiment. Each advice is a tuple consisting of <advice identifier, advice text>. The "advice identifier" uniquely identifies the advices and "advice text" contains the list of steps which have to be followed by the farmer to improve the farm productivity. These sentences are written in English language and consists of words from both English literature and agricultural literature. In addition, the advice text also includes "transliteration words" representing the phonological words in Telugu literature. The agricultural experts use transliteration words to help the coordinators for an easy interpretation and translation of the advice to the farmers.

The data set is noisy. It contains typographical errors, numerical and special characters. At first, all numerical and special characters were removed. To correct typographical errors the following two step procedure is followed.

1. Traditional words from English literature are spell corrected using Roget's Thesaurus [9].
2. Agricultural word repository was built by selecting the words related to names of pest, diseases, symptoms and agricultural practices. Based on these words we have corrected the spelling mistakes related to agricultural area.

After correcting the errors, we have extracted the words by applying text processing methods. The rank-frequency graph was built on the extracted words... After forming rank-frequency graph of advice data set, we have removed high frequency words and low frequency words. The remaining words are taken as keywords which are used to build advice vector matrix $(m \times n)$, where m indicates the number of advices and n indicates the number of key words. The entry $V_{(i,j)}$ indicates that frequency of j'th keyword in i'th advice.

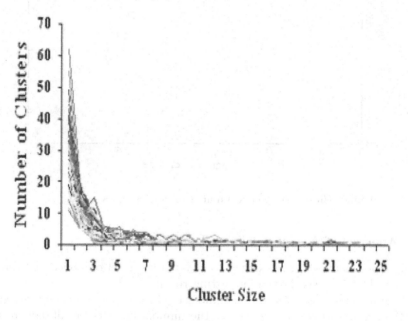

Fig. 3. Cluster size versus number of clusters

4   Experiment Results

In the experiments, we have employed cosine similarity criteria with threshold value as 80% to compare similarity between two advices. Let q and r be the advice vectors. The similarity between two advices is calculated using Equation 1.

$$cos(q,r) = \left(\frac{\sum_y q(y)r(y)}{\sqrt{\sum_y q(y)^2 \sum_y r(y)^2}} \right) \tag{1}$$

With advice-vector matrix as input, we have carried out two experiments. In the first experiment we have performed cluster analysis on advice data set. In the second experiment, we have analyzed the behavior of farms that belong to the same cluster over consecutive weeks.

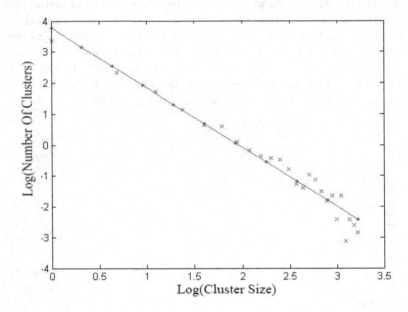

Fig. 4. Cluster size versus number of clusters on a log-log scale

4.1 Analysis of Advices Delivered on Each Day

In this experiment, we have extracted clusters on the set of advices delivered on each day. The analysis showed interesting results.

The graphs of cluster-size verses number of clusters for all the days are shown in 3. The graphs of cluster size versus the number clusters for all days resulted into an exponential curve. Figure 4 shows a curve of *cluster size* versus *number of clusters* on a log-log scale by considering the mean values of all days. It resulted into a straight line with slope of -1.6. The equation of corresponding exponential curve comes to $\left(\frac{1}{2e^{1.6i}} \right)$.

A sample of clustering results for four days is shown in Table 3. It can be observed that as cluster size increases the number of clusters decrease sharply.

Table 3. Sample results of number of clusters and number of farms at different cluster sizes on four days (year = 2004)

Cluster Size (a)	27-August		20-September		20-October		23-November	
	Clusters (b)	Farms (c=a*b)	Clusters (d)	Farms (e=a*d)	Clusters (f)	Farms (g=a*f)	Clusters (h)	Farms (i =a*h)
1	22	22	33	33	49	49	40	40
2	5	10	8	16	10	20	8	16
3	6	18	6	18	15	45	6	18
4	4	16	6	24	3	12	4	16
5	3	15	5	25	1	5	2	10
6	1	6	4	24	2	12	1	6
7	1	7	0	0	1	7	0	0
8	0	0	1	8	1	8	1	8
9	0	0	0	0	2	18	1	9
10	2	20	1	10	0	0	1	10
11	1	11	1	11	1	11	0	0
12	0	0	1	12	1	12	0	0
13	0	0	2	26	0	0	0	0
14	0	0	0	0	0	0	0	0
15	0	0	1	15	0	0	0	0
16	1	16	0	0	0	0	0	0
17	1	17	0	0	0	0	0	0
18	0	0	1	18	0	0	0	0
19	0	0	1	19	0	0	0	0
20	0	0	0	0	0	0	0	0
21	0	0	1	21	0	0	2	24

However, as cluster-size increases, the number of farms in a cluster increases. Table 4 shows number of farms in the clusters of size equal to 1, 2 to 5, and above 5 for the data in Table 3. Figure 5 displays the percentage of farms in the clusters of size equal to 1, 2 to 5, and above 5.

If the cluster size is one, the number of singleton clusters are equal to about 20 percent of farm advices on that day. It means that about 20 percent crops are facing distinct problems. At the same time, about 40 percent of farm advices have formed medium size clusters of size ranging from 2 to 5. The remaining advices have formed few clusters of large size varying from 5 to 25.

The results show an interesting phenomena. Significant number of farms are receiving distinct advices and the remaining farms are forming clusters of different sizes and each cluster of farms is receiving the same advice (or facing same problems). This shows that it is necessary to design a personalized agricultural advisory systems to the improve efficiency of farming. The results also show that several farms are facing same problems which provides an opportunity to improve the performance of eSagu system.

Table 4. Number of farms in clusters of different sizes for the data of Table 3 (year = 2004)

Cluster Size	27-August	20-September	20-October	23-November
1	22	33	49	40
2 to 5	59	83	82	60
above 5	77	143	110	58

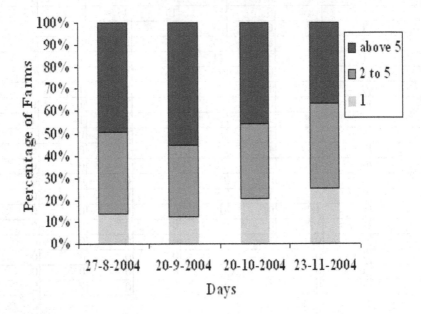

Fig. 5. Percentage of farms in different cluster sizes for the data of Table 4

4.2 Dynamics of Farm Problems over Weeks

In this experiment we have selected a sample cluster of farms which received the same advice in a particular week and examined the advices received by these farms in subsequent and preceding weeks. (Note that, in eSagu prototype, every crop has received advice once in a week.) Table 4 shows the sample results. The cluster number 18 contains four farms and have received the same advice in week number 17 (July first week is considered as week number 1). The farm identifiers are shown in the first column. It can be observed that even though these four farms belong to one cluster in week number 17, they fall in different clusters in the subsequent weeks and also in the preceding weeks. We found similar phenomenon for the farms of various other clusters.

This indicates that farm conditions are dynamic as several factors like the stage of the crop, the cultural practises followed by the farmer, soil properties, farm history, and irrigation facility are influencing the condition of the crop. So, information requirement is unique to each farm and varies dynamically.

Table 5. Variation of farm problems in different weeks in the year 2004. (July first week is considered as number one week).

Farms	15th Week	16th Week	17th Week	18th Week	19th Week	20th Week
far10761_0	13	18	18	58	92	48
far10875_0	31	11	18	51	31	54
far10220_0	31	11	18	92	88	65
far10521_0	31	67	18	82	88	65

5   Conclusions

The eSagu prototype was implemented for 1051 cotton farms by delivering personalized agricultural advice to each farm once in a week. In this paper we have analyzed the corresponding advice data set by applying text analysis methods. The analysis shows that the farm problems are complex, dynamic and influenced by several factors. The results show evidence that farming community requires personalized agricultural advice service in a regular manner for an efficient farming.

Acknowledgements

The eSagu system has been developed by IIIT-H, Hyderabad, India and Media Lab Asia. The initial part of the project was supported by Ministry of Communications & Information Technology, New Delhi.

References

1. Silberschatz, A., Zdonik, S.B.: Strategic Directions in Database Systems - Breaking Out of the Box. In: ACM Computing Surveys, pp. 764–778. ACM Press, New York (1996)
2. Pine II, B.J.: Mass Customization. In: Hardward Business School Press, Boston, Massachusetts (1993)
3. Han, J., Kamber, M. (eds.): Data Mining Concepts and Techniques. Morgan Kaufmann, San Francisco (1988)
4. Sharma, R.: Reforms in Agricultural extension: New policy framework. In: Economic and Political Weekly, pp. 3124–3131 (July 27, 2002)
5. Krishna Reddy, P., Ankaiah, R.: A Framework of information technology based agricultural information dissemination system to improve crop productivity. Current Science 88(12), 1905–1913 (2005)
6. Ratnam, B.V., Krishna Reddy, P., Reddy, G.S.: eSagu: An IT based personalized agricultural extension system prototype - Analysis of 51 farmers' case studies. International Journal of Education and Development using ICT (IJEDICT) 2(1) (2006)
7. Krishna Reddy, P., Ramaraju, G.V., Reddy, G.S.: $eSagu^{TM}$: A Data Warehouse Enabled Personalized Agricultural Advisory System. In: Proceedings of the 2007 ACM SIGMOD Conference, ACM Press, New York (2007)

8. Krishna Reddy, P., Sudharshan Reddy, A., Venkateswar Rao, B., Reddy, G.S.: eSagu: Web-based Agricultural Expert Advice Dissemination System (2004-2005). Final Completion Report of Research Project. Submitted to Ministry of Communications and Information Technology (2005)
9. Roget, P.M.: MICRA Inc. An electronic thesaurus derived from the version of Roger's Thesaurus (published in 1911), http://www.infomotions.com/etexts/gutenberg/dirs/etext91/roget15a.htm

Adaptive Query Interface for Mining Crime Data

B. Chandra, Manish Gupta, and M.P. Gupta

Indian Institute of Technology, Delhi
Hauz Khas, New Delhi., India 110 016
bchandra104@yahoo.co.in

Abstract. In present day scenario, law enforcement agencies are looked upon not only to control crime but also to analyze the crime so that future occurrences of similar incidents can be overcome. There is need for user interactive interfaces based on current technologies to meet and fulfill the new emerging responsibilities and tasks of the Police. The paper proposes adaptive query interface to assist police activities. The significance of such interface for police is to adapt interactive behavior of system with consideration of individual needs of the police and altering conditions within an application environment. The proposed interface is used to extract useful information, find crime hot spots and predict crime trends for the crime hot spots based on crime data using data mining techniques. The effectiveness of the proposed adaptive query interface has been illustrated on Indian crime records. A query interface tool has been designed for this purpose.

1 Introduction

The law enforcement agencies are constantly need of obtaining hidden and useful information from crime data. One major challenge for law enforcement agencies is the difficulty of analyzing large volume of crime data. Crime database consists of various relational tables which contains the information about crime details in a region under various crime heads such as murder, rape etc. at different time points. Advanced analytical methods are required to extract useful information from large amount of crime data. Data mining is looked as a solution to such problems.

Data mining is defined as the process of discovering, extracting and analyzing meaningful patterns, structure, models, and rules from large quantities of data. Data mining is emerging as one of the tools for crime detection, clustering of crime location for finding crime hot spots, criminal profiling, predictions of crime trends and many other related applications. Major objectives of crime data mining is to group crimes and to predict trends and models in crime. The main techniques of the crime data mining are clustering[7], association rule mining[1], classification[5] and sequential pattern mining[2]. Clustering technique is used to locate the crime clusters in the area of interest, which in turn will help in identifying the crime hot spots.

S. Bhalla (Ed.): DNIS 2007, LNCS 4777, pp. 285–296, 2007.

The user is interested to identify crime hot spots of a particular region on certain crime types for a specific period. In order to fulfill such a requirement, a user interactive query interface is needed. Kumar et al[10] has presented an interactive media system with capability to adapt to various conditions from user preferences and terminal capabilities to network constraints. Newsome et. al.[13] has proposed HyperSQL as a web-based query interfaces for biological databases. The design of query interfaces to biological database has also been presented by Che. et. al.[3]. However, no online adaptive query interface has been designed for mining crime data. The purpose of the paper is to design an adaptive query interface for mining crime data or similar kind of problems. The proposed query interface provides a tool for making an online query and helps in identifying crime hot spots, predict crime trends for the crime hot spots based on the query.

The overview of clustering algorithms are given in the next section of the paper. Section 3 of the paper describes the crime database structure. The proposed query interface has been described in details in section 4 of the paper. Section 5 presents the results and analysis of query interface on Indian crime records. Concluding remarks has been given in the last section of the paper.

2   Overview of Clustering Algorithms

In this section, some of the widely known clustering algorithms like K-means clustering, Hierarchical clustering and Self Organizing Map (SOM) have been described in brief.

K-means[11] is one of the most popular clustering algorithms. K-means is a partitioning method, which creates initial partitioning and then uses iterative relocation technique that attempts to improve the partitioning by moving objects from one group to another. The algorithm is used to classify a given data set into fixed number of clusters (K). K-means uses the concept of centroid, where a centroid represents the center of a cluster. In the process, K centroids, one for each cluster is defined apriori. Then each object of the data set is assigned a group with the closest centroid. The positions of k centroids are recomputed when all objects have been assigned to any of the clusters. The process is repeated until the centroids are no longer move.

Hierarchical clustering[8,6] groups the data objects by creating a cluster tree called dendrogram. Groups are then formed by either agglomerative approach or divisive approach. The agglomerative approach is also called the bottom-up approach. It starts with each object forming a separate group. Groups, which are close to each other, are then gradually merged until finally all objects are in a single group. The divisive approach is also called as top-down approach. It begins with a single group containing all data objects. Single group is then split into two groups, which are further split and so on until all data objects are in groups of their own. The drawback of Hierarchical clustering is that once a step of merge or split is done it can never be undone.

SOM[9] is a neural network based unsupervised clustering. It maps high dimensional data into a discrete one or two-dimensional space. SOM performs clustering through a competitive learning mechanism. In the process, several units compete for the current object and the unit whose weight vector is closest to the current object becomes the winning or active unit. Only the winning unit and its nearest neighbors participate in the learning process using Mexican Hat function.

Clustering techniques described above deal only with static data. In a data stream, data is coming constantly as in the case of crime data. Data stream clustering is required to identify crime hot spots periodically. Data streams and techniques of data stream clustering are described briefly in the next subsection.

2.1   Data Streams and Two Phase Clustering

Data Streams. In recent years, advances in hardware technology have allowed us to automatically record transactions of everyday life at a rapid rate. Such processes lead to large amounts of data which grow at an unlimited rate. These data processes are referred to as data streams. Data stream consists of a set of multi-dimensional records $X_1, X_2 \ldots X_n$ arriving at time stamps $T_1, T_2, \ldots T_n$. Each X_i is a multi- dimensional record containing d dimensions which are denoted by $X_i = (X_{i1}, X_{i2} \ldots X_{id})$.

Two Phase Clustering. Data stream clustering[4] can be done using one pass data stream clustering or two phase clustering methods. In a one-pass clustering algorithm over an entire data streams is dominated by the outdated history of the data streams. Therefore, it does not suit to the requirement of crime data mining since recent trend should not be dominated by the past crime trends. Thus, two phase clustering methods such as micro-clustering phase method and macro-clustering is employed in order to identify crime hot spots.

Micro-clustering phase method deals with online statistical data collection. This process is not dependent on any user input such as the time horizon or the required granularity of the clustering process. The aim is to maintain statistics at a sufficiently high level of (temporal and spatial) granularity. For a set $X_{i1}, X_{i2} \ldots X_{id}$ with time stamps $T_{i1}, T_{i2} \ldots T_{in}$, micro-cluster is given by tuple $(CF2^x; CF1^x; CF2^t; CF1^t; n)$, wherein cluster feature $CF2^x$, and $CF1^x$ each correspond to a vector of d entries. The pth entry of $CF2^x$ is equal to $\Sigma_{i=1}^{n}(X_{ij}^p)^2$ and p^{th} entry of $CF1^x$ is equal to $\Sigma_{i=1}^{n}(X_{ij}^p)$. A cluster feature is used to store data summary and time stamp maintained the elapsed time. Snapshots at different time stamp are taken to favor the most recent data. A snapshot contains a pre-specified number of micro-clusters. Any new data points is assigned to one of the micro-clusters in previous snapshot if it falls within the maximum boundary of that micro-cluster. If a new data points fails to fit into any of the existing micro-cluster, a new micro-cluster is created and simultaneously an existing micro-cluster is deleted or two micro-clusters merged. A cluster is removed if the average time-stamp is least recent.

Macro-clustering is an offline component, in which a user has the flexibility to explore stream clusters over different horizons. The macro-clustering process does not use the data stream, but the compactly stored summary statistics of the micro-clusters. Therefore, it is not constrained by one-pass requirements. The analyst gives a wide variety of inputs (such as time or number of clusters) in order to provide a quick understanding of the broad clusters in the data stream.

Two phase data stream clustering has been applied on Indian crime records. For crime data, a cluster feature is used to store the summary of crime records. Snapshots of crime data are taken for every fortnight. Initially, snapshots containing micro clusters are maintained as different crime zones. Micro-Clustering of crime data is given below in Fig. 1. In case 1 of Fig. 1, new crime record falls

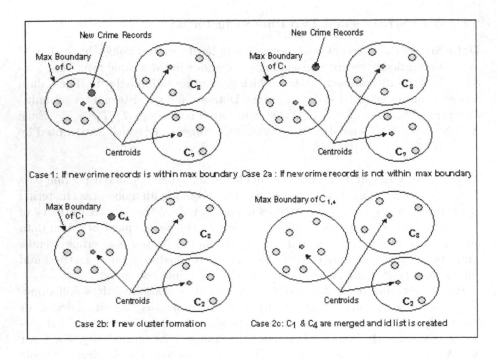

Fig. 1. Micro-Clustering of Crime Data

within the boundary of existing cluster. In case 2a of Fig. 1, new crime record does not fall within the boundary of any of the micro clusters. Case 2b of Fig. 1 shows the formation of new cluster whereas case 2c of Fig. 1 shows that clusters C_1 and C_4 are merged. If a new cluster is created, then crime hot spot is further identified by selecting the cluster with maximum average crime density among all the clusters. A cluster is deleted only if the average time-stamp is least recent by absorbing m new crime records.

3 Crime Database

Data is collected from various sources such as passport department, transport, and crime records. Raw data sources are stored in three kinds of database i.e. entity based database, crime statistics database and lost vehicle database. Entity based database contains the information about relationship among entities involved in crime. Crime statistics database contains the information of overall crime in India. Lost vehicle database are the collection of records about lost vehicles. Crime database consists of all of these databases. The structure of crime database is given below in Fig. 2.

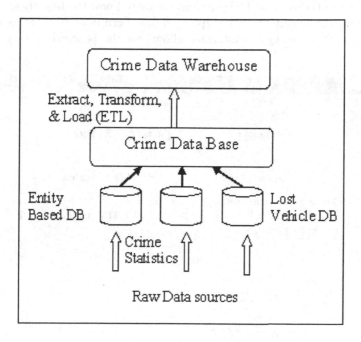

Fig. 2. Structure of Crime Database

The crime database consists of many tables(relations) which contains the information about each and every crime occurring in India. The tables of crime database contain the information related to the location, crime types, date of occurrence etc. Every record has information in coded form i.e. location code, major head key, police station code, district code are given in place of their description. The code descriptions are given in separate tables. e.g. Location table contains the codes and description of different places in India. Major Head Key table contains the information about crime type such as murder, rape, kidnapping etc. and their codes. The information in the relational tables are in very abstract form. Mining of crime data requires the information in the aggregated form. Cluster analysis is used for identification of crime hot spots in the area of location.

4   Proposed Query Interface for Mining Crime Data

Adaptive interfaces[3,10,12,13,14] are required to overcome the problems due to increasing complexity of human computer interfaces. Adaptive query interface interacts with the users in a variety of way. The proposed adaptive query interface provides a user-friendly interface between the user and knowledge acquisition system. The objective of query base interface is to shift functionality as possible from the user to the system. The user only needs to know the basic tasks to carry out for crime data analysis. The type of desired crime analysis is ascertained through this interface. Query interface allows for the processing of continuous

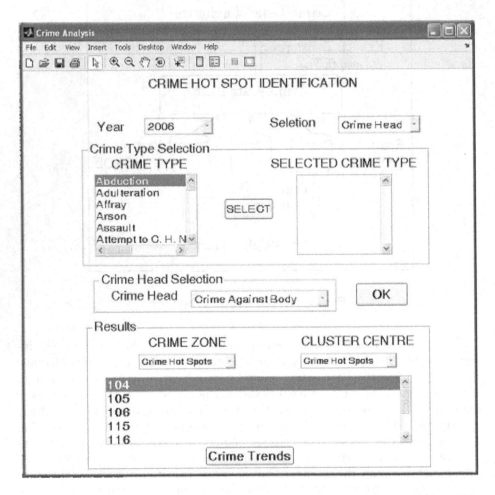

Fig. 3. A query about Selection Type as Crime Head:Crime Against Body for 2006

and snapshot queries over crime data. Any query interface is required to extract the information from the crime relational tables.

The traditional way of executing the queries is the extraction of records and aggregate them for every execution. This is more time consuming process because of rescanning of database every time. The proposed online query interface eliminates the rescanning of the database for every new query. It provides the user an interactive and fast way to carry out process of identification of crime hot spots and crime zones. Interaction of many relational tables is required for analyzing crime data. The proposed query interface extracts the records from these tables and aggregates them for further online querying. The crime analysis software as online query interface has been designed on Indian crime records.

Database connection is established by a database connection program, which requires the database name, username and password as inputs from the user. The extraction of records is also carried out through the database connection program, which also generates some text files for further use. It is a one-time effort to run the program. Prior to executing database connection program, an ODBC data source is required to create from the original database.

The online adaptive query interface, Crime Analysis GUI(See Fig. 3) is executed after the execution of database connection program. This interface provides a tool for making an online query and based on the query crime hot spots and crime zones of a particular region on certain crime types for specific period are identified. The user can select year, selection type from crime head or crime type from the option given in the GUI.

For crime head selection, user can select one of the major crime heads i.e. crime against body, crime against women, crime against property and kidnapping and abduction. The crime heads and their crime types are as follows, crime against body includes abduction, attempt to murder, hurt, kidnapping, murder; crime against women consists of cruelty by husband or husband s relative(s), deaths-dowry, molestation, rape, sexual harassment (eve teasing) and crime against property includes dacoity, robbery, theft. Kidnapping and abduction is also considered as one of the crime heads. Crime heads help the laymen users for analyzing crime hot spots and crime zones of general type of crimes without going into the specific details.

For crime type selection, the user can select the desired list of crime type from the crime type list box options of the GUI. More than one crime type can also be selected.

The results of query consist of crime hot spots, high crime zones, moderate crime zone and low crime zones based on the average density of these crime. Cluster center as average density of various crime zones are also given in the results of the query interface. The results obtained using proposed adaptive query interface will be helpful in identifying the crime hot spots, predicting crime trends for the crime hot spots which will ultimately help in controlling the crime. The details of results and analysis is given in the next section of the paper.

5   Results and Analysis

In this section, some of the results obtained using the proposed adaptive query interface along with analysis has been presented. A query about crime head selection for crime against body for year 2006 is given in the Fig. 3. Districts 104, 105, 106, 115, 116, 118, 123 and 125 have been identified as crime hot spots. 5.

It is highly important to know crime trends for those districts, which have been designated as crime hot spots. Figure 4 presents the crime trends of district 106 with respect to crime type murder(homicide) from year 2000-2005. Figure 4 shows decreasing crime rate for district 106 from 2000-2005 except 2003. District 106 is still in the crime hot spots even though it is showing decreasing crime

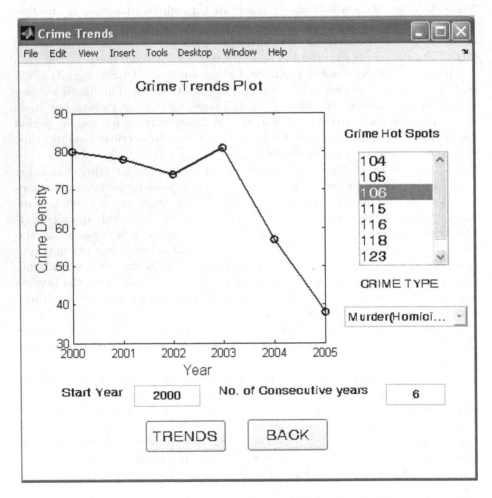

Fig. 4. Crime Trends Plot for District '106' under Murder(Homicide) from year 2000-2005

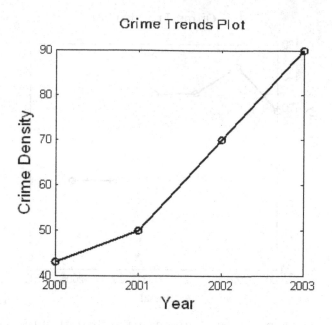

Fig. 5. Crime Trends Plot for District '105' under Dacoity from year 2000-2003

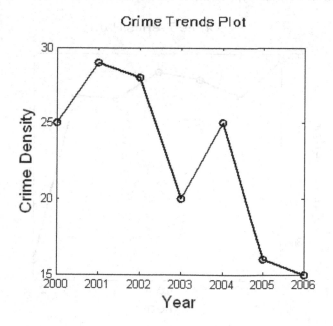

Fig. 6. Crime Trends Plot for Crime Hot Spot District '105' with respect to crime against women under crime type as rape from year 2000-2006

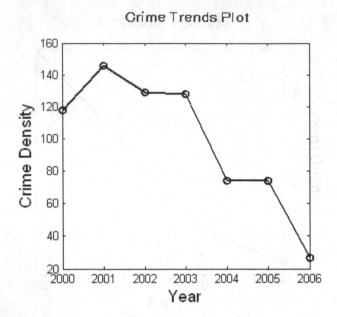

Fig. 7. Crime Trends Plot for Crime Hot Spot District '105' with respect to crime against women under crime type as molestation from year 2000-2006

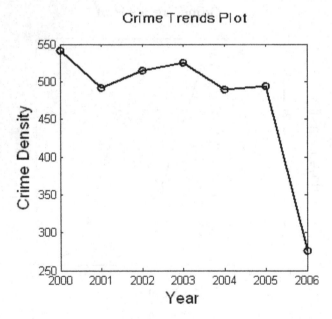

Fig. 8. Crime Trends Plot for Crime Hot Spot District '105' with respect to crime against women under crime type as cruelty by husband from year 2000-2006

trends since the average crime density of the district 106 is greater than the average density of districts in the other crime zones. Whereas, crime rate of district 105 under dacoity crime is increasing from 2000-2003 as shown in Fig.5.

In response to another query, district 105 is the only district identified as crime hot spot with respect to crime against women for year 2006. Crime trends plots for identified crime hot spot district '105' with respect to crime against women from the period 2000-2006 under crime type as rape, molestation, cruelty by husband are shown in Fig. 6, Fig.7 and Fig. 8. It is seen that all three figures show highly fluctuating trends during the above mentioned period. It may also be noted although district '105' shows low crime density for every women related crime during the year 2006, it is still designated as crime hot spot since average crime density is greater than any of the other districts.

Similarly, crime trends of other crime hot spots with respect to individual crime types can be obtained using crime trends interface.The proposed interface can also be utilized for mining of crime data at the police station level. This tool not only mines the crime data in user interactive manner but also enhances the compatibility of analyzing crime data. By analyzing the crime trends, user can predict the crime density for a location and also take preventive measures for future occurrences of crime in those areas.

6   Concluding Remarks

In this paper, an online adaptive query interface has been proposed for mining of crime data. The interface identifies crime hot spots and crime zones of a particular region on certain crime types for period specified by the user. The proposed interface further analyze the crime trends with respect to the individual crime type of crime hot spots. The two phase clustering method has been applied for identification of crime hot spots and crime zones. The effectiveness of the proposed adaptive query interface has been illustrated on crime records of India using a developed software tool. The proposed interface will be helpful for analyzing the crime and controlling the crime. The interface can play an important role for wider variety of similar problems.

References

1. Agrawal, R., Imielinski, T., Swami, A.N: Mining association rules between sets of items in large databases. In: Proceedings of the ACM SIGMOD International Conference on Management of Data (1993)
2. Agrawal, R., Srikant, R.: Mining sequentiel motifs. In: 11th Int'l Conf. on Data Engineering (1995)
3. Che, D., Aberer, K., Chen, Y.: The design of query interfaces to the GPCRDB biological database. In: Proceedings of User Interfaces to Data Intensive Systems (1999)
4. Guha, S., Mishra, N., Motwani, R., Callaghan, L.O.: Clustering Data Streams. IEEE FOCS Conference (2000)

5. Han, J., Kamber, M.: Data mining: concepts and techniques. Morgan Kaufmann, San Francisco (2001)
6. Hartigan, J.A.: Clustering Algorithms. John Wiley and Sons, Inc, New York (1975)
7. Jain, A.K., Murty, M.N., Flynn, P.J.: Data clustering: a review. ACM Computing Surveys 31(3), 264–323 (1999)
8. Johnson, S.C.: Hierarchical clustering schemes. Psychometrika 32(3), 241–254 (1967)
9. Kohonen, T.: The Self Organizing Map. Proc. IEEE 78, 1464–1480 (1990)
10. Kumar, M., Gupta, A., Saha, S.: Approach to Adaptive User Interfaces using Interactive Media Systems. In: Proceedings of the 11th international conference on Intelligent user interfaces (2006)
11. McQueen, J.: Some methods for classification and analysis of multivariate observations. In: Proc. Symp. Math. Statist. And Probability, 5th, Berkeley, vol. 1, pp. 281–298 (1967)
12. Michelson, M., Knoblock, C.A.: Phoebus: A System for Extracting and Integrating Data from Unstructured and Ungrammatical Sources. In: Proc AAAI- (2006)
13. Newsome, M., Pancake, C., Hanus, J.: HyperSQL: web-based query interfaces for biological databases. In: Proceedings of the Thirtieth Hawaii International Conference on System Sciences (1997)
14. Tuchinda, R., Szekely, P., Knoblock, C.A.: Building Data Integration Queries by Demonstration. In: Proceedings of the 12th international conference on Intelligent user interfaces (2007)

Aizu-BUS: Need-Based Book Recommendation Using Web Reviews and Web Services

Takanori Kuroiwa and Subhash Bhalla

Graduate School of Computer Science and Engineering,
University of Aizu, Aizu-Wakamatsu, Fukushima 965-8580, Japan
{m5101202,bhalla}@u-aizu.ac.jp

Abstract. Presently, there are three approaches that constitute recommender systems: collaborative filtering, content-based approach and a hybrid system. This paper proposes a complementary recommendation methodology, focusing on book recommendation. By retrieving web reviews of books using existing Web Services, an infrastructure has been developed for need-based book recommendation system. Implementation results shows that our book recommendation allows a user to eliminate irrelevant books and presents the desired books to the user from given book set. The proposed book recommender is one of the first systems in terms of focusing on meeting individuals' needs rather than calculating similarity or preferences automatically, which is adopted by the traditional recommender system.

Keywords: Book recommendation, Web review retrieval, Web Services.

1 Introduction

Recommendation systems provide users with products that they might want. However, current recommender systems such as Amazon.com's book recommender do not take each user's need into consideration. In fact, many times a user does not decide to purchase unknown recommended books promptly. This may be because in most cases, the user checks his/her own needs. For example, he/she may think "Which book is my favorite author's/publisher's?", "Which book is good for beginner?" or "If I purchase the book, are there any benefits for me?" One of the ways to resolve such questions is to check customer reviews provided by Amazon.com. Another way is to search the Web to find effective web reviews. Recently information transmissions through Blogs or virtual-community logs have been very active. Many people write book reviews on their Blogs. Therefore, if a recommender system has a facility for collecting and checking web reviews, the recommender system would be able to recommend books that meet individuals' needs.

The purpose of this study is to build an infrastructure for a need-based recommendation that supports such a facility by collecting web reviews though existing Web Service integration. Also, we presents the effectiveness of the proposed recommendation approach by providing implementation results.

S. Bhalla (Ed.): DNIS 2007, LNCS 4777, pp. 297–308, 2007.

The rest of the paper is organized as follows. The background information is described in Section 2. In Section 3, we introduce an infrastructure of our book recommendation. Section 4 discusses needs-based recommendation. In Section 5, results are discussed. Finally, Section 6 presents the summary and conclusions.

2   Background

Presently, there are three recommendation approaches that constitute recommender systems: collaborative filtering, content-based approach and their hybrid system [3]. Each approach has some drawbacks and advantages. In addition to these traditional recommendation methodologies, a new approach has been proposed. In this section, we briefly describe them.

2.1   Collaborative Filtering Approaches

The collaborative filtering approaches recommend books based on similarity between the books the user preferred in the past and the ones other users have purchased. Amazon.com's recommender system adapts this approach [7][10]. The collaborative approaches can provide books of unknown category. However, such recommender systems have to learn a user's preference from the user's previous purchases. In addition, the recommender systems require a sufficient number of other user's ratings before making a recommendation.

2.2   Content-Based Approaches

The content-based approaches recommend books based on a user's preferences [9][11][12]. The content-based approaches are appropriate for recommending books by extracting the user's preferences from the user's ratings. One of the drawbacks of such recommender systems is that they require a sufficient number of the user's ratings to make an accurate recommendation. Also, this approach may cause a loss of information problem in the course of extracting featured terms to calculate a user's interests. For example, when selecting featured terms from text, terms regarded as insignificant are eliminated. However, in terms of context, such terms may not be ignored.

2.3   Hybrid Approaches

The hybrid approaches combine the collaborative filtering and the content-based approaches [4][13]. The hybrid approaches are able to make more accurate recommendation than the above two approaches by overcoming weakness of pure approaches. However, recommended books still depends on users' own effort.

2.4   Review-Based Approach

Aciar et al. propose a new recommendation approach [1][2]. It is based on consumer product reviews written in free-form text. In this approach, each review
comment is mapped into the ontology's information structure and classified into
three categories: "good", "bad" and "quality" to decide whether a product is
good or bad taking each commenter's skill level into consideration. The drawback
is that this approach cannot be applied to all review comments. For example,
some long, complicated sentences cannot be classified into any category.

2.5   Human-to-Human Approach

We have been proposed a human-to-human recommendation approach [8]. In
this approach, users share a Knowledge Base (KB) and each user recommends
his/her chosen books to other users like real-world borrower or lender through
the web-based system. The purpose of this approach is to enhance utilization of
existing books among users. In this approach, to visualize individuals' preference,
featured keywords are extracted from the KB. However, to make an accurate
featured keyword extraction, a sufficient number of user's ratings are required.

2.6   The Proposal

Both the collaborative filtering approach and the content-based approach have
a common drawback: these require a sufficient number of users' efforts before
making a recommendation. Similarly, the proposed review-based approach cannot work well in a few cases. The human-to-human approach also have some
drawbacks. This may be because these approaches tend to focus on automatically processing something (such as extracting users' interests, calculating the
similarities between users, analyzing review comments) rather than reflecting
users' direct needs. Thus, making a recommendation based on these approaches
may cause a gap between the recommended books and the ones users really
want. This is because the approaches ignore users' current interests and needs.
For this reason, we propose a recommender system that supports a facility for
checking if recommended books meet the users' needs or not.

3   Aizu Book Utilization System (Aizu-BUS)

In this Section, we introduce an infrastructure for the proposed book recommendation system called Aizu Book Utilization System (Aizu-BUS).

3.1   System Architecture

Figure 1 depicts an architecture of Aizu-BUS. The Aizu-BUS is composed of
Web User Interface (WUI), Web Service handler, XML DB handler, external
resources and a Knowledge Base (KB). The WUI receives users' requests such
as gathering and searching operation for books and generates query to the Web

Services. The Web Service handler searches the external resources according to the request and give the results as store/query operation. The XML DB handler stores XML data or retrieves XML data from the XML database called Xpriori [16]. The KB is used for book utilization activities: book search, book recommendation and book sharing among users. In this study, we focus on book recommendation activity.

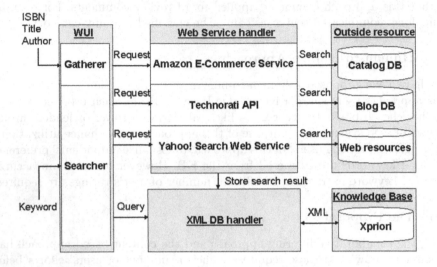

Fig. 1. System Architecture

3.2   Building Knowledge Base

Initially, data about books (available in store) is entered. When a user enters an ISBN (as shown in Figure 1), the system generates a request to Amazon E-Commerce Service [14] and searches book catalog to retrieve basic book information. The basic book information includes: image URL; ISBN; title; author name; product page URL; price; publication date; publisher name; and the number of pages and Amazon.com's sales rank. The basic book information is stored in an XML database called Xpriori. The XML database serves as a KB for the proposed book recommendation system. Technorati API [15] and Yahoo! Search Web Service [17] are applied for Web and Blog information retrieval.

3.3   Book Rating

Aizu-BUS provides a book rating method considering the fact that different users may have different rating scale [5][6]. The book rating approach given by the following formula depends not on the score of each book but on the context of each book review.

$$r = \frac{N_p - N_n}{N_p + N_n} \times (r_{ave} - 1) + r_{ave} \tag{1}$$

Table 1. Book Rating

r	Evaluation
1	Poor
$1 < r \leq 2$	Bad
$2 < r \leq 3$	So-so
$3 < r \leq 4$	Nice
$4 < r \leq 5$	Excellent

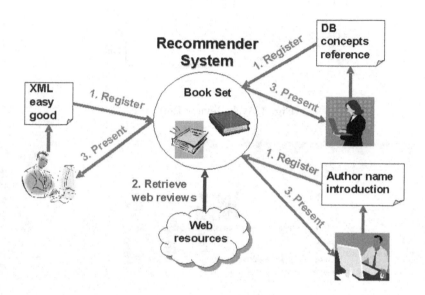

Fig. 2. Need-Based Book Recommendation

where r_{ave} is the average rating, N_p and N_n represents the positive review count and the negative review count respectively. The value of r ranging from 1 to 5 represents book ratings. The larger the value of r is, the better evaluation is. Table 1 shows the definition of book rating in this study. For example, if the number of the positive reviews of a book (for example, review comments are saying "good" or "easy") equals to 3 and the number of the negative reviews of the book (for example review comments is saying "bad" or "difficult") equals to 1, r will be 4, which represents the book is "Nice". Figure 4 and 5 depict the proposed book rating. Suppose that each book review is collected manually. In case two positive reviews (that contain "excellent" and "useful" in the review comments) are added to Knowledge Base, the book rating (r) become excellent (5) as shown in Figure 4. However, if one negative review (that contains "worst") is added, the book review (r) is reduced to "nice" (3.66) as shown in Figure 5.

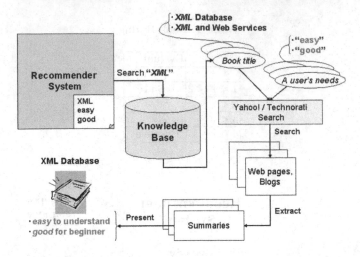

Fig. 3. Web Review Retrieval

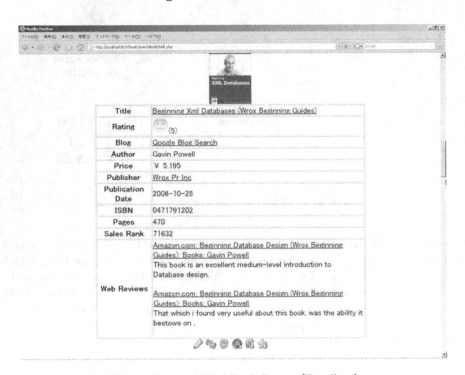

Fig. 4. Review-based Book Rating [Excellent]

4   Need-Based Book Recommendation

In this Section, we propose a need-based recommendation as a facility of the Aizu-BUS. The need-based recommendation approach corresponds users' direct needs

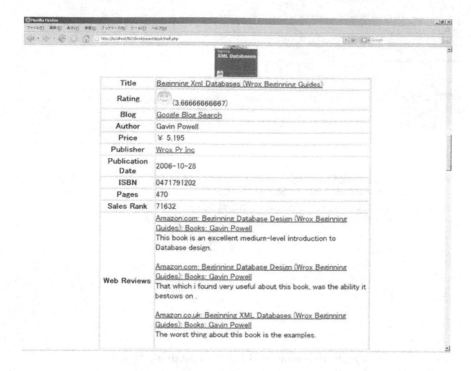

Fig. 5. Review-based Book Rating [Nice]

with book reviews available on the Web using the existing Web Services. As shown in Figure 2, the need-based book recommendation follows Step 1 to 3 below.

1. Needs registration;
2. Web review retrieval; and
3. Book recommendation

4.1   Needs Registration

First, each user registers his/her needs into the recommender system. The format of contents is as follows 1) a representative keyword of his/her interests (such as an interested category name, a favorite author/publisher name), 2) his/her first need (such as "easy", "concepts" or "introduction"), and 3) his/her second need (such as "good" or "reference"). These registered needs are used as keywords for narrowing down the search result when searching the Web/Blogs using Yahoo!/Technorati Web Services.

4.2   Web Review Retrieval

Once the needs are registered, the recommender system automatically assembles queries and searches the Web/Blogs using Yahoo!/Technorati Web Services.

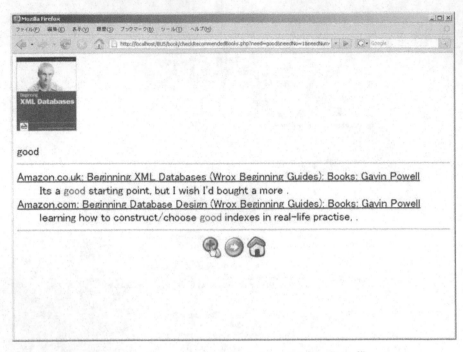

Fig. 6. Recommended book 1 containing "good"

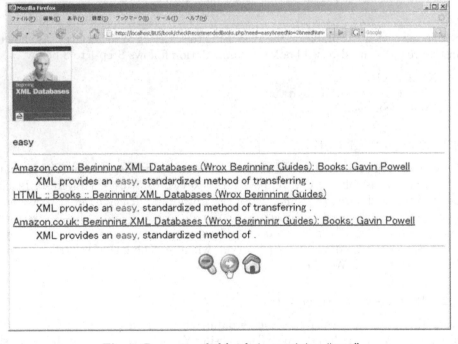

Fig. 7. Recommended book 1 containing "easy"

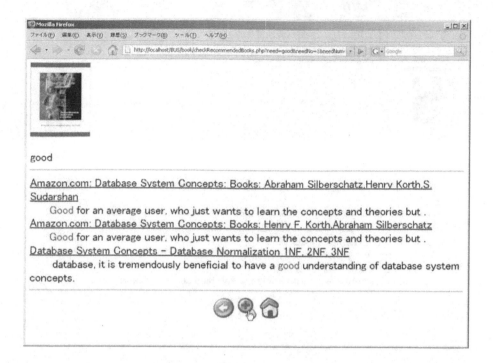

Fig. 8. Recommended book 2 containing "good"

Figure 3 shows the procedures of Web review retrieval. First, recommender system searches the Knowledge Base by using a user's first need (that is, for example, a representative keyword such as "XML") and book titles containing the keyword are retrieved. Then, the system automatically assembles the query to search web reviews by combining a selected book title and the user's next need "easy". By using Yahoo!/Technorati Web Service, short summaries containing the assembled query are extracted from retrieved Web pages/Blogs. Finally, the extracted summaries are presented to the user.

4.3   Book Recommendation

After retrieving the web reviews according to the above method, our book recommender system presents recommended books to a user one by one according to Amazon.com's sales rank. Currently, the web review retrieval and recommended book presentation are performed in real-time. In other words, a user can trigger those two activities. When a user requests to browse recommended books from given book set (such as other user's library) , the recommender system searches the web reviews using Web Services and presents the selected books (that meet his/her needs) to him/herself. The implementation details and results are discussed in the next section.

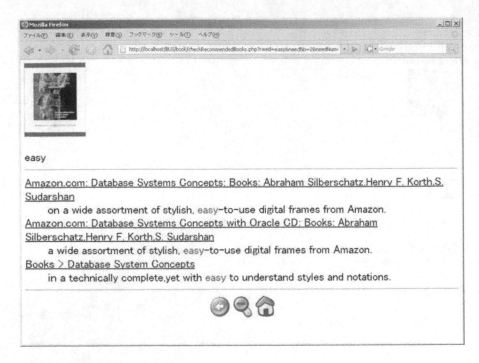

Fig. 9. Recommended book 2 containing "easy"

5   Implementation Details

A prototype system was implemented on Windows 2000 using Apache 2.0.5 and PHP 5.1.4. To test our book recommendation facility, we collected 100 books from Amazon.com. Assume that a user has registered his/her needs like "Database", "good" and "easy". Also, note that the top three pages of Yahoo!/Technorati search are retrieved as web reviews of a selected book.

Figure 6 to 9 show the recommended books and the review comments retrieved by Yahoo! Web Service. Most reviews are from Amazon.com related sites. This is because Blog search results have not been taken into account. Technorati API (that is used for retrieving book reviews from Blogs) is unsuitable for retrieving web contents written in space delimited language like English because it does not support a perfect matching search (at this time). However, it is appropriate for retrieving Japanese contents. Therefore, we apply Technorati API to Japanese book review retrievals.

6   Summary and Conclusions

This paper proposed a need-based book recommendation using Web reviews. Implementation results show that existing Web Service integration contributes to retrieving book reviews available on the Web and corresponding the web reviews with each user's needs. This recommendation method is not an alternative approach to traditional approaches such as the collaborative filtering and the content-based approaches, but a complementary approach to such approaches. We believe the proposed book recommender is one of the novel systems in terms of taking users' needs directly into consideration and focusing on meeting the needs rather than automatically predicating recommended books, as in the case of the traditional recommender systems.

For future work, we will investigate the following things: coordination with traditional recommendation approaches and web review clustering to automatically classify web reviews into the positive ones and the negative ones.

References

1. Aciar, S., Zhang, D., Simoff, S., Debenham, J.: Recommender System Based on Consumer Product Reviews. In: Proceedings of the 2006 IEEE/WIC/ACM International Conference on Web Intelligence, pp. 719–723 (2006)
2. Aciar, S., Zhang, D., Simoff, S., Debenham, J.: Informed recommender: basing recommendations on consumer product reviews. IEEE Intelligent Systems 22(3), 39–47 (2007)
3. Adomavicius, G., Tuzhilin, A.: Toward the next generation of recommender systems: A survey of state-of-the-art and possible extensions. IEEE Transactions on Knowledge and Data Engineering 17(6), 734–749 (2005)
4. Balabanovic, M., Shoham, Y.: Fab: Content-Based, Collaborative Recommendation. Communications of ACM 40(3), 66–72 (1997)
5. Jin, R., Si, L., Zhai, C.: Preference-based Graphic Models for Collaborative Filtering. In: Proceedings of Nineteenth Conference on Uncertainty in Artificial Intelligence (August 2003)
6. Jin, R., Si, L., Zhai, C., Callan, J.: Collaborative Filtering with Decoupled Models for Preferences and Ratings. In: Proceedings of the Twelfth International Conference on Information and Knowledge Management, pp. 309–316 (November 2003)
7. Konstan, J.A., Miller, B.N., Maltz, D., Herlocker, J.L., Gordon, L.R., Riedl, J.: GroupLens: Applying Collaborative Filtering to Usenet News. Communications of the ACM 40(3), 77–87 (1997)
8. Kuroiwa, T., Bhalla, S.: Dynamic Personalization for Book Recommendation System using Web Services and Virtual Library Enhancements. In: CIT 2007. IEEE 7th International Conference on Computer and Information Technology, University of Aizu, Japan (2007)
9. Lang, K.: NewsWeeder: Learning to Filter Netnews. In: Proceedings of the 12th International Conference on Machine Learning, pp. 331–339 (1995)
10. Linden, G., Smith, B., York, J.: Amazon.com recommendations: item-to-item collaborative filtering. IEEE Internet Computing, 76–80 (January/February 2003)
11. Mooney, R.J., Roy, L.: Content-based Book Recommending Using Learning for Text Categorization. In: Proceedings of the 5th ACM conference on Digital Libraries, pp. 195–204. ACM Press, New York (2000)
12. Pazzani, M., Billsus, D.: Learning and Revising User Profiles: The Identification of Interesting Web Sites. Machine Learning 27, 313–331 (1997)
13. Pazzani, M.: A Framework for Collaborative, Content-Based and Demographic Filtering. Artificial Intelligence Review , 393–408 (December 1999)

14. Amazon Web Service. http://aws.amazon.com/
15. Technorati API. http://technorati.com/developers/api/
16. Xpriori (XMS), http://www.xpriori.com/
17. Yahoo! Search Web Service, http://developer.yahoo.com/search/

Search in Virtual and Study in Real: A Ubiquitous Query/Answer System[*]

Lei Jing[1,2], Zixue Cheng[1,3], and Tongjun Huang[1,4]

[1] The University of Aizu, Aizu-Wakamatsu City, Fukushima Preference, 965-8580, Japan
[2] The graduate school of computer science and engineering, The University of Aizu
[3] The school of computer science and engineering, The University of Aizu
[4] Information systems and technology center, The University of Aizu
{d8071202,z-cheng,t-huang}@u-aizu.ac.jp

Abstract. U-Learning has emerged with the development of UbiComp. In the precomputer era, people can only learn from surrounding environment or other people. Given a subject, it is difficult to find out the concerning learning contents in the real environment. In the era of CAI and M-Learning, people can choice to learn in the virtual world and rapidly find out the related virtual learning contents. But they still could not rapidly find the right learning contents in the real world which has confine the learner's learning practice. U-Learning, by making use of digital artifacts in the ubiquitous environment, has provided the possibility for a learner to learn from more extensive learning contents including both of the virtual contents and real contents, and bridge the virtual and real contents seamlessly. In such a system, people learn not only from books or web pages, but also they can explore their knowledge in the real world as well. In this paper, the architecture of a ubiquitous query and answer system for pupils is presented to show the possibility of integration of searching and providing real and virtual learning contents in a ubiquitous environment. And a matching algorithm is proposed for find out the proper learning contents by matching the learning contents database with the ubiquitous learning environment database. Finally, the scene of learning by such a system is shown through a concrete example.

Keywords: Ubiquitous Computing, Ubiquitous Learning, Digital Artifacts, Question and Answer, Mapping between virtual and real worlds.

1 Introduction

With the progress of computing technologies, the E-Learning (Electrical Learning) has developed from Distance Learning to M-Learning (Mobile Learning), and further to U-Learning (Ubiquitous Learning).

So far, Distance Learning on the network by a PC like terminal has become a widely accepted learning style. Key advantages of this style are flexibility,

---

[*] This research is partially supported by a funding of Fukushima Prefecture and a funding from JSPS (Japan Society for the Promotion of Science) No. 18650251.

S. Bhalla (Ed.): DNIS 2007, LNCS 4777, pp. 309–321, 2007.

convenience and the ability to work at any place where an internet connection is available and at one's own pace. But the difficulty to operate such a system has confine the users to adult people who would like to spend some time on learning how to operate PC and the instruction software itself.

With the development of portable technology, M-Learning has come into people's horizon. In nowadays society, not every one has a PC, but most of people have at least one portable device like cell phone or PDA. It is especially true for the young adult. M-Learning could enable them to learn in various places like school, workplace, museum, rural areas, and so on. Some novel learning styles have been investigated such as "blended learning" (as part of formal courses) and "location-depended learning" (during visit to museums etc.). However, the portable attribute of the user terminal is a double edged sward. On one hand, portability means people can learn at anywhere and anytime. On the other hand, it confines the size of the portable device include the input part and screen. Then, how to interact with small devices and display useful content in small-screen devices have been the key problems for M-Learning.

Ubiquitous computing is a new trend of information and communication technologies. Computer-augmentation and proactive service are two main characteristics of ubiquitous computing system. Computer-augmentation means embedded tiny computing nodes into everyday artifacts to yield digital artifacts. Each node is equipped with the sensors and/or actuators to interact with our living environment, and with communication functions to exchanging their data. The smallest node could be NRFID (Networked RFID)[12]. When embedded such nodes into the normal artifacts, we get the computer-augmented artifact or digital artifacts. Research groups at MIT Media Lab have produced a wide range of computer-augmented artifacts [1]. MediaCups [2] have embedded sensors and communication modules into the bottom of the normal size coffee cups, so that these cups can judge their states (full or empty, held or put down) and give a hint to the user. U-schoolbag could be another example research which integrated the tiny computer into the normal schoolbag to provide some proactive services like object reminder service and timely review service [7]. Moreover, the digital artifacts like the nerves in the human body. When the environment are filled with the digital artifacts, the environment becomes can "see", "smell", "hear", "touch", etc. In another word, the environment could to some extent understand the situation or context of the people. Thus, a ubiquitous computing system could provide proactive service based on the information about the context.

Ubiquitous learning is supported by ubiquitous computing and represents the next step of E-Learning. So far, a few new learning styles have been developed. A new concept ubiquitous museum is proposed in [9], which means an environment that can provide museum-like visiting experience. ENLACE project proposed a pedagogical and technical approach to support the flow of learning activities outside of school and in class [13][14]. Perkam project has concerned on the collaborate learning in the ubiquitous environment [3][4]. For example when a learner runs into some problems, how to find out the proper helper based on the learner's situation. In paper [11], the possibility of building ubiquitous learning environment by only attaching NRFID to objects and places is presented.

The novel learning styles introduced above are inspiring and to some extend give us a vision of future direction of U-Learning. However, the potential possibility of U-Learning has not been fully explored, so that it is difficult for the current U-Learning researches to different from M-learning researches. The target deployment environment of a U-Learning system is a ubiquitous computing environment. Then from ubiquitous computing point of view, let's have a look at whether U-Learning can do something different from the M-Learning or Distance Learning. Since digital artifacts have linked the real and virtual world, it has given us a chance to learn in the real world by interacting with these digital artifacts. However, so far, most of the researches adopting the digital artifacts into U-Learning have treated each digital artifact independently. In this paper, we will show the potential merits taken by linking the digital artifacts and setting up the mapping mechanism between the virtual learning contents and the linked digital artifacts -----real learning contents.

Questions are one of the important motivations of learning. When people run into some questions, they will eager to find out the answers. The process of finding out the answers is a process of learning, but how to find out the answers and where to find out the real learning contents have always been a problem. In the era of pre-computer, people can ask other people or consult on some books if they can find out the proper people or proper books. Obviously, this process is lack of instruction and time consuming. As a result, it will reduce the learner's curiosity and confidence. In the era of E-Learning, people can find out the answers from the knowledge repositories with little effort. But the virtual learning contents in the knowledge repositories have separated from the real world. It has disobeyed learning by doing rules. Thus the learners tend to lose the curiosity on the question since they get the answers too easily with no practical experience. In the era of Ubicomp, the digital artifacts can act as a bridge between virtual learning contents and real world. When people have any questions, a U-Learning system could help the people to find out the answers by exploring not only the virtual world but also the real world. In this way, people can not only find out the answers but also learn how to find out the answers.

To show the feasibility of real environment learning with the diverse digital artifacts, in this paper a ubiquitous query and answer system for pupils will be presented. When pupils have any questions, they can input their questions through some digital artifacts like U-Pens, and these questions are parsed by a U-Learning Agent in the system. Then the agent queries the database to find out the proper and available learning contents according to pupils' ability and location. Finally, the agent helps the pupils to find out the answers by interacting with real objects (Digital Artifacts) nearby.

2   Model

Different pupils have different interests and different knowledge levels, and can run into various questions in various kinds of situations. Even a pupil run into a given question, under different situations, the ways to provide the help and the help contents should be adjusted by the system automatically.

A model of the ubiquitous query and answer system is shown in Fig. 1. Supposing in a Ubiquitous Learning Environment (ULE), each object, from a bean to a house, is attached with a NRFID. U-Pen integrated with multiple kinds of UMI (User Machine Interface), such as handwriting and voice reorganization, RFID reader etc., is used as user input interface [5] [8]. U-Schoolbag [7] with a unique ID and it is supposed that the schoolbag is combined with the pupil. Then the U-Schoolbags can be used to identify the pupils. U-Schoolbag with LCD displayer can be used as the output interface as well. The user information server analyses and extracts the pupils' knowledge levels and interests from the history records and storage them into the user information DB. The virtual learning contents are storage in the learning contents DB. The information of the real learning contents---digital artifacts, like position, size, owner, and material etc., storage in the ULE DB.

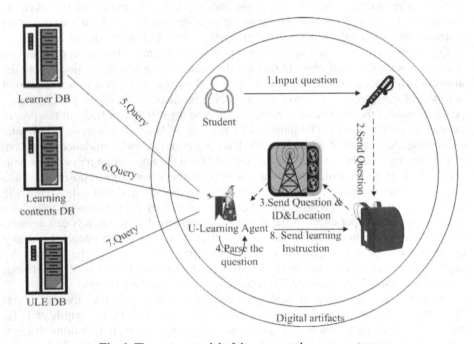

Fig. 1. The system model of the query and answer system

①When a pupil runs into a question, he or she can input the question by U-Pen in a natural way based on the situation. For example, when he is reading a book and the keywords of the question is in the book, he can get the keywords by using the U-Pen to scan the keywords. When he is watching TV, he can tell U-Pen the question in word. When he is on the way to school and have question on an object, he can scan the RFID tag on the object by U-Pen. ②Now the question has been informed to the system. Then U-Pen sends the question to the U-Schoolbag via the wireless communication module. ③The U-Schoolbag resends the question together with the ID and location to the remote U-Learning Agent via the GPRS (General Packet Radio Service). ④Since some times the keywords on the question can not be gotten directly. For example, a sentence may be inputted in by voice, a series of binary number may be inputted in by RFID reader. So at first the U-Learning Agent has to parse the question to get the keywords. ⑤To provide the understandable contents for the learner, the U-Learning Agent access the learner DB to get the pupil's knowledge level on the subject. ⑥The U-Learning Agent queries the Learning contents database on the keywords and knowledge level to get the spare learning contents for the pupil. ⑦ Since some contents maybe not available to the current pupil's situation, the U-Learning Agent will matching the spare learning contents and pupils current location with the ULE DB to get the available real learning contents. ⑧These learning contents and suggestions will be send back to the pupil under the control of the U-Learning Agent.

The model has shown a whole image on the system elements and their interactions. It is a big system including with many problems to be solved. Since the limit of the paper, we will concentrate on the database design and the matching algorithm to find out the proper learning material surrounding learners.

3 Matching Between Real and Virtual

In this section, how to find out the proper learning task based on the learner's situation will be discussed. At first, some key concepts will be presented. Then the database schemas and data exchange mechanism will be given out.

3.1 Only NRFID Is Not Enough

NRFID means networked RFID tag. An object with a NRFID tag is the simplest Digital Artifact. Some research institute including Ubiquitous ID Center and EPCglobal (formerly Auto ID Lab) have set up the standards to set up an ID platform so that the information can be shared by different purpose applications. Some important application fields including supply chain management, library management and vision-impaired people guiding system have been pointed out and implemented [10]. By adopting NRFID into these applications, the productivity is improved. But to trace the objects, human interventions are still necessary in these applications which have shown that only NRFID is not enough for most of ubiquitous applications.

3.2 Digital Artifact

Digital Artifact is a normal artifact augmented with controller modules, sensor modules and/or communication modules. NRFID can be deemed as the smallest digital artifacts. On one hand, these digital artifacts can provide usual usage. For example, using a digital cup, one can drink, digital tiles can be used as normal tiles and a digital schoolbag can take books. On the other hand, these digital artifacts can provide some other function by augmenting tiny computing node on them. For example, a digital cup [2] can know the states (full or empty, held or put down) and give a hint to systems or users. Digital tiles [6] can know what is on them. A digital schoolbag [7] can know at what time, what objects have been put in or taken out of it and give a hint to the systems or users.

3.3 Knowledge Level

To provide understandable and enough learning suggestion to the pupils, a knowledge level management system is necessary. Knowledge level is assigned to each of the record in the learning contents DB to mark the pre-requirement of the corresponding record. In this paper, the pupils' grade is adopted as a standard to evaluate learner's knowledge level.

3.4 Data Model

3.4.1 Architecture of the Data Model

If the digital artifacts are around every corner of the world, one computer would not enough to deal with all of the operation required. So the distributed databases have been adopted into this model. For example, the earth can be divided into many parts by 1' longitude and 1' latitude and the square of each part is about $3.4km^2$ (the distance of 1' longitude and 1' latitude is about 1852m). Each ULE DB point response for one part and the address list of all ULE DB points are storage on the gateway to find out the according database point.

The database organization architecture is shown in Fig.2. Learner DB is used to store the learners' information. Learning contents DB is used to store the learning contents including both of instructions and learning materials. ULE DB is used to store the location and other information about the physical materials. The agent can query these databases through the database service.

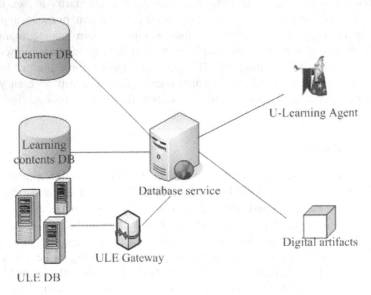

Fig. 2. The database organization architecture

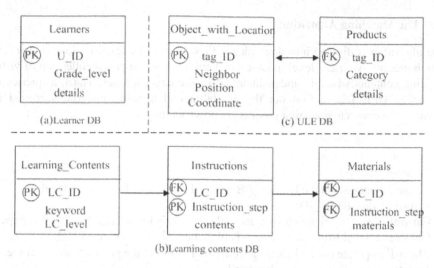

Fig. 3. Schemas of database instance

3.4.2 Learner DB

As shown in Fig.3 (a), Leaners Table is used to place the information about the learners. "U_ID" and "tag_ID" are 128 bit binary code which store in the RFID tag. "Grade_level" is the grade number of a pupil, which is used as a representation of the pupil's knowledge level. "details" is a extensible attribute, which is application dependent.

3.4.3 Learning Contents DB

As shown in Fig.3 (b), three tables are in the DB. Generally, concerning one learning point, multiple experiments can be found in the learning contents DB. And each experiment contains multiple steps. Each of steps needs multiple experiment materials. Subsequently, one Learning _Contents records can have several Instructions and one Instructions can have several Materials.

3.4.4 Ubiquitous Learning Environment DB

As shown in Fig.3 (c), Object_with_Location Table is used to store the relative position (neighbor and position) and absolute position (coordinate) of the objects. "Neighbor" functions as a reference about which object has detects this object. For example, a desk detects that a cup is on it, and then the cup's neighbor is the desk. "Position" is the spatial relative relation between an object and its neighbor. "coordinate" is the global position which is presented by longitude and latitude. Product Table keeps the record about every product. We suppose the data about each product has been input into the table in the production factory. "Tag_ID" is a unique 128 bit number which can identify the product on the world. "Category" is the category of the product. "details" is a extensible attribute which can be used to describe any attribute about the products.

3.5 The Matching Algorithm

The algorithm is divided into three phase. First, it gets the learner's knowledge level which present by Grade_level in Learners Table. Second, it gets a collection of virtual learning contents which is understandable for the target learner. Third, it queries the according ULE DB to find out the available real learning material surround the learner. The concrete matching process is shown as follow.

> **Matching Algorithm:**
> Learning_contents_set←∅
> Material_set←∅
> **Phase I :** operate on the Learner DB to get the learner's knowledge level
> **Input:** learner ID (U_ID)
> **Output:** Grade_level which is equal to the learner's knowledge level in this paper
> Get Grade_level from the Learners Table where U_ID = Learners.U_ID
> **Phase II :** operate on the Learning Contents DB to get the proper learning contents
> **Input:** U_ID, Grade_level, a keyword
> **Output:** the learning_contents_set which is understandable to the learner
> **For each** record in Learning_Contents Table
>   **If** keyword = Learning_Contents.keyword **and**
>   (Grade_Level -1) < LC_level<(Grade_Level +1)
>   **Then** add the LC_ID to the Learning_contents_set
> **Phase III:** operate on the ULE DB and Learning contents DB to find out the proper physical learning material
> **Input:** the current location of the learner (U_Location), learning_contents_set
> **Output:** the proper learning suggestions which the learning material is available to the learner
> /** 1-4: Get the ULE DB address which the learner is located in from the ULE gate way list**/
> 1.**For each** ULE DB on the ULE gate way list
> 2.   **If** U_Location ∈ [longitude$_{min}$, longitude$_{max}$] **and**
> 3.   U_Location ∈ [latitude$_{min}$, latitude$_{max}$]
> 4.   **Then** ULE_DB_address = gateway_list.ULE_DB_address
> /** 5-18: Iterate the learning_contents_set to find out whether **/
> /**any experiments are available for the learner**/
> 5.**For each** LC_ID in learning_contents_set
> /**6-7: Get the materials list for a given experiment**/
> 6.   **If** LC_ID = Materials.LC_ID
> 7.   **Then** add materials to the Material_set
> /**8-18: Judge on whether the all learning materials are surround the learner**/
> 8.   material_matched = FALSE
> 9.   location_matched = FALSE
> 10.   **For each** material in Material_set
> 12.   **If** material matches any Products.category
> 13.   **Then** material_matched = TRUE and
> 14.   Get tag_ID of the category and
> 15.   Location_matched = neighbor (tag_ID, U_Locatioin)

16.   **Else** material_matched = FALSE
17.   **If** material_matched = TRUE and location_matched = TRUE
18.   **Then**   add   the   tag_ID   of   each   material   in   material_set   into available_contents_list

3.6 Find Out the Coordinate of Digital Artifacts

As shown in Fig.4, supposing a ubiquitous environment is filled with various kinds of Digital Artifacts. The tracing of physical objects in an automatic way will become possible. And the relative positions of the artifacts can be detected automatically. For example, in Fig.4, a smart house can know U-tiles in the house. U-tiles can know a bookshelf, a desk and a schoolbag is on it. The desk can know a cup and some paper is on it. Moreover, the space coordinate can be detected by two ways. One is the spatial database which records the coordinate of most of the building. Another is GPS which has been integrated onto various devices like mobile phones, cars etc. As a result, in the example, the space coordinate of these digital artifacts could be mined out from the ULE DB.

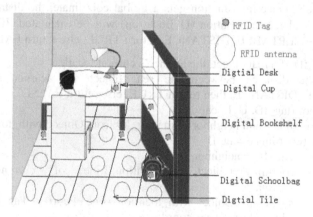

Fig. 4. An example space with Digital Artifacts

In this section, we will show the process of finding out the coordinate by an example and give out a function which has been applied into the Matching Algorithm. An instance of ULE DB is shown in Table 1 and Table 2. It can be deemed as a description of Fig.4.

Table 1. An instance of Products Table

Tag_ID	Category	details	
		Material	producer
01	cup	Glass	SUN
02	chair	null	
03	desk	wood	null
04	house	null	
05	tiles	null	

Table 2. An instance of Object_with_Location Table

Tag_ID	neighbor	position	coordinate
01	Desk (03)	on	null
02	Tiles (05)	on	null
03	Tiles (05)	on	null
04	null	null	(XXXX, YYYY)
05	House (04)	in	null

Now given a Tag_ID, for example, 01 for a cup, then we can get the coordinate of the cup by literately checking whether the neighbor has a global coordinate. Here, the process of finding is 01->03->05->04.

The definition of function neighbor (tag_ID, U_Location) is shown as follow. The ACCEPTABLE_DISTANCE is a constant used to mark if the object is easy to get for the learner. The ACCEPTABLE _DISTANCE can be defined by users or developers. In this function, the function iteratively traces whether an object neighbor to it has a global coordinate. Once the function gets a global coordinate, the distance between the coordinate and learner location (U_Location) will be calculated. If the result is less than the ACCEPTABLE_DISTANCE, return TRUE, else return FALSE.

Input: the tag_ID of a real object, the learner's current location
Output: if the distance between the real object and learner is less than ACCEPTABLE_DISTANCE, then return TRUE, else return FALSE

```
Bool neighbor (tag_ID, U_Location)
    Get coordinate and neighbor of the tag_ID from Object_with_location Table
    While (coordinate = null)
        Tag_ID = neighbor
        coordinate=Object_with_location.(Line    of    current    Tag_ID).
        coordiante
        neighbor = Object_with_location. (line of current Tag_ID).neighbor
        Object_Location = coordinate
    If (abs (U_Location, Object_Location) < ACCEPTABLE_DISTANCE)
        Return TRUE
    Else
        Return FALSE
```

4 An Example

As shown in the left part of Fig.5, Eric is a pupil of grate 3. When he is watching TV at his home, he heard the word "Atmospheric pressure". He shows strong curiosity on the term. So he asks the U-Pen "1.What is atmospheric pressure?" The U-Pen interacts with the ubiquitous query and answer system and firstly an easy understanding explanation about the concept is shown on the LCD displayer of the schoolbag as follow "2.it is the weight of air". Eric can not imagine that air could have weight. Then he wants to know more about the concept, "3.Tell me more". So the U-Schoolbag tells him that he can do an experiment on the concept. He can use the glass cup and papers on his desk in his sleeping room as the experiment materials. That is "Please take the cup on your desk in your sleeping room". After Eric gets the materials he tells the U-Pen "5. I am ready". Then U-Pen helps Eric to complete the experiment step by step as follow.

Step one: U-Pen: Put the cap fill with water.
        Eric: OK!
Step two: U-Pen: cover the cup with a piece of paper, be careful do not let air in it
        Eric: OK!
Step three: U-Pen: Reverse the cup, be careful do not let air in it
        Eric: OK!
Step four: U-Pen: now keep the position and draw out the paper
        Eric: OK!

Since there are more of wonderful experiments on the question, such a process could be continued if Eric still has interest on this question.

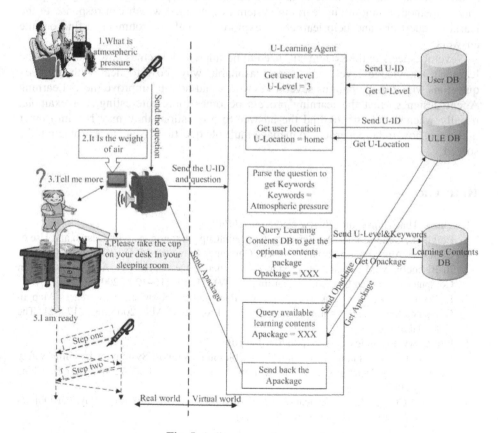

Fig. 5. A diagram on the example

In the virtual world shown in the right part of Fig.5, begin with the query by Eric, the data process in the system is shown in Fig.5. The question together with the U-ID is send to the U-Learning Agent. Then the U-Learning Agent queries the User DB and ULE DB to get the U-Level and U-Location respectively. Then the agent will parse on the question to get the keywords. Then it queries the Learning Contents DB to get the optional learning package (Opackage). Although the contents in the Opackage is understandable for the learners, the necessary learning materials to perform the learning contents maybe not nearby the learners. So the agent will query the ULE DB to find out which learning contents are available for the current learners and get the available package (Apackage). Then it sends back the Apackage to the U-Schoolbag and U-Pen to interactive with the learner.

5 Conclusion and Future Work

In this paper, the brick of ubiquitous environment ----the concept of Digital Artifacts is proposed. In an environment deployed with digital artifacts, the method which can find out the spatial coordinate of any digital artifacts is presented. And by adopting such a method, a ubiquitous learning system is presented which can response to the learner's questions and help learners to explore the real environment to find out the answers.

Several aspects in the system still need to be improved, such as how to evaluate the learner's knowledge level in a more reasonable way, how to deal with complex questions which may contains several keywords and how to improve the U-Learning Agent design so that the learning process becomes more interesting. For example, usually when people try to find the answer to a question, they may run into other questions, how to help learners to manage multiple questions is another problem to be solved.

Reference

1. Things That Thinks, http://www.media.mit.edu/ttt/
2. Beigl, M., Gellersen, H.-W., Schmidt, A.: MediaCups: Experience with Design and Use of Computer-Augmented Everyday Objects. Computer Networks 35(4), 401–409 (2001)
3. El-Bishouty, M.M., Ogata, H., Yano, Y.: Learner-Space Knowledge Awareness Map in Computer Supported Ubiquitous Learning. In: WMTE pp. 116–120 (2006)
4. El-Bishouty, M.M., Ogata, H., Yano, Y.: Personalized Knowledge awareness Map in Computer Supported Ubiquitous Learning. In: Proc. of ICALT 2006, pp. 817–821, The Netherlands (2006)
5. http://www.manningsrfid.com/pdfs/IDBlue.pdf
6. Han, Q., Jing, L., Huang, T., Cheng, Z.: A Ubiquitous Support System for Learning Safety Knowledge by Detecting Kids' Dangerous Situations. In: ICITA2007, Harbin, China (January 2007)
7. Jing, L., Cheng, Z.: An Educational Schoolbag System for Providing Physical Objects Reminding Service. IPSJ 48, 656–666 (2007)

8. Kurihara, K., Goto, M., Ogata, J., Igarashi, T.: Speech pen: predictive handwriting based on ambient multimodal recognition. In: Proceedings of the SIGCHI conference on Human Factors in computing systems CHI, pp. 851–860 (2006)

9. Liu, T.-Y., Tan, T.-H., Chu, Y.-L.: The Ubiquitous Museum Learning Environment: Concept, Design, Implementation, and a Case Study. ICALT,: 989-991 (2006)

10. Liu, Y., Bacon, J., Wilson-Hinds, R.: On Smart-Care Services: Studies of Visually Impaired Users in Living Contexts. First International Conference on the Digital Society (ICDS), pp:32-37 (2007)

11. Sakamura, K., Koshizuka, N.: Ubiquitous Computing Technologies for Ubiquitous Learning. In: IEEE International Workshop on Wireless and Mobile Technologies in Education (WMTE), pp: 11-20 (2005)

12. YRP Unbiquitous Networking Laboratory: Philosophy of Networked RFID Technologies and Applications. Asia-Pacific Telecommunity Standardization Program 2005 (ASTAP) (October 2005)

13. Verdejo, M.F., Celorrio, C., Lorenzo, E., Sastre-Toral, T.: An Educational Networking Infrastructure Supporting Ubiquitous Learning for School Students. In: ICALT, pp. 174–178 (2006)

14. Verdejo, M.F., Celorrio, C., Lorenzo, E.: Improving Learning Object Description Mechanisms to Support an Integrated Framework for Ubiquitous Learning Scenarios. In: WMTE, pp. 93–97 (2006)

Building a Scalable Web Query System

Meichun Hsu and Yuhong Xiong

Hewlett-Packard Laboratories, 1501 Page Mill Road, Bldg. 1U, Palo Alto,
CA 94304, USA
{meichun.hsu,yuhong.xiong}@hp.com

Abstract. Nowadays, the dominant way to find information on the web is
through search. General search engines are very effective, but search phrases
and results are unstructured and that limits a user's ability to further automate
the processing of the search results. In recent years, we have seen efforts to
build systems that support more precise query on the web for certain content
verticals. We describe the general problems for building an extensible web
query system and report some of our work in this area.

Keywords: Focused Crawling, Web Page Classification, Metadata Extraction.

1 Introduction

It is well known that the Web contains a huge amount of useful information.
Currently, the dominant way to obtain information from the web is through search.
General search engines are very effective, but search phrases and results are
unstructured and that limits a user's ability to further automate the processing of the
search results. For example, a user generally cannot embed the search results in an
application program the way one would embed the results from querying a
conventional relational database. Ideally, we want to be able to *query* the web using a
more structured query language. For example, in finding product information from
HP's web site, one might wish to say:

SELECT product_name FROM hp.com WHERE product_type = PC

and get a reasonably accurate and complete result.

In recent years, progress in several areas, including web content discovery,
analysis, and information extraction, have moved us closer to this goal. Some research
and industry groups have built systems that support more precise query on certain
content verticals, such as academic publications and products [1][2]. Such systems are
referred to as vertical search, entity search, object-level search, or content aggregation
systems. Some people also observe such efforts as evidence of a trend towards a
convergence of database systems and information retrieval systems [3].

At HP Labs, we are conducting research that aims at developing technologies that
enable domain-specific content discovery on the web that are generally extensible to
multiple content domains. This paper describes the general problems for building an
extensible web query system and reports our recent findings.

S. Bhalla (Ed.): DNIS 2007, LNCS 4777, pp. 322–328, 2007.

2 General Problems

To be able to query the web like a relational database, we need to find the relevant pages on the web, categorize them, extract key information, feed the extracted information into a database, then serve the information through a database query interface, or a web portal.

2.1 Focused Crawling

Web search is the most popular way to discover information on the web. However, as mentioned above, search result is often incomplete and imprecise. Another way to discover web contents is to use a *focused crawler* [4][5], which is a crawler that tries to identify and download relevant pages, and filters out irrelevant ones while crawling the web. One major problem in focused crawling is to determine how likely a link will lead to target pages, so as to decide whether to follow that link.

2.2 Web Content Categorization

After we have obtained a set of web pages, which hopefully include mostly the relevant pages, but inevitably some irrelevant ones, we need to separate out the relevant ones from the irrelevant ones. This is a binary classification problem. Further more, we want to classify the relevant pages into finer categories. For online courses for example, we want to divide the course pages into CS, biology, history, etc. Web page classification can be done using the URL features [6], the page content [7], and the link structure [8]. Related to classification, main page finding is also an important problem. In many websites, an entity is described by more than one page, with one main page among them. For example, a movie site may have one main page for each movie, which links to a review page, a trailer page, a ticket info page, etc. Identifying the main page can greatly help classification and information extraction.

2.3 Information Extraction

After the web pages are categorized, we need to extract key information before importing it to a relational database. This key information, or metadata, is domain specific. Many techniques have been developed for metadata extraction [9], including template based method [10], rule based method, matching based method, generative models like Hidden Markov Model (HMM) [11]. Some researchers also reported success with recently developed methods like Conditional Random Field (CRF) [12].

Crawling, classification, and metadata extraction do not have to be performed in sequence. In face, information obtained at one stage can be used to improve the performance of other stages. For example, extracted metadata can help improve classification and crawling accuracy.

2.4 Human Knowledge Injection and Training Data Collections

From system point of view, there are several practical problems we need to solve. When working on a specific content domain, human knowledge can be very useful. For example, if we work in the movie domain, we know that concepts like movie, director, actor, actress, cast, release, theater, trailer, box office, and critics are important, so how do we inject this knowledge into a classifier that picks out movie pages? Of course, we can use machine learning to try to automatically discover these terms as features, but providing an interface for human to directly inject this knowledge may greatly speed up system development, or improve performance.

Another problem people face when working in a specific domain is the lack of training data for their classifiers and metadata extractors. One way to alleviate this problem is to use techniques that leverage human's domain knowledge to reduce the need for training data. The other is to develop tools that enable human to quickly create a relatively large amount of training data.

2.5 Computing and Storage Infrastructure

Reducing infrastructure cost is a general problem in any crawling and content analysis system. This problem can happen at different scales. An enterprise may want to develop efficient crawling algorithms to reduce the size of the crawling cluster, and an individual developing a personalized system may need to fit all data into a PC.

In the next section, we will describe efforts at HP Labs in addressing the above problems.

3 The Fusion Project at HP Labs

We chose to focus on on-line courseware as an initial vertical domain to develop our research.

We have developed a focused crawling algorithm that is able to quickly find course pages from school web sites. As mentioned before, one major problem in focused crawling is determining how to measure the likelihood that a page will lead to target pages. To address this problem, we developed a measurement of this likelihood, called Navigational Rank (NR). The intuition behind NR is that each page is rewarded by pointing to pages with high relevance and penalized by pointing to pages with low relevance. We have used NR in two ways for focused crawling, based on the passive and active learning models, respectively. In the passive model, we download several entire websites and compute the NR of each page. We then learn the correlation between the NR scores and the URL features of a webpage. For a new site, the learnt results are used to guess the NR of each page encountered in the crawling process. In the active model, we dynamically update the NR of each downloaded page while we crawl a website. As the sub-graph expands, the NR score becomes an increasingly accurate reflection of the likelihood that a page will lead to target pages. In order to test our method, we conduct large-scale experiments on the focused crawling of course materials from university web sites. Our experimental results show that our method significantly outperforms several other link-based methods. The resulting focused crawler can download about 90% of the target pages by crawling only 30% of the entire website.

In deciding whether a page is a course page or not, a binary classifier is needed. To build a classifier using standard methods like SVM requires availability of a large quantity of training data. Instead, we explored a method to leverage human knowledge to quickly develop prototype classifiers. We call our method *Weighted*

Search Tree (WST). The basic idea is to use search-based techniques to build a classifier. When doing a search, a user submits a set of keywords to a search engine, and gets a ranked list of web pages. The list is ranked by a certain relevance score. If we use a threshold on the score to separate the pages into two classes, we obtain a binary classifier. The key problem here is how to compute the relevance score. WST is a way for a human to express his knowledge about a content domain, in the form of keywords and their weights, together with the method to compute page relevance based on those weights. Figure 1 shows an example of a WST. According to this WST, if the word "resume" or "vita" appears in the URL, the page gets a negative score of -1. If the word "homework" appears in page content, the page accumulates a score of 0.6. The keywords do not have to be complete and the weights do not have to be fine-tuned for the classifier to work. We have tested using the WST technique to build classifiers for course pages. In addition, we tested the technique in a different content domain, the on-line pages about movies; we were able to build the classifier for the movie domain in just half an hour, by picking some keywords from movie site, and the classification result was surprisingly good by subject tests.

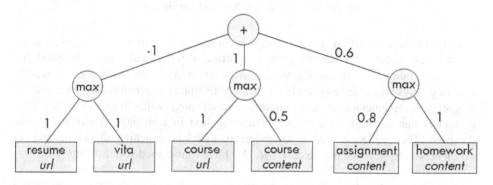

Fig. 1. An example Weighted Search Tree (WST)

In addition to studying ways to build a classifier by directly injecting human knowledge, we also explored technologies to enable the generation of large amount of training and testing data. We designed and implemented a labeling tool to help human classify web pages. In our tool, we use the structural information, the URL tree and the link structure, as well as the content information of a web page, to help identify important pages. We call our tool SALT - Structure-Assisted Labeling Tool. Our tool is web based so people from different locations can label the same dataset at the same time. Figure 2 shows a screen shot. The left part of the window shows the URLs in a tree structure. Any branch of the tree can be expanded or shrank. The leaf nodes of the tree are URLs. When one of them is clicked, the web page is shown on the right part of the window. On the left window, any sub-tree can be labeled all at once. To help the user identify important pages, the tool also computes the WST scores of each page, and use heuristics to identify potential main pages. Pages with high WST scores and main page candidates are marked to give user hints. Using this tool, five colleagues are able to label more than 220,000 pages from one school website, as either course page or not, in about four hours. This averages to about 3 pages per second.

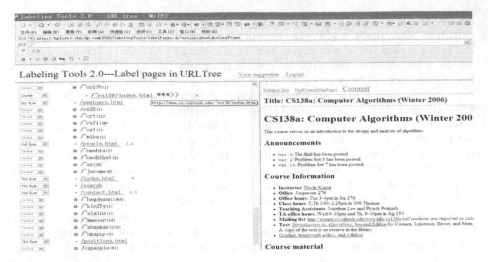

Fig. 2. SALT – Structure-Assisted Labeling Tool

For metadata extraction, we have implemented a system to extract metadata from course web pages. Our current implementation is rule-based, which is good for expressing human intuitions. Figure 3 shows an example result. We are currently studying ways to combine the rules with CRF to improve performance. In a related project, we implemented another system to extract movie titles from web pages. This system is match-based. That is, we match the text in web pages against a movie database. The challenge here is to design an efficient matching algorithm and an efficient way to store the movie database. We have found good solutions for both.

Course Page Metadata Extraction Results (Course Pages Number=328 Runtime=34.348s)

No.	URL	Course ID	Course Name	Course Time	Teacher Name	Teacher Email	Literature	All Valid Metadata
	Precision	92.16%	76.83%	65.61%	64.95%	75.93%	37.93%	68.27%
	Recall	85.98%	76.83%	72.03%	55.75%	63.08%	36.26%	43.29%
	F1 score	88.96%	76.83%	68.67%	60.00%	68.91%	37.08%	52.99%
1	data\coursePages\caltech\1.html	+	+	+	+	+	+	6/6
2	data\coursePages\caltech\102.html	+	+	null	+	null	null	3/3
3	data\coursePages\caltech\103.html	+	+	null	null	null	null	2/2
4	data\coursePages\caltech\104.html	+	+	null	+	null	null	3/3

Fig. 3. Example metadata extraction result from course web pages

4 Summary

In this extended abstract we described a framework and recent findings in developing technologies to enable more precise query of web contents. Such web content discovery and analysis techniques have been an active research area. The technologies can be used to support structured queries over contents extracted and aggregated from the Web. They are also foundational to personalization, by offering more insights into the web content of interest to particular users. While such technologies require existence of human knowledge to frame the metadata of the contents in a specific domain, they in turn are tools for discovering or evolving the metadata or ontology with which we could further improve or personalize such systems. This would be the promise of the convergence of database technology and information retrieval in the era of the Web.

Acknowledgments. We would like to thank our Fusion project colleagues for the work reported in this paper: Li Zhang at HPL Palo Alto, Shi-cong Feng, Baoyao Zhou, Yong Zhao, Yu Yang, Demiao Lin, Weichun Wang, and Zhiyao Luo at HPL China, and our collaborators at Peking University.

References

1. Castellanos, M., Chen, Q., Dayal, U., Hsu, M., Lemon, M., Siegel, P., Stinger, J.: Component Advisor: A tool for automatically extracting electronic component data from Web datasheets. In: Proceedings of the Workshop on Reuse of Web-based Information, 7th International World Wide Web Conference (WWW7), Brisbane, Australia (1998)
2. Nie, Z., Wen, J., Ma, W.: Object-level Vertical Search. In: Proceedings of Conf. on Innovative Data Systems Research, Pacific Grove, California (2007)
3. Weikum, G.: DB&IR: both sides now. In: Proceedings of the 2007 ACM SIGMOD international conference on Management of data, Beijing, China, pp. 25–30 (2007)
4. Chakrabarti, S., van den Berg, M., Dom, B.: Focused Crawling: A New Approach to Topic-Specific Web Resource Discovery. Computer Networks 31(11-16), 1623–1640 (1999)
5. Diligenti, M., Coetzee, F., Lawrence, S., Giles, C.L., Gori, M.: Focused Crawling Using Context Graphs. In: Proceedings of 26th Int. Conf. on Very Large Databases (VLDB), Cairo, Egypt, pp. 527–534 (2000)
6. Kan, M., Thi, H.: Fast Webpage Classification Using URL Features. In: Proceedings of the 14th Int. Conf., Bremen, Germany (2005)
7. Dumais, S., Chen, H.: Hierarchical Classification of Web Content. In: Proceedings of the 23rd ACM Int. Conf. on Research and Development in Information Retrieval (SIGIR-00), Athens, Greece (2000)
8. Calado, P., Cristo, M., Moura, E., Ziviani, N., Ribeiro-Neto, B., Gonalves, M.: Combining Link-Based and Content-Based Methods for Web Document Classification. In: CIKM 2003. Proceedings or the 12th Int. Conf. on Information and Knowledge Management, New Orleans, Louisiana (2003)
9. McCallum, A.: Information Extraction: Distilling Structured Data from Unstructured Text. In: ACM QUEUE, pp. 49–57 (November 2005)

10. Arasu, A., Garcia-Molina, H.: Extracting Structured Data from Web Pages. In: Proceedings of the 2003 ACM SIGMOD Int. Conf., San Diego, California (2003)
11. Yin, P., Zhang, M., Deng, Z., Yang, D.: Metadata Extraction from Bibliographies Using Bigram HMM. In: Chen, Z., Chen, H., Miao, Q., Fu, Y., Fox, E., Lim, E.-p. (eds.) ICADL 2004. LNCS, vol. 3334, pp. 310–319. Springer, Heidelberg (2004)
12. Lafferty, J., McCallum, A., Pereira, F.: Conditional Random Fields: Probabilistic Models for Segmenting and Labeling Sequence Data. In: Proceedings of the 18th Int. Conf. on Machine Learning, pp. 282–289. Morgan Kaufmann, San Francisco, CA (2001)

Author Index

Lecture Notes in Computer Science

Sublibrary 3: Information Systems and Application, incl. Internet/Web and HCI

For information about Vols. 1– 4295
please contact your bookseller or Springer

Vol. 4560: N. Aykin (Ed.), Usability and Internationalization, Part II. XVIII, 576 pages. 2007.

Vol. 4559: N. Aykin (Ed.), Usability and Internationalization, Part I. XVIII, 661 pages. 2007.

Vol. 4558: M.J. Smith, G. Salvendy (Eds.), Human Interface and the Management of Information, Part II. XXIII, 1162 pages. 2007.

Vol. 4557: M.J. Smith, G. Salvendy (Eds.), Human Interface and the Management of Information, Part I. XXII, 1030 pages. 2007.

Vol. 4541: T. Okadome, T. Yamazaki, M. Makhtari (Eds.), Pervasive Computing for Quality of Life Enhancement. IX, 248 pages. 2007.

Vol. 4537: K.C.-C. Chang, W. Wang, L. Chen, C.A. Ellis, C.-H. Hsu, A.C. Tsoi, H. Wang (Eds.), Advances in Web and Network Technologies, and Information Management. XXIII, 707 pages. 2007.

Vol. 4531: J. Indulska, K. Raymond (Eds.), Distributed Applications and Interoperable Systems. XI, 337 pages. 2007.

Vol. 4526: M. Malek, M. Reitenspieß, A. van Moorsel (Eds.), Service Availability. X, 155 pages. 2007.

Vol. 4524: M. Marchiori, J.Z. Pan, C.d.S. Marie (Eds.), Web Reasoning and Rule Systems. XI, 382 pages. 2007.

Vol. 4519: E. Franconi, M. Kifer, W. May (Eds.), The Semantic Web: Research and Applications. XVIII, 830 pages. 2007.

Vol. 4518: N. Fuhr, M. Lalmas, A. Trotman (Eds.), Comparative Evaluation of XML Information Retrieval Systems. XII, 554 pages. 2007.

Vol. 4508: M.-Y. Kao, X.-Y. Li (Eds.), Algorithmic Aspects in Information and Management. VIII, 428 pages. 2007.

Vol. 4506: D. Zeng, I. Gotham, K. Komatsu, C. Lynch, M. Thurmond, D. Madigan, B. Lober, J. Kvach, H. Chen (Eds.), Intelligence and Security Informatics: Biosurveillance. XI, 234 pages. 2007.

Vol. 4505: G. Dong, X. Lin, W. Wang, Y. Yang, J.X. Yu (Eds.), Advances in Data and Web Management. XXII, 896 pages. 2007.

Vol. 4504: J. Huang, R. Kowalczyk, Z. Maamar, D. Martin, I. Müller, S. Stoutenburg, K.P. Sycara (Eds.), Service-Oriented Computing: Agents, Semantics, and Engineering. X, 175 pages. 2007.

Vol. 4500: N.A. Streitz, A.D. Kameas, I. Mavrommati (Eds.), The Disappearing Computer. XVIII, 304 pages. 2007.

Vol. 4495: J. Krogstie, A. Opdahl, G. Sindre (Eds.), Advanced Information Systems Engineering. XVI, 606 pages. 2007.

Vol. 4480: A. LaMarca, M. Langheinrich, K.N. Truong (Eds.), Pervasive Computing. XIII, 369 pages. 2007.

Vol. 4471: P. Cesar, K. Chorianopoulos, J.F. Jensen (Eds.), Interactive TV: A Shared Experience. XIII, 236 pages. 2007.

Vol. 4469: K.-c. Hui, Z. Pan, R.C.-k. Chung, C.C.L. Wang, X. Jin, S. Göbel, E.C.-L. Li (Eds.), Technologies for E-Learning and Digital Entertainment. XVIII, 974 pages. 2007.

Vol. 4443: R. Kotagiri, P. Radha Krishna, M. Mohania, E. Nantajeewarawat (Eds.), Advances in Databases: Concepts, Systems and Applications. XXI, 1126 pages. 2007.

Vol. 4439: W. Abramowicz (Ed.), Business Information Systems. XV, 654 pages. 2007.

Vol. 4430: C.C. Yang, D. Zeng, M. Chau, K. Chang, Q. Yang, X. Cheng, J. Wang, F.-Y. Wang, H. Chen (Eds.), Intelligence and Security Informatics. XII, 330 pages. 2007.

Vol. 4425: G. Amati, C. Carpineto, G. Romano (Eds.), Advances in Information Retrieval. XIX, 759 pages. 2007.

Vol. 4412: F. Stajano, H.J. Kim, J.-S. Chae, S.-D. Kim (Eds.), Ubiquitous Convergence Technology. XI, 302 pages. 2007.

Vol. 4402: W. Shen, J.-Z. Luo, Z. Lin, J.-P.A. Barthès, Q. Hao (Eds.), Computer Supported Cooperative Work in Design III. XV, 763 pages. 2007.

Vol. 4398: S. Marchand-Maillet, E. Bruno, A. Nürnberger, M. Detyniecki (Eds.), Adaptive Multimedia Retrieval: User, Context, and Feedback. XI, 269 pages. 2007.

Vol. 4397: C. Stephanidis, M. Pieper (Eds.), Universal Access in Ambient Intelligence Environments. XV, 467 pages. 2007.

Vol. 4380: S. Spaccapietra, P. Atzeni, F. Fages, M.-S. Hacid, M. Kifer, J. Mylopoulos, B. Pernici, P. Shvaiko, J. Trujillo, I. Zaihrayeu (Eds.), Journal on Data Semantics VIII. XV, 219 pages. 2007.

Vol. 4365: C.J. Bussler, M. Castellanos, U. Dayal, S. Navathe (Eds.), Business Intelligence for the Real-Time Enterprises. IX, 157 pages. 2007.

Vol. 4353: T. Schwentick, D. Suciu (Eds.), Database Theory – ICDT 2007. XI, 419 pages. 2006.

Vol. 4352: T.-J. Cham, J. Cai, C. Dorai, D. Rajan, T.-S. Chua, L.-T. Chia (Eds.), Advances in Multimedia Modeling, Part II. XVIII, 743 pages. 2006.

Vol. 4351: T.-J. Cham, J. Cai, C. Dorai, D. Rajan, T.-S. Chua, L.-T. Chia (Eds.), Advances in Multimedia Modeling, Part I. XIX, 797 pages. 2006.

Vol. 4328: D. Penkler, M. Reitenspiess, F. Tam (Eds.), Service Availability. X, 289 pages. 2006.

Vol. 4321: P. Brusilovsky, A. Kobsa, W. Nejdl (Eds.), The Adaptive Web. XII, 763 pages. 2007.

Vol. 4317: S.K. Madria, K.T. Claypool, R. Kannan, P. Uppuluri, M.M. Gore (Eds.), Distributed Computing and Internet Technology. XIX, 466 pages. 2006.

Vol. 4312: S. Sugimoto, J. Hunter, A. Rauber, A. Morishima (Eds.), Digital Libraries: Achievements, Challenges and Opportunities. XVIII, 571 pages. 2006.

Vol. 4306: Y. Avrithis, Y. Kompatsiaris, S. Staab, N.E. O'Connor (Eds.), Semantic Multimedia. XII, 241 pages. 2006.

Vol. 4302: J. Domingo-Ferrer, L. Franconi (Eds.), Privacy in Statistical Databases. XI, 383 pages. 2006.

Vol. 4299: S. Renals, S. Bengio, J.G. Fiscus (Eds.), Machine Learning for Multimodal Interaction. XII, 470 pages. 2006.